Field Guide to
Environmental Engineering
for Development Workers

Other Titles of Interest

Advances in Water and Wastewater Treatment, **edited by Rao Y. Surampalli and K. D. Tyagi.** (ASCE Committee Report, 2004). Describes the application of innovative technologies for water and wastewater treatment with an emphasis on the scientific principles for pollutant or pathogen removal. (ISBN 0-7844-0741-X)

Appropriative Rights Model Water Code, **edited by Joseph W. Dellapenna.** (ASCE Committee Report, 2007). Presents a legal framework that balances management of water with social, economic, political, and administrative concerns. (ISBN 978-0-7844-0887-2)

Climate Variations, Climate Change, and Water Resources Engineering, **by Jurgen D. Garbrecht and Thomas C. Piechota.** (ASCE Committee Reports, 2006). Highlights current knowledge about climate variations and change and their impact on water resources systems. (ISBN 978-0-7844-0824-7)

Sharing Water in Times of Scarcity: Guidelines and Procedures in the Development of Effective Agreements to Share Water Across Political Boundaries, **edited by Stephen E. Draper.** (ASCE Committee Report, 2006). Offers narrative guidelines and procedures for formulating a water sharing agreement. (ISBN 0-7844-0846-7)

Sustainable Engineering: An Introduction, **by the Committee on Sustainability of the Technical Activities Committee.** (ASCE Committee Report, 2004). Provides a broad, fundamental understanding of sustainability principles and their application to engineering work. (ISBN 0-7844-0750-9)

Field Guide to Environmental Engineering for Development Workers

Water, Sanitation, and Indoor Air

James R. Mihelcic, Ph.D.
Lauren M. Fry
Elizabeth A. Myre
Linda D. Phillips, P.E.
Brian D. Barkdoll, Ph.D., P.E.

Library of Congress Cataloging-in-Publication Data

Field guide to environmental engineering for development workers : water, sanitation, and indoor air / James R. Mihelcic ... [et al.].

   p. cm.

 Includes bibliographical references and index.

 ISBN 978-0-7844-0985-5

  1.   Sanitary engineering--Developing countries. 2.   Water-supply--Developing countries. 3. Indoor air quality--Developing countries. I. Mihelcic, James R.

 TD127.F54 2009

 628.09172'4--dc22

2009022603

Published by American Society of Civil Engineers
1801 Alexander Bell Drive
Reston, Virginia 20191
www.pubs.asce.org

The cover shows the construction of a 10,000-gallon, reinforced brick water storage tank built for the Los Chaguites community in southern Honduras. The system serves 150 homes and consists of a catchment structure feeding the storage tank, 5,400 m of conduction line, two break pressure tanks, and 1,000 m of distribution piping. Watercolor batik on Japanese rice paper by Linda D. Phillips.

Contents

Wastewater Treatment

Solid Waste

Air Quality

Appendix

Foreword

Peace falls under the general umbrella of many things: freedom, democracy, human rights, the alleviation of suffering, and the improvement of environmental quality and health. Yet it is global health that connects us all.

Global health can be found in a small child in Mali learning to wash her hands to prevent the spread of diarrheal disease. It appears in the work of a community erecting a water storage tank in the mountains of western Honduras or village members pouring concrete slabs used for latrines in Timor-Leste. Global health is also present in groups of women constructing more efficient cooking stoves in Asia that not only reduce indoor air pollution but also make better use of scarce tree resources.

From these seemingly simple beginnings, global health broadens its reach to affect the prosperity and stability of whole nations—whether empowering communities to better educate and provide for themselves, stopping a rapid outbreak of illness, preventing famine, or eliminating socially and economically devastating diseases.

Having access to the basic needs of water, sanitation, hygiene, and shelter are issues of human rights, not just issues of development and engineering. Providing these basic rights in a sustainable manner is critical if we are to improve the health among poor people afflicted with disease, people who are often isolated, forgotten, ignored, and without hope.

If you examine engineering closely, it is about solving problems. Engineering is also tied to eradicating poverty and disease, just as much as it is tied to planning, design, and construction. One key, though, is to provide engineers and others with resources so they understand the beneficial use of appropriate technology. This type of technology is developed and deployed with ongoing input from local communities to accommodate economic, social, environmental, and cultural conditions unique to each locale. Such an approach holds greater promise of fostering healthy communities and reducing gender inequalities.

One principle of the Carter Center is that people can improve their lives when provided with the necessary skills, knowledge, and access to resources. What is special about this book is that it provides all of these so that people can improve the lives of their families, their communities, and their countries.

I am proud to see that much of the knowledge in this book was created by those who served in the U.S. Peace Corps as water-sanitation engineers. Many do not know that my mother also served in the Peace Corps as a health volunteer. Like the child in Mali, she has been an inspiration to me. Only after global health is improved will we all be connected in a world of peace and equality.

President Jimmy Carter

Preface

Welcome to our vision, a world where all have access to sanitation, potable water, and safe indoor air; where all children are able to learn in well built classrooms; where families no longer suffer from disease, starvation, and poverty; where renewable energy has replaced fossil fuels—and engineers are part of the solution.

For over a decade, Linda Phillips and I have directed undergraduate and graduate programs that allowed hundreds of engineering students to come to a robust understanding of their discipline and its relationship to social needs, the environment, and civic responsibility. In the process, students created knowledge as they applied and researched appropriate technological solutions to a wide variety of global challenges with the added constraints of the world's many economic, social, and environmental limitations.

Interestingly, both programs grew independently, but both were born in the late 1990s. During those early years, Linda and I had little idea of what the other was doing until one day, when I was on sabbatical and Linda was back in her office, we connected through several e-mails and realized the characteristics that were similar to both programs. Looking back on the past, we both clearly see how things that developed at the grassroots level fit into larger issues of eliminating poverty, improving global health, stemming the tide of environmental destruction, reforming engineering education, and attracting new students to our discipline.

As an educator and engineer, I have a personal and professional goal to nurture and educate young people to value and implement our field guide's vision of a just and better world. Unfortunately, few have the in-country experience to implement our collective vision of a better world. As you are probably aware, women and children unfairly bear the greatest burden of environmental risk associated with poor water and sanitation and polluted indoor air environments. The information contained in this book fits with global efforts to build capacity to solve these global problems of poverty, shelter, justice, gender inequalities, health, water supply, sewerage, solid waste, and indoor air. This book integrates design and construction of appropriate technology with critically important issues of health, project management, and community participation. We also included detailed chapters on the use of common engineering materials and, importantly, on proper construction techniques that will be encountered in the field.

Our field guide shows an engineer how to apply professional skills they have learned to serve humanity. The book is concise and, thanks to Linda Phillips's multiple talents as an engineer and artist, it is beautifully illustrated. One goal we have is that this book allow students, engineering professionals, development workers, government officials, and communities to evaluate, design, size, plan, construct, operate, and maintain technology that is culturally, economically, and geographically appropriate. We are in debt to many others around the world who have developed similar materials, especially those focused in great detail on a particular topic. Accordingly, at the end of each chapter we provide readers with more focused readings so they can expand their understanding of a particular topic.

The book is also designed such that non-engineers can use the book's information to solve problems in a sustainable manner.

I am especially grateful for the long and mutually beneficial partnership I have had with the Peace Corps through the extensive committed support provided by the Master's International program. This partnership allows students at over 60 universities to incorporate 11 weeks of training and 2 years of service in the Peace Corps with a graduate degree program that is in a scarce skill area (see http://www.peacecorps.gov/masters). The Master's International program fits with the vision of President John F. Kennedy, who in 1961 established the Peace Corps to promote world peace and friendship. The Peace Corps mission has three simple goals, all which fit into our vision of engineering education and service:

1. Helping the people of interested countries in meeting their need for trained men and women.
2. Helping promote a better understanding of Americans on the part of the peoples served.
3. Helping promote a better understanding of other peoples on the part of Americans.

I am also very thankful to the financial and professional support provided to me by the University of South Florida through the State of Florida 21st Century World Class Scholar Program and the Dr. Kiran C. Patel Center for Global Solutions. The Patel Center's mission is to promote and support nonpartisan, independent applied research that leads to the discovery, dissemination, and application of new knowledge about the sources of and solutions to problems of global concern. In addition to research programs, the Center supports educational programs for training current and future leaders and educators and for expanding public awareness about global concerns (see http://www.patelcenter.usf.edu).

A large amount of the knowledge presented in this book was generated by engineering students who are graduates of a Master's International program in civil and environmental engineering that I founded and of the International Capstone Design course that Linda Phillips and Dennis Magolan founded. Their contribution and the change they have effected on me and other engineering faculty through their Peace Corps service, undergraduate and graduate education and research, overseas internships, and professional service has been invaluable.

I thank Betsy Kulamer and the staff at ASCE Press, who embraced this book's concept and vision at an early stage, and Tony Marjoram, who is head of engineering sciences at UNESCO, for his support of this project. Special thanks go to the dedicated staff of the Peace Corps and the hundreds of communities we have partnered with, for their support, efforts, and passion for effective and just change.

I especially thank again the many students who have taken the leap of faith with me. Thanks also to Karen. You have all enriched my life. Some have called what we do "doing the right thing." I would emphasize, it is about doing the right thing, the right way.

James R. Mihelcic
Tampa, Florida

Contributors

Lead Authors

James R. Mihelcic, Ph.D., BCEEM
Civil & Environmental Engineering
Patel Center for Global Solutions
University of South Florida

Lauren M. Fry
Civil & Environmental Engineering
Michigan Technological University

Elizabeth A. Myre
International Water and Sanitation Program Manager

Linda D. Phillips, P.E., P.M.P, C.D.T.
Civil & Environmental Engineering
Patel Center for Global Solutions
University of South Florida

Brian D. Barkdoll, Ph.D., P.E., D.WRE, F.ASCE
Civil & Environmental Engineering
Michigan Technological University

Illustrations by:
Linda D. Phillips, P.E., P.M.P, C.D.T.
Civil & Environmental Engineering
Patel Center for Global Solutions
University of South Florida

This book is based on the collective knowledge, contributions, and authorship of many individuals, most of whom are returned Peace Corps volunteers (RPCV) who were part of the Master's International program, some of whom participated in International Capstone Design (ICD), and others who have worked in the developing world for other education, research, and service programs. This book is special because of their vision, their commitment to service, their willingness to share information with others, and especially their important contributions to this effort. These other authors and contributors are:

Dennis J. Magolan, P.E.
Erlande Omisca
Maya A. Trotz, Ph.D.
Qiong Zhang, Ph.D.
Civil & Environmental Engineering
Patel Center for Global Solutions
University of South Florida

Blair Orr,[1] Ph.D. (RPCV, Lesotho, Somalia)
John S. Gierke,[2] Ph.D., P.E.
David W. Watkins Jr.,[3] Ph.D.
Kurt G. Paterson,[3] Ph.D., P.E.
Thomas Van Dam,[3] Ph.D., P.E. (RPCV, Tanzania)
[1]School of Forestry & Environmental Science
[2]Geological Engineering & Sciences
[3]Civil & Environmental Engineering
Michigan Technological University

Except for Marilyn M. Phillips (International Capstone Design on-site coordinator), the following contributors are all environmental and civil engineering students from Michigan Technological University. They are all either returned Peace Corps volunteers and graduates of the Master's International program in civil and environmental engineering, participants in International Capstone Design, and/or engaged in other campus sustainability efforts.

Brooke Tyndell Ahrens (RPCV, Mali)
Jonathan E. Annis (RPCV, Madagascar)
Matthew D. Babcock (RPCV, Panama)
Joshua R. Cowden, Ph.D.
James Dumpert (RPCV, Ghana)
Valerie J. Fuchs (EWB project, Honduras)
Christopher G. Gilbertson (EWB project, Guatemala)
Stephen P. Good (RPCV, Dominican Republic)
Meghan E. Housewright (RPCV, Mali)
Jason N. Huart (RPCV, Honduras)
Daniel M. Hurtado (RPCV, Panama)
Milagros JeanCharles (RPCV, Mali)
Josephine Kaiser (RPCV, Panama)
Paul M. Kennedy (RPCV, Kenya)
Kraig Lothe (RPCV, Nepal, Honduras)
Jennifer R. McConville (RPCV, Mali)
Jessica A. Mehl (RPCV, Panama)
Matthew A. Niskanen (RPCV, Dominican Republic)
Daniel Nover (RPCV, Phillipines)
Emily L. Owens (RPCV, East Timor, Palau)

Marilyn M. Phillips (ICD on-site coordinator, Bolivia, Dominican Republic)
Marc F. Plotkin (ICD, Bolivia)
Jennifer L. Post (RPCV, Uzbekistan, Jamaica)
Nathan Reents (RPCV, Honduras, and NGO, Bolivia, Thailand)
Ryan W. Schweitzer (RPCV, Dominican Republic)
John D. Simpson (RPCV, Honduras)
Kelly L. Stanforth (RPCV, Jamaica)
Ed Stewart (RPCV, Jamaica)
Lyle J. Stone (RPCV, Dominican Republic)
Ryu Suzuki (RPCV, Panama)
Eric Tawney (RPCV, Vanuatu)
Alexis M. Troschinetz
Glenn A. Vorhes (RPCV, Dominican Republic)

Introduction

1

Engineering a Better World

Environmental engineers use the fundamentals of mathematics, physics, chemistry, biology, engineering science, sustainability science and engineering, economics, and social science to protect human health and the environment. The field has its origins in public health. Present-day environmental engineers continue this tradition by addressing issues including waste management, indoor and outdoor air pollution, contaminant transport, water resources and environmental management, municipal and on-site wastewater treatment, and drinking water quality, green engineering, ecological restoration, and green building design.

Although this knowledge and technology have greatly improved public health in developed nations, much of the developing world still experiences public health crises that could be significantly mitigated by engineering. Polluted water affects the health of more than 1 billion people, and UNICEF reports that 1.5 million children die every year from unsafe water and sanitation. Furthermore, some 2.5 billion people lack access to adequate sanitation, including 1 billion who are children. Infants, often carried on their mothers' backs or kept close to the warm hearth, can spend many hours breathing indoor smoke during their first year of life, when their developing airways make them particularly vulnerable. As a result, 56% of all indoor air pollution-attributable deaths occur in children under 5 years of age.

Other global health issues are HIV/AIDS, tuberculosis, and malaria, which are among the world's largest killers. They have their greatest effect on developing nations, interact in ways that make their combined effect worse, and create an enormous economic burden on families and communities, especially where economic livelihood depends on good health (UNESA 2004). Engineering activities can provide significant improvements for people living with such diseases, as well as indirect or direct barriers to transmission.

Figure 1-1 shows the risk factors that contribute the most environmental burden of disease around the world. These risks lead to a great loss in disability-free days of a person's life, especially in the developing world. Almost one half of the risk is associated with poor access to drinking water and sanitation, and much of the other half results from exposure to indoor and urban air pollution. Furthermore, the global effects on the water cycle from human development stressors of changes in land cover, urbanization, and water resource development may surpass effects from recent or anticipated climate change.

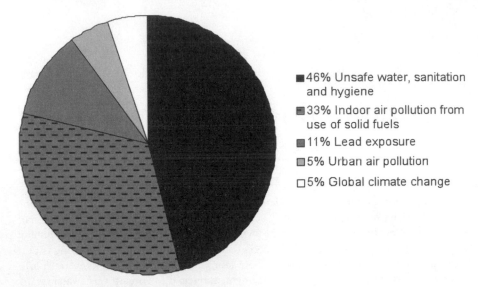

Figure 1-1. Attributable Fraction of Factors Contributing to Environmental Risk, Measured by Disability Adjusted Life Years (DALYs).

Source: Data from Ezzati et al. 2004.

▶▶▶ *"Health is both a resource for, as well as an outcome of, sustainable development. The goals of sustainable development cannot be achieved when there is a high prevalence of debilitating illness and poverty, and the health of a population cannot be maintained without a responsive health system and a healthy environment. Environmental degradation, misman-agement of natural resources, and unhealthy consumption patterns and lifestyles impact health. Ill-health, in turn, hampers poverty alleviation and economic development."*

World Health Organization (2002)

For people living in poverty, illness and disability translate directly into loss of income. The effects of ill health have significant ramifications at the macroeconomic scale as well. For instance, a significant portion of Africa's economic shortfall may be attributed to climate and disease burden (WHO 2001). Environmental degradation also has a direct effect on household income. The income derived from ecosystems (i.e., environmental income) is recognized to provide a "fundamental stepping stone in the economic empow-erment of the rural poor" (WRI 2005). This "natural capital" is the environmental stock that yields the flow of natural resources.

Health, the environment, and livelihood are intertwined in ways that are impossible to separate. On a practical level, maintaining a healthy ecosystem is crucial to sustainable development. This fact is true at all levels of income, but the feedback loop is generally shorter for communities living closer to the survival level. The effects of soil degradation, loss of biodiversity, water source pollution, or deforestation are often felt first by developing countries and low-income communities within high-income countries. As population increases, so does the urgency of maintaining the environments on which people depend for social and economic prosperity.

Worldwide, numerous discussions have yielded considerable contributions to the concept of sustainability. For example, the United Nations (U.N.) Conference on the Human Environment in Stockholm in 1972 was significant because for the first time, it added the environment to the list of global problems (Box 1-1). This conference also resulted in the creation of the U.N. Environment Programme (UNEP). Principle 2 of the 1972 Stockholm Conference on the Human Environment states, "The natural resources of the earth, including air, water, land, flora, and fauna and especially representative samples of natural ecosystems, must be safeguarded for the benefit of present and future generations through careful planning and management, as appropriate."

The United Nations Environment Programme (UNEP 2002) lists 10 existing or emerging environmental issues:

1. globalization, trade, and development;
2. coping with climate change and variability;
3. the growth of megacities;
4. human vulnerability to climate change;
5. freshwater depletion and degradation;
6. marine and coastal degradation;
7. population growth;
8. rising consumption in developing countries;
9. biodiversity depletion; and
10. biosecurity.

Clearly, engineering has an important role to play in addressing all of these topics.

Two items in this list that show the integration of these global issues are the growth of megacities and marine and coastal degradation. Today, 21 of the world's 33 megacities are located in coastal areas, and most of the 3 billion people added over the next century will be in urban areas (Fig. 1-2). Yet large segments of the world's population depend on fishing for their livelihood, and fisheries are critically dependent on coastal water quality.

⟫⟩ Box 1-1. Mission of the United Nations Environment Programme

The stated mission of the UNEP is "to provide leadership and encourage partnership in caring for the environment by inspiring, informing, and enabling nations and peoples to improve their quality of life without compromising that of future generations." The UNEP has many initiatives related to sustainability; these initiatives include sustainable industrial development, understanding the forces that drive consumption patterns, and environmental consequences of energy production.

Source: www.unep.org.

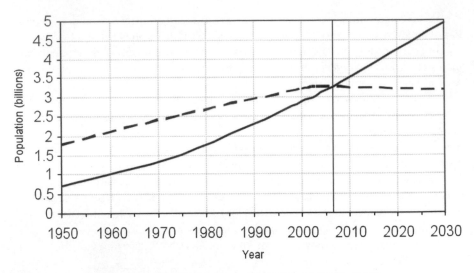

Figure 1-2. Urban and Rural Populations of the World: 1950–2030.

Note: The world's rural population in 2003 was 3.26 billion and is predicted to decrease to 3.19 billion by 2030. In the year 2007, for the first time in human history, the human urban population exceeded the rural population, and by 2030, 60.8% of the global population is expected to be living in urban areas. The world's urban population growth rate is predicted to be 1.83% between now and then, nearly double the world's total population growth rate.

Source: UNESA 2004.

Also, when urban infrastructure is not able to develop at the same rate as the population, urbanization can be a source of health problems.

Another important existing and emerging trend is rising consumption. Worldwide, resource consumption per capita is on the rise (Box 1-2). Many current engineering activities are also based on increasing resource consumption on a per capita level. A paradigm shift is required that will allow the quality of life of billions of people to be improved while simultaneously pursuing a path of development that is more equitable and less resource-intensive.

At the 2002 World Summit on Sustainable Development in Johannesburg, world leaders reaffirmed the principles of sustainable development[1] adopted at the Earth Sum-

---

[1]In 1987, *Our Common Future* (World Commission on Environment and Development) was released by the U.N. (also referred to as the Brundtland Commission report). Gro Harlem Brundtland, a former prime minister of Norway, chaired the commission. This influential document not only adopted the concept of sustainable development but also provided the stimulus for the 1992 U.N. Conference on Environment and Development (i.e., Earth Summit). It defined sustainable development as "development which meets the needs of the present without compromising the ability of the future to meets its needs."

Sustainable design has been defined as the design of human and industrial systems to ensure that humankind's use of natural resources and cycles do not lead to diminished quality of life due either to losses in future economic opportunities or to adverse effects on social conditions, human health, and the environment (Mihelcic et al. 2003).

))) Box 1-2. Global State of Consumption

"By virtually any measure—household expenditures, number of consumers, extraction of raw materials—consumption of goods and services has risen steadily in industrial nations for decades, and it is growing rapidly in many developing countries.

By one calculation, there are now more than 1.7 billion members of "the consumer class"—nearly half of them in the developing world. A lifestyle and culture that became common in the twentieth century is going global in the twenty-first.

Worldwide, private consumption expenditures—the amount spent on goods and services at the household level—topped $20 trillion in 2000, a four-fold increase over 1960 (in 1995 dollars).

The 12 percent of the world's population that lives in North America and Western Europe accounts for 60 percent of private consumption spending, while the one-third living in South Asia and sub-Saharan Africa accounts for only 3.2 percent."

Source: Worldwatch Research Institute, http://www.worldwatch.org/node/810.

mit 10 years earlier. One outcome was the development of millennium development goals (MDGs), shown in Table 1-1. The MDGs provide a global vision of development in which health and education are equal pillars of importance. The eight MDGs represent commitments to reduce poverty, hunger, ill health, gender inequality, lack of access to clean water, and environmental degradation. For each goal, one or more targets have been set, most for 2015, using 1990 as a benchmark. At the 56th World Health Assembly, the World Health Organization reports that 3 out of 8 goals, 8 of the 16 targets, and 18 of the 48 indicators relate directly to health.

Engineers who want to address water, sanitation, and air quality on a global scale must understand the interconnectedness of environmental issues listed in UNEP 2002 and also the MDGs (this interconnectedness is shown in Table 1-1).

Despite the great promise of engineering, engineering projects do not always have their intended effects on people's lives. The world is littered with broken-down wells and unused latrines. Large-scale dam projects have displaced hundreds of thousands of people, with devastating environmental and societal effects. These projects may be constructed with the best of intentions, but nontechnical considerations are not always taken into account.

Often, community preferences and habits are ignored, and communities have only a token role in decision-making processes. In other situations, initial funds for construction are provided by outside agencies, but no realistic plans for paying for operation and maintenance are put into place. The technology chosen might be beyond the capabilities of the communities to repair when needed, with replacement parts not easily obtainable. Issues of governance and financing are thus as important as technology because "the widespread diffusion of sustainable technologies greatly depends on suitable financial vehicles as well as robust governance institutions that are able to manage and maintain technologies and infrastructures over the long term, continuously adapting human systems to their natural environment" (Mihelcic et al. 2007).

Table 1-1. The Millennium Development Goals (MDGs), with Background, Targets, and Relationship to Environmental Engineering

Millennium Development Goal (Background)	Target(s)	Example Relationship(s) with Environmental Engineering Activities
1. *Eradicate extreme poverty and hunger.* (More than 1 billion people still live on less than US$1 a day. Sub-Saharan Africa, Latin America and the Caribbean, and parts of Europe and Central Asia are falling short of the poverty target.)	*Target for 2015:* Halve the proportion of people living on less than a dollar a day and those who suffer from hunger.	Malnutrition is exacerbated by diarrhea, which is mostly attributed to lack of water, sanitation, and hygiene. Additionally, food production requires 3,000 L of water per day per capita, 10% of which is supplied by irrigation.
2. *Achieve universal primary education.* (As many as 113 million children do not attend school, but the target is within reach. India, e.g., should have 95 percent of its children in school by 2005.)	*Target for 2015:* Ensure that all boys and girls complete primary school.	Girls are more likely to attend schools where adequate sanitation is available and the sanitation technology that is provided is separated for boys and girls so that girls have their own facilities.
3. *Promote gender equality and empower women.* (Two-thirds of illiterates are women, and the rate of employment among women is two-thirds that of men. The proportion of seats in parliaments held by women is increasing, reaching about one-third in Argentina, Mozambique, and South Africa.)	*Targets for 2005 and 2015:* Eliminate gender disparities in primary and secondary education, preferably by 2005, and at all levels by 2015.	Women and children are more vulnerable to harassment and assault when they are forced to walk away from their homes for excreta disposal. About 40–45% of the burden of chronic obstructive pulmonary disease experienced by women in the developing world is caused by indoor air pollution from solid fuel use.
4. *Reduce child mortality.* (Every year, almost 11 million young children die before their 5th birthday, mainly from preventable illnesses, but that number is down from 15 million in 1980.)	*Target for 2015:* Reduce by two-thirds the mortality rate among children under 5.	The estimated disease burden from childhood and maternal undernutrition; unsafe water, sanitation, and hygiene; and global climate change is responsible for 85% of disease burden in the under-5 age group, most of which is improved, at least in part, by increasing water and sanitation coverage. For children under the age of 5, acute respiratory infections caused by exposure to indoor air pollution are a major category of deaths and DALYs.

Table 1-1. The Millennium Development Goals (MDGs), with Background, Targets, and Relationship to Environmental Engineering (*Continued*)

Millennium Development Goal (Background)	Target(s)	Example Relationship(s) with Environmental Engineering Activities
5. *Improve maternal health.* (In the developing world, the risk of dying in childbirth is one in 48, but virtually all countries now have safe motherhood programs.)	*Target for 2015:* Reduce by three-quarters the ratio of women dying in childbirth.	Water, sanitation, and hygiene decrease the risk of diarrhea, which would exacerbate maternal undernutrition.
6. *Combat HIV/AIDS, malaria, and other diseases.* (40 million people are living with HIV, including 5 million newly infected in 2001. Countries like Brazil, Senegal, Thailand, and Uganda have shown that the spread of HIV can be stemmed.)	*Target for 2015:* Halt and begin to reverse the spread of HIV/AIDS and the incidence of malaria and other major diseases.	Water supply projects save time for women and children by decreasing the distance they walk to get water. In addition to the time-saving effects that could reduce the risk of exposure to HIV/AIDS by allowing children more time for school and mothers more time for caretaking, environmental engineering technologies can improve the standard of living of people living with HIV/AIDS. Indoor air pollution is implicated in the deaths of more than 2 million people every year, and it is responsible for 2.6% of global ill health.
7. *Ensure environmental sustainability.* (More than 1 billion people lack access to safe drinking water, and more than 2 billion lack sanitation. During the 1990s, however, nearly 1 billion people gained access to safe water and the same number to sanitation.)	Integrate the principles of sustainable development into country policies and programs and reverse the loss of environmental resources. *By 2015:* Reduce by half the proportion of people without sustainable access to safe drinking water and basic sanitation. *By 2020:* Achieve significant improvement in the lives of at least 100 million slum dwellers.	Water and sanitation projects improve the overall sustainability. Additionally, improvements in water and sanitation in urban slums will improve slum-dwellers' lives by reducing disease and improving aesthetics.

(continued)

Table 1-1. The Millennium Development Goals (MDGs), with Background, Targets, and Relationship to Environmental Engineering (*Continued*)

Millennium Development Goal (Background)	Target(s)	Example Relationship(s) with Environmental Engineering Activities
8. *Develop a global partnership for development.*	Develop further an open, rule-based, predictable, nondiscriminatory trading and financial system. Address the special needs of the least developed countries. Address the special needs of landlocked countries and small island developing states. Deal comprehensively with the debt problems of developing countries through national and international measures to make debt sustainable in the long term. In cooperation with developing countries, develop and implement strategies for decent and productive work for youth. In cooperation with pharmaceutical companies, provide access to affordable, essential drugs in developing countries. In cooperation with the private sector, make available the benefits of new technologies, especially information and communication.	Environmental engineering projects provide opportunities for cooperation with the private sector. Innovative technologies, mass production, and social marketing are required to rapidly improve sanitation coverage.

Note: The MDGs are an ambitious agenda embraced by the world community for reducing poverty and improving lives in the global community. Learn more at http://www.un.org/millenniumgoals/.

One factor preventing engineering from reaching its full potential is the traditional approach to engineering education. Typical curricula in the United States include several years of science, math, and technical subjects, but touch only lightly on the humanities, economics, and social sciences. Foreign language requirements, ecology, and public health are often nonexistent. Whereas engineering students can opt to take general education courses in world cultures, history, or geography, a more direct education in international topics would include a significant international experience. However, of the almost

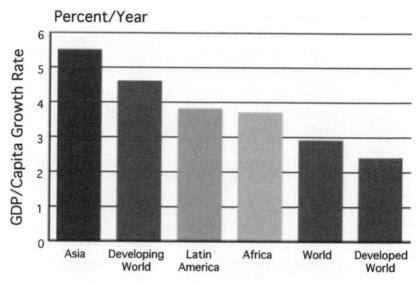

Figure 1-3. Estimated Percent Increase in Economic Growth in Different Parts of the World (as % GDP/Capita per Year).

175,000 U.S. students who studied abroad in 2003, only 3% were engineering students, which is only 1% of the U.S. students enrolled in engineering programs.

Solving the world's problems requires technical as well as nontechnical skills, and an understanding of the dynamic between society, the economy, and the environment (Fuchs and Mihelcic 2006). Engineering practice for sustainable development must also consider nontraditional principles of sufficiency, equity, gender, and efficiency. The ideal operational model for research and education is a global partnership, enhanced by integrating the best and most appropriate knowledge, methodologies, techniques, principles, and practices from both the developed and developing worlds. Unfortunately, conventional engineering education still does not do an adequate job of integrating technological development with development that is compatible with society and the environment (Hokanson et al. 2007).

Having an awareness of social and economic issues will improve engineers' abilities to interact with organizations from other sectors and to assume leadership roles in alliances formed. Engineers of the future will need to be able to partner effectively with major groups, such as women, farmers, the scientific and technological community, children and youth, indigenous peoples and their communities, workers and trade unions, business and industry, nongovernmental organizations, and local authorities.

>>> *"By now we know that peace and prosperity cannot be achieved without partnerships involving governments, international organizations, the business community and civil society."*
Former United Nations Secretary General Kofi Annan

In the chapters that follow, we fill in some of the gaps in the conventional approach to engineering. We present the fundamental scientific and technical information necessary to evaluate, design, and construct a project, as well as the broader sustainability picture required to put the project in its proper context and ensure its long-term success. The relationship of public health to engineering is a major theme of this book. Community approaches to project design and management are discussed, and appropriate technology[2] is emphasized. Though the solutions in this book are community-focused and primarily useful for rural areas, the principles discussed can be expanded to address issues of periurban and urban regions.

Reaching the tangible objectives of the MDGs requires the application of science and problem definition and solution, which form the core definition of engineering. In support of the vision that engineers can assist in solving many of world's problems, the World Federation of Engineering Organizations (WFEO) has stated that "engineers play a crucial role in improving living standards throughout the world. As a result, engineers can have a significant impact on progress towards sustainable development."

In the upcoming decades, engineers will have the opportunity to play a critical role in the eradication of global poverty and hunger; the facilitation of sustainable development, appropriate technology, and beneficial infrastructure; and the promotion of change that is environmentally and socially just. Our hope is that this book will assist engineers, other professionals, and lay people who serve the global community and, like us, envision a better world.

⫸ References

Ezzati, M., Rodgers, A., and Lopez, A. D. (2004). "Mortality and burden of disease attributable to individual risk factors." *Comparative quantification of health risks: Global and regional burden of disease attributable to selected major risk factors*, Vol. 2, M. Ezzati and A. D. Lopez, eds., World Health Organization, Geneva, Switzerland.

Fuchs, V. J., and Mihelcic, J. R. (2006). "Engineering education for international sustainability: Curriculum design under the sustainable futures model." *Proc., 5th annual ASEE global colloquium on engineering education*, ASEE, Washington, D.C.

Hokanson, D. R., Mihelcic, J. R., and Phillips, L. D. (2007). "Educating undergraduate and graduate engineers to achieve a more sustainable future: Education and diversity initiatives with a global perspective." *International J. Engineering Education*, 23(2), 254–265.

Mihelcic, J. R., Crittenden, J. C., Small, M. J., Shonnard, D. R., Hokanson, D. R., Zhang, Q., Chen, H., Sorby, S. A., James, V. U., Sutherland, J. W., and Schnoor, J. L. (2003). "Sustainability science and engineering: Emergence of a new metadiscipline." *Envir. Sci. and Technol.*, 37(23), 5314–5324.

Mihelcic, J. R., Zimmerman, J. B., and Ramaswami, A. (2007). "Integrating developed and developing world knowledge into global discussions and strategies for sustainability. Part 1: Science and technology." *Envir. Sci. and Technol.*, 41(10), 3415–3421.

United Nations Economic and Social Affairs (UNESA). (2004). *World urbanization prospects: The 2003 revision.* UN Sale No. E.04.XIII.6. United Nations, New York.

United Nations Environment Programme (UNEP). (2002). *Global environmental outlook 3*, Earthscan Publications Ltd., London.

---

[2]*Appropriate technology* is defined here as the use of materials and technology that are culturally, economically, and socially suitable to the area in which they are implemented.

World Health Organization (WHO). (2001). *Report of the commission on macroeconomics and health*, World Health Organization, Geneva, Switzerland.

World Health Organization (WHO). (2002). *World summit on sustainable development.* Johannesburg, South Africa, August 26–September 4, World Health Organization, Geneva, Switzerland, <http://www.who.int/wssd/en/> (June 10, 2005).

World Resources Institute (WRI) in collaboration with United Nations Development Programme, United Nations Environment Programme, and World Bank. (2005). *World resources 2005: The wealth of the poor—Managing ecosystems to fight poverty.* World Resources Institute, Washington, D.C.

))) Further Reading

Fry, L. M., Mihelcic, J. R., and Watkins, D. W. (2008). "Water and non-water-related challenges of achieving global sanitation coverage." *Envir. Sci. and Technol.*, 42(4), 4298–4304.

Mihelcic, J. R., and Zimmerman, J. B. (2009). *Environmental engineering: Fundamentals, sustainability, design*, John Wiley & Sons, New York.

United Nations Economic and Social Affairs (UNESA) Population Division, United Nations, New York.

Zimmerman, J. B., Mihelcic, J. R., and Smith, J. A. (2008). "Global stressors on water quality and quantity: Sustainability, technology selection, and governance in a dynamic world." *Envir. Sci. and Technol.*, 42(4), 4247–4254.

2

Project Motivation: Public Health and the Role of Engineers

Health, standard of living, and environmental conditions are not independent of one another. It is the engineer's responsibility to work with communities to reduce the disease burden caused by modifiable environmental factors. Engineers and other extension agents must take time to understand the actual health problems faced by a particular community. This work may involve health surveys, community mapping, community meetings, collaborating with social scientists or health workers, and informal discussions with members of the community, all of which are discussed in Chapter 3. The better understanding of health issues and cultural practices and beliefs gained from these activities can maximize the health benefits of an engineered solution.

To fully understand the potential effect of environmental engineering technologies, a basic understanding of public health concerns is necessary. Box 2-1 lists definitions for common terms used to describe public health.

According to the World Health Organization (2004), at the global level, infectious and parasitic diseases account for 23.5% of the burden of disease. The largest contributors to this category are tuberculosis, HIV and AIDS, diarrheal diseases, childhood cluster diseases (e.g., pertussis, poliomyelitis, diphtheria, measles, and tetanus), and malaria. Infectious and parasitic diseases are followed by neuropsychiatric conditions (12.9%), cardiovascular disease (9.9%), unintentional injuries (8.9%), and respiratory infections (6.5%).

Table 2-1 shows the distribution of burden of disease according to risk factor for two economic regions, according to the World Health Organization's *World Health Report* (2002). Three of the leading risk factors in high-mortality developing countries—underweight (14.9% of all DALYs); unsafe water, sanitation, and hygiene (5.5%); and indoor smoke from solid fuels (3.7%)—can all be directly improved by environmental engineering activities.

Other leading risk factors may also be indirectly affected by improvements in the environment. For example, the risk factor unsafe sex (10.2%) can be influenced by access to water and sanitation. Education level affects a woman's likelihood of becoming a sex worker, and girls are less likely to attend school if girls and boys do not have separate sanitation facilities. Zinc, iron, and vitamin A deficiencies can all be exacerbated by infections resulting from unsafe drinking water and lack of sanitation (WHO 2002).

>>> **Box 2-1. Common Definitions Used to Describe Public Health**

Morbidity: Incidence of illness.

Mortality: "Mortality" and "death" are used interchangeably.

Risk: "Risk" can have various meanings. In this book, it is defined using the following two definitions: (1) Probability of an adverse health outcome or (2) A factor that raises the probability of an adverse health outcome (WHO 2002).

Relative Risk: A ratio of the risk of morbidity or mortality for a population receiving an intervention to the risk for a population not receiving the intervention.

Population Attributable Risk or Fraction: The fraction of a disease or death that is attributable to a particular risk factor in a population.

Disability Adjusted Life Years (DALYs): A measure of burden of disease. One DALY is equal to the loss of one healthy life year due to death or the inability to work because of illness. In 2001, there were almost 1.5 billion DALYs, or roughly 0.24 DALY per person globally (WHO 2002).

Attributable Burden of Disease: The fraction of a disease or injury burden that results from past exposure to a risk (WHO 2002).

Avoidable Burden of Disease: The proportion of future disease or injury burden that is avoidable if exposure levels are reduced to an alternative distribution (WHO 2002).

Table 2-1. Ten Leading Risk Factors Contributing to Burden of Disease in 2000, According to Economic Region

Developing Countries		Developed Countries	
Risk Factor	% DALYs	Risk Factor	% DALYs
Underweight	14.9	Tobacco	9.4
Unsafe sex	10.2	Blood pressure	7.2
Unsafe water, sanitation, and hygiene	5.5	Alcohol	6
Indoor smoke from solid fuels	3.7	Cholesterol	3.5
Zinc deficiency	3.2	Overweight	3
Iron deficiency	3.1	Low fruit and vegetable intake	2.8
Vitamin A deficiency	3	Physical inactivity	2.6
Blood pressure	2.5	Illicit drugs	2.5
Tobacco	2	Unsafe sex	2.5
Cholesterol	1.9	Iron deficiency	2.4

Note: The elimination of the leading 5 risk factors would increase life expectancy by 14 years in African nations with high child and very high adult mortality, and 11 years in African nations with high child and high adult mortality.

Source: WHO 2002.

A study by Prüss-Üstün and Corvalán (2006) used the World Health Organization's 2002 comparative risk assessment (Ezzati et al. 2004), along with expert opinion, to estimate that 24% of the global burden of disease (in terms of DALYs) and 23% of all deaths can be attributed to modifiable environmental factors. Table 2-2 lists the diseases with the largest absolute burden attributable to modifiable environmental factors. Also included in this table are the environmental causes of the diseases and examples of engineering activities that can modify the environment to decrease the burden of disease.

Causes of death and disease are different in developed versus developing regions. Lopez et al. (2006) showed that the percentage of deaths attributable to communicable diseases is five times greater in low- and middle-income countries than in high-income countries. Figure 2-1 shows the 10 leading causes of burden of disease by broad income group.

Sustainable development requires that engineers implementing solutions to current problems consider how risks in the community will change over time (e.g., with increasing income). Many noncommunicable diseases that are dominant in developed countries are, at least in part, attributable to environmental factors. The risk factors overweight and physical inactivity (Table 2-1) are examples of risks that can be improved by sustainable development (e.g., incorporating pedestrian-friendly routes into community development plans).

In the comparative risk assessment in Lopez et al. (2006), environmental risks were divided into five groups: unsafe water, sanitation, and hygiene; indoor air pollution from use of solid fuels; lead exposure; urban air pollution; and global climate change (Fig. 1-1 from Chapter 1). The engineering solutions discussed in this book directly reduce exposure to the risk factors unsafe water, sanitation, and hygiene; indoor air pollution from use of solid fuels; and urban air pollution. As a result, the interventions are appropriate for public health goals to reduce water-related diseases, respiratory diseases, and diseases that result from burning of open dumps because of poor solid waste management.

Additional indirect benefits may also result from these interventions. For example, a water project that reduces the time required for drawing water may allow mothers more time for food preparation and breastfeeding, which might lead to improved child nutrition. More time may also allow women to pursue income-generating activities to improve household economics (Esrey 1996). Environmental conditions can also have a significant indirect effect on other important disease groups, such as malnutrition, HIV, and AIDS, as well as on other risk factors, such as climate change.

))) 2.1 Water-Related Diseases

The risk factor water, sanitation, and hygiene is responsible for 5.5% of all DALYs in low- and middle-income countries (WHO 2002). In fact, 88% of all diarrhea is caused by unsafe water, sanitation, and hygiene (Prüss-Üstün and Corvalán 2006).

Diseases that can be directly affected by water, sanitation, and hygiene improvements fall under two broad disease transmission categories: water-related and fecal–oral transmission. Water-related diseases fall into four categories, according to their transmission routes: (1) water-borne, (2) water-washed, (3) water-based, and (4) insect–vector (Table 2-3). Fecal–oral diseases are considered water-related, but they are diseases that fall under

Table 2-2. Diseases with the Largest Absolute Burden of Disease Attributable to Modifiable Environmental Factors and Potential Environmental Engineering Activities to Lower the Environmental Burden

Disease or Disease Group	Global Burden of Disease (% of all DALYs)[a]	Percent of Disease Burden Attributable to Modifiable Environmental Factors[b]		Environmental Factors Contributing to Disease Burden[b]	Engineering Activities to Lower the Burden of Disease
Diarrhea	4.3%	World	94%	Water, sanitation, and hygiene (88%) of all cases	Improved water and sanitation; mitigation of climate change
		Developing countries	94%		
		Developed countries	90%		
Lower respiratory infections	6.3%	World	41%	Indoor air pollution from solid fuel use; outdoor air pollution; tobacco smoke; housing conditions; hygiene	Improved cook stoves; air pollution controls; improved housing conditions
		Developing countries	42%		
		Developed countries	20%		
Other unintentional injuries	3.2%	World	44%	Contact with heavy machinery or sports equipment; off-road transportation accidents; animal bites and venomous plants; exposure to ionizing radiation or electric currents; suffocation; natural forces; contact with hot substances; complications from medical and surgical care	Improved safety during construction; improved transportation; mitigation of climate change; flood control; improved land use patterns; improved building materials
		Developing countries	45%		
		Developed countries	30%		
Malaria	2.3%	World	42%	Stagnant or slowly moving freshwater	Improved drainage; modified house design, including gutters and roof drains; wastewater management; improved irrigation; vegetation management; improved domestic water storage; solid waste management; maintenance of urban water supply and sanitation systems; mitigation of climate change
		Americas	64%		
		Eastern Mediterranean	36%		
		Europe	50%		
		Southeast Asia	42%		
		Sub-Saharan Africa	42%		
		Western Pacific	40%		

Note: Data are representative of conditions in 2002.

[a]WHO 2004.

[b]Prüss-Üstür and Corvalán 2006.

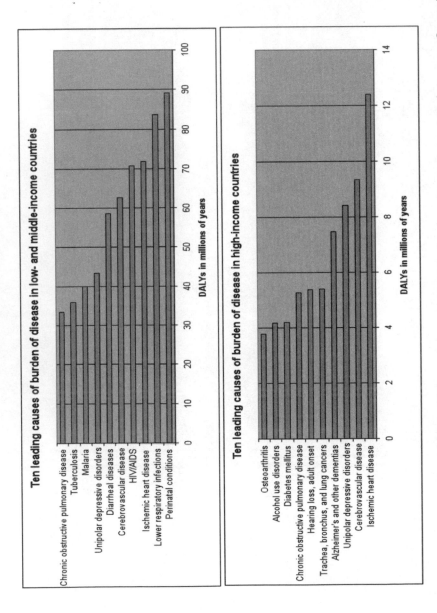

Figure 2-1. The 10 Leading Causes of Disease Burden in Low- and Middle-Income Countries and High-Income Countries.

Note: Low- and middle-income countries represent about 85% of the global population.

Source: Lopez et al. 2006.

Table 2-3. Water-Related Diseases, According to Transmission Route

Transmission Route	Definition	Preventive Strategies	Examples of Diseases
Water-borne	Transmitted when a person or animal ingests water containing the pathogen	Improve drinking water quality; prevent casual use of unprotected sources of water	Cholera; typhoid; infectious hepatitis; diarrhea; dysentery
Water-washed	Transmitted because of inadequate amounts of water used for personal hygiene	Increase water quantity; improve accessibility; improve reliability; improve hygiene	Cholera; bacillary dysentery; bacterial skin sepsis; scabies; fungal skin infections; louse-borne epidemic typhus; louse-borne relapsing fever; trachoma
Water-based	Transmitted by a pathogen that spends part of its life cycle in a water snail or other aquatic animal. All pathogens causing these diseases are parasitic worms (helminthes).	Reduce the need for contact with infected water; control snail populations; reduce contamination of surface waters	Schistosomiasis; Guinea worm
Insect–vector	Transmitted by insects breeding or biting in or near water	Improve surface water management; destroy breeding sites; reduce the need to visit breeding sites; use mosquito netting	Malaria; yellow fever; dengue; onchocerciasis

Note: The transmission routes described for diseases may not be the only means of transmission (see Fig. 2-2 for an example of the complicated nature of disease transmission).

Source: Cairncross and Feachem 1993.

both water-borne and water-washed, such as cholera and bacillary dysentery. In other words, fecal–oral diseases are transmitted by both direct ingestion of contaminated water and inadequate hygiene (Cairncross and Feachem 1993).

Fecal–oral transmission routes are often described by the *F-diagram*, so called because transmission is related to fingers, flies, fluids (i.e., water), food, and fields (Fig. 2-2). The F-diagram shows how diseases are transmitted and how engineering activities and personal hygiene act as barriers to the routes of transmission. Without sanitation and hygiene, human waste can directly enter surface water and wash onto fields, flies can land on uncovered feces, and fecal pathogens can be transmitted directly by human contact. In some cultures, fields may be considered the appropriate place for defecating. Humans can then ingest fecal pathogens through direct ingestion of contaminated water, eating

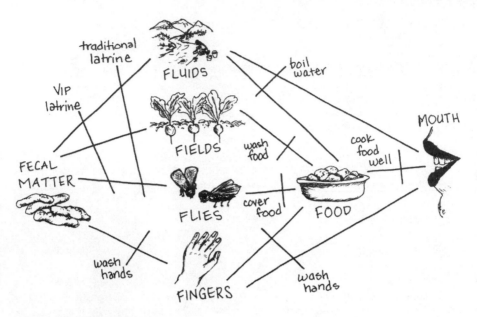

Figure 2-2. The F-Diagram.

Note: Fecal–oral transmission results from contamination of fluids (water), fields, flies, food, and fingers. A combination of engineering activities and hygiene education, such as the interventions depicted on this diagram, can effectively act as barriers to all routes of transmission. There are a greater number of hygiene interventions than engineered barriers (e.g., latrines) to ensure 100% prevention.

Source: Adapted from Wagner and Lanoix 1958.

contaminated foods, through contact with other infected people, or when people, especially children, put fingers in their mouths. Contamination of food results from washing or preparing it with fecally infected water; harvesting food from fields irrigated with contaminated water, polluted by runoff, or used for defecation; or through preparation of food with unwashed hands.

A combination of engineering activities and personal hygiene can act as barriers to every route of fecal–oral transmission. For example, traditional latrines prevent feces from contaminating surface water and fields. If properly sited, they also prevent feces from contaminating groundwater. Ventilated improved pit (VIP) latrines go a step further and prevent flies from entering or leaving the pit. Treating drinking water prevents ingestion of contaminated water as well as contamination of foods during preparation. The F-diagram also shows the great importance of hygiene education because even if engineered barriers are in place to prevent transmission through water, fields, or flies, a major transmission route remains if hand washing does not take place after defecation, before preparing foods, or before and after personal contact with others.

Engineering activities that are strategies to prevent water-related diseases are discussed throughout this book. These activities include water supply and treatment technologies, as well as latrines and other small-scale wastewater treatment solutions. The question of which type of intervention to implement is discussed in Box 2-2.

))) Box 2-2. Water, Sanitation, or Hygiene?

The Historical Perspective

In the 1980s and 1990s, Steven A. Esrey published several studies evaluating the reduction in diarrheal morbidity following water, sanitation, and hygiene interventions (Esrey et al. 1985, 1991; Esrey 1996). His work showed that improvements in water quality alone resulted in a smaller reduction in diarrhea than sanitation or hygiene improvement. For example, the expected reductions in diarrheal disease morbidity from a multicountry study in 1991 are listed in the following table (table courtesy of Thomas Clasen, London School of Hygiene and Tropical Medicine).

	All Studies		Rigorous Studies	
Route	No. of Studies	Reduction	No. of Studies	Reduction
Water and sanitation	7	20%	2	30%
Sanitation	11	22%	5	36%
Water Quality and Quantity	22	16%	2	17%
Water Quality	7	17%	4	15%
Water Quantity	7	27%	5	20%
Hygiene	6	33%	6	33%

The meta-analyses conducted by Esrey (Esrey et al. 1985, 1991; Esrey 1996) provided a clear message: sanitation and hygiene interventions are more important than water quality interventions, and water quality improvements should be accompanied by sanitation and hygiene improvements. Further research, however, suggests that the implications of Esrey's results may not be as simple as these conclusions imply.

The Question Reexamined

More recently, there has been increased focus on point-of-use treatment of drinking water (treating drinking water at the home, rather than at the source), which was not included in Esrey's studies. A recent review showed reductions in diarrhea morbidity between 35% and 46% following point-of-use water quality interventions (Clasen and Cairncross 2004). Unfortunately, these studies did not include other water, sanitation, or hygiene interventions to compare their effect or their combined effects. It should also be noted that the point-of-use studies mentioned above did not consider effects on water-washed diseases, such as trachoma. Point-of-use systems have only been evaluated for their effectiveness in reducing diarrhea, which is usually water-borne.

In 2005, Lorna Fewtrell and others conducted a meta-analysis that included point-of-use treatment, improved water supply, improved hygiene, improved sanitation, and multiple interventions. Although all interventions resulted in significant reductions in morbidity from diarrhea, the surprising result of their study was that multiple interventions did not have a greater effect than a single intervention implemented alone, but they suggest that this result may be a result of "piecemeal implementation of more ambitious intervention programs." Fewtrell et al. (2005) advise that a better method might involve phased implementation of multiple projects, giving each component sufficient attention.

⟩⟩⟩ 2.2 Diseases Related to Air Pollution

Among the leading 10 disease groups contributing to the global burden of disease are lower respiratory infections, ischemic heart disease, stroke, and chronic obstructive pulmonary disease. For all of these diseases, urban air pollution and indoor air pollution from use of solid fuels have an appreciable population attributable fraction. Table 2-4 shows these fractions, which demonstrate how disease burden changes with development.

In developed countries, indoor air pollution from use of solid fuels is not an important risk factor, and lower respiratory infections do not appear as one of the 10 leading disease groups. However, trachea, bronchus, and lung cancers are among the top 10 disease groups for developed countries, and 3% of these diseases are attributable to urban air pollution (Lopez et al. 2006). Urban air pollution is a major environmental justice concern in developed countries and thus should be a consideration for sustainable development.

Engineers can play a role in preventing diseases resulting from indoor smoke from solid fuels and urban air pollution. For example, engineers can improve urban air quality in developing countries through solid waste management, which reduces the need

Table 2-4. Percent Attributable Fractions for the Risk Factors "Indoor Air Pollution from Use of Solid Fuels" and "Urban Air Pollution" for Four of the 10 Leading Diseases

Percent Attributable Fraction	Lower Respiratory Infections	Ischemic Heart Disease	Stroke	Chronic Obstructive Pulmonary Disease
Global				
Indoor air pollution from use of solid fuels	35%			30%
Urban air pollution	1%	2%	3%	2%
Low- and middle- Income countries				
Indoor air pollution from use of solid fuels	36%			35%
Urban air pollution	4%	2%	4%	2%
Developed countries				
Indoor air pollution from use of solid fuels	NA			
Urban air pollution	NA	1%	1%	1%

Notes: In developed countries, lower respiratory infections are not among the 10 leading diseases. Not listed on this table, however, is "trachea, bronchus, and lung cancers," which is among the top 10 disease groups in developed countries. Urban air pollution is accountable for 3% of these diseases in developed countries, which suggests that as countries develop, they experience a shift in the diseases that are important. This notion should be a consideration for sustainable development. NA is not applicable.

Source: Lopez et al. 2006.

for burning of open dumps. Chapter 24 discusses solid waste management technologies and practices.

In Chapter 25, improved cook stoves are presented as an appropriate technology for regions where households' primary source of cooking fuel has been firewood. By improving the efficiency of burning and including ventilation of smoke away from the kitchen, women and children can greatly reduce their exposure to indoor smoke from solid fuels. An additional advantage of reducing the amount of required fuel wood is reduced deforestation, an important public health topic that relates to climate change risks, malnutrition, water resource and land management, and traditional lifestyles.

⟩⟩⟩ 2.3 Diseases and Risks Affected Indirectly by Environmental Engineering Activities

Although they are not the main focus of this book, health outcomes that are indirect results of environmental engineering activities do provide further motivation for environmental engineers to be engaged in public health development. Examples of diseases and risks that can be indirectly affected by the environmental engineering activities presented in this book include malnutrition, HIV and AIDS, and climate change risks.

2.3.1 Malnutrition

Despite the fact that food production increased by 25% during the second half of the 20th century (UN-Water/WWAP 2006), nutritional deficiencies still accounted for 2.3% of all DALYs worldwide in 2002 (WHO 2004). The risk factor underweight accounts for 14.9% of all DALYs in high-mortality developing countries and 3.1% of DALYs in low-mortality developing countries. The fact that 3% of all DALYs in developed nations is attributable to overweight is yet another striking contrast showing how disease burden and risks shift as a result of development (Table 2-1). Malnutrition can be improved in several ways, including improved food production and improved nutrient uptake through the reduction of infectious diseases.

On average, worldwide food production requires 3,000 L of water per day per capita, 10% of which is supplied by irrigation (UN-WATER/WWAP 2006). Engineers can increase food production by improving water resource management and irrigation technology. Related engineering activities are watershed management and water supply systems.

Malnutrition is exacerbated by diarrhea, which reduces dietary intake; increases fecal and nutritional loss; reduces intestinal transit time, resulting in reduced absorption of nutrients; and increases protein catabolism caused by acceleration in the basal metabolic rate. This fact is especially true for children under the age of two, as they are being weaned from breast milk to solid foods (Braghetta 2006). Parasitic infections such as hookworm, giardia, and schistosomiasis also influence malnutrition by causing diminished appetite, nutrient loss, and decreased nutrient absorption (Stephenson et al. 2000). By improving water supply, sanitation, and hygiene, the resulting reduction in diarrhea will improve child nutritional status.

2.3.2 HIV and AIDS

HIV and AIDS is the fourth leading cause of DALYs in low- and middle-income countries (Lopez et al. 2006). According to the Joint United Nations Programme on HIV/AIDS (United Nations Programme on HIV/AIDS 2006a), sub-Saharan Africa, Asia, and Latin America accounted for 34.4 million of the 38.6 million people worldwide who were living with HIV in 2005. In the *2006 Report on the Global AIDS Epidemic*, United Nations Programme on HIV/AIDS (2006b) states that "mitigation efforts need to address the root causes of child labor, including poverty, illiteracy and food shortages."

Engineering activities can have an indirect effect on HIV and AIDS by improving education, family economics, and food consumption. Water supply projects and improved cook stoves not only reduce the burden of diarrheal disease and respiratory infections, respectively, but also save time by decreasing the distance to collect water and reducing the need for firewood collection for solid fuel burning (United Nations Programme on HIV/AIDS 2006b). In addition, such technologies can improve the standard of living of people living with HIV and AIDS.

It is widely understood that people infected with HIV and AIDS are at greater risk for malaria and that bouts of malaria are more severe for people living with HIV and AIDS. Recent studies have also shown that malaria actually increases the risk of transmission of the AIDS virus. A study of HIV-infected women in Uganda found that mother-to-child transmission of HIV was significantly associated with both the AIDS viral load and placental malaria infection (Brahmbhatt et al. 2003). Additional research in Kenya has shown that a person infected with both HIV and the malaria parasite has an increased viral load during malaria episodes, which results in increased likelihood of HIV transmission (Abu-Raddad et al. 2006). Therefore, engineering efforts that reduce the transmission of malaria, such as surface water management and projects that destroy mosquito breeding grounds, may also have an effect on transmission of the AIDS virus.

2.3.3 Climate Change Risks

The risk factor global climate change is defined by McMichael et al. (2004) as "current and future changes in global climate attributable to increasing atmospheric concentration of greenhouse gases." Climate change is expected to bring some health benefits, such as lower cold-related mortality and improved crop yields in temperate regions. However, many adverse health outcomes are also expected as a result of climate change, and these outcomes will outweigh the benefits (McMichael et al. 2004). Even today, climate change is considered to be an important risk: climate change risks are accountable for 5% of all environmental risk factors (refer to Fig. 1-1 in Chapter 1).

Climate change affects many health outcomes, including malnutrition, diarrhea, malaria, injuries and deaths due to floods, and cardiovascular diseases. For example, in seasonally dry and tropical regions at lower latitudes, food production is expected to decrease with even small increases in local temperature (IPCC 2007). The analysis by McMichael et al. (2004) suggests that climate change affects the burden of disease from malnutrition, diarrhea, and vector-borne diseases significantly more than it affects the burden from flooding or deaths due to thermal extremes. These negative effects will likely be the most pronounced among poorer populations at low latitudes, where climate-sensitive health outcomes are already common and for whom vulnerability is greatest.

Studies have been conducted to predict how climate change will affect water supply and use (e.g., Alcamo et al. 1997; Arnell 1999; Meigh et al. 1999). Although in many areas precipitation and runoff will increase as a result of climate change, other regions will experience a decrease. For example, Alcamo et al. (1997) predicted that under the best estimate, 75% of the earth's land area will experience increases in runoff from 1995 levels, but 25% will experience decreases. This study indicated that decreases in runoff of more than 50 mm/year were possible outcomes for northern Brazil, Chile, Taiwan, and the Indian west coast. Countries such as Cyprus, Israel, Jordan, and Morocco, which are already experiencing water scarcity, will see an even further intensification of water scarcity.

Whereas it increases water availability, increased precipitation in the other regions will also create public health challenges. The expected increase in temperature, precipitation, and runoff in most regions is expected to increase the geographical range of vector-borne diseases, the frequency of contamination of drinking water, and the frequency of weather disasters (McMichael et al. 2004).

Engineers are already working to mitigate emissions of greenhouse gases, which in the long run will decrease the health risks related to global climate change. Although limiting greenhouse gas emissions is the most important step in reducing health risks resulting from climate change, such actions must be accompanied by development of adaptation strategies to cope with the damage that has already been done and the irreversible warming trends that have been initiated.

The African Development Bank and other development organizations have come to the agreement that integrating adaptation responses into development planning is an important way to address climate change effects on the poor. Projects to improve water, sanitation, and air quality should, therefore, incorporate adaptation planning and assessment into project design. For example, is a sanitary sewer an appropriate technology in a city that will move into water scarcity by 2025? Sewers require up to 75 L/capita-day, whereas sanitation technologies also exist that require no water to convey waste (Gleick 1996). If a sewer project is deemed appropriate, what should the community do to prepare for the effects of climate change?

Traditional knowledge and coping strategies already exist in many places to deal with the effect of changing climates, and interventions should take these strategies into account (UNDP et al. 2003). Engineers and communities interested in developing climate change adaptation strategies can learn more from the UNDP's *Adaptation Policy Frameworks for Climate Change* (Lim and Spanger-Siegfried 2004).

))) References

Abu-Raddad, L. J., Patnaik, P., and Kublin, J. G. (2006). "Dual infection with HIV and malaria fuels the spread of both diseases in sub-Saharan Africa." *Science,* 314(5805), 1603–1606.

Alcamo, J., Doll, P., et al. (1997). *Global change and global scenarios of water use and availability: An application of WaterGAP1.0.* Center for Environmental Systems Research (CESR), University of Kassel, Germany.

Arnell, N. W. (1999). "Climate change and global water resources." *Global Environmental Change,* 9, S31–S49.

Braghetta, A. (2006). "Drawing the connection between malnutrition and lack of safe drinking water in Guatemala." *Journal AWWA,* 98(5), 97–106.

Brahmbhatt, H., Kigozi, G., et al. (2003). "The effects of placental malaria on mother-to-child HIV transmission in Rakai, Uganda." *AIDS,* 17, 2539–2541.

Cairncross, S., and Feachem, R. (1993). *Environmental health engineering in the tropics: An introductory text,* 2nd ed. John Wiley & Sons, New York.

Clasen, T. F., and Cairncross, S. (2004). "Household water management: Refining the dominant paradigm." *Tropical Medicine and International Health,* 9(2), 187–191.

Esrey, S. A. (1996). "Water, waste, and well-being: A multicountry study." *American J. Epidemiology,* 143(6), 608–622.

Esrey, S. A., Feachem, R. G., and Hughes, J. M. (1985). "Interventions for the control of diarrhoeal diseases among young children: Improving water supplies and excreta disposal facilities." *Bulletin of the World Health Organization,* 63(4), 757–772.

Esrey, S. A., Potash, J. B., Roberts, L., and Shiff, C. (1991). "Effects of improved water supply and sanitation on ascariasis, diarrhoea, dracunculiasis, hookworm infection, schistosomiasis, and trachoma." *Bulletin of the World Health Organization,* 69(5), 609–621.

Ezzati, M., Lopez, A. D., Rodgers, A., and Murray, C. J. L., eds. (2004). *Comparative quantification of health risks: Global and regional burden of disease attributable to selected major risk factors,* Vol. 2. World Health Organization, Geneva, Switzerland.

Fewtrell, L., Kaufmann, R., Kay, D., et al. (2005). "Water, sanitation, and hygiene interventions to reduce diarrhoea in less developed countries: A systematic review and meta-analysis." *Lancet Infect. Dis.,* 5, 42–52.

Gleick, P. H. (1996). "Basic water requirements for human activities: Meeting basic needs." *Water International,* 21, 83–92.

Intergovernmental Panel on Climate Change (IPCC). (2007). "Summary for policymakers." *Climate change 2007: Impacts, adaptation and vulnerability. Contribution of Working Group II to the Fourth Assessment Report of the Intergovernmental Panel on Climate Change,* M. L. Parry, O.F. Canziani, J. P. Palutikof, P. J. van der Linden, and C. E. Hanson, eds., Cambridge University Press, Cambridge, U.K. <http://www.ipcc.ch/ipccreports/ar4-wg2.htm> (Jan. 11, 2008).

Lim, B., and Spanger-Siegfried, E., eds. (2004). *Adaptation policy frameworks for climate change: Developing strategies, policies and measures.* United Nations Development Programme and Cambridge University Press, Cambridge, U.K.

Lopez, A. D., Mathers, C. D., Ezzati, M., Jamison, D. T., and Murray, C. J. L. (2006). *Global burden of disease and risk factors.* The International Bank for Reconstruction and Development/The World Bank and Oxford University Press, New York.

McMichael, A. J., Campbell-Lendrum, D., Kovats, S., et al. (2004). "Chapter 20: Global climate change." *Comparative quantification of health risks: Global and regional burden of disease attributable to selected major risk factors,* Vol. 2, M. Ezzati, A. D. Lopez, et al., eds., World Health Organization, Geneva, Switzerland.

Meigh, J. R., McKenzie, A. A., and Sene, K. J. (1999). "A grid-based approach to water scarcity estimates for eastern and southern Africa." *Water Resour. Mgmt.,* 13, 85–115.

Prüss-Üstün, A., and Corvalán, C. (2006). *Preventing disease through healthy environments: Towards an estimate of the environmental burden of disease.* WHO Press, Geneva, Switzerland.

Stephenson, L. S., Latham, M. C., and Ottesen, E. A. (2000). "Malnutrition and parasitic helminth infections." *Parasitology,* 121, S23–S38.

United Nations Development Programme (UNDP) et al. (2003). *Poverty and climate change: Reducing the vulnerability of the poor through adaptation.* Published by UNDP and eight other international groups. UNDP, New York.

United Nations Programme on HIV/AIDS. (2006a). *Global facts and figures.* Joint United Nations Programme on HIV/AIDS. <http://www.who.int/hiv/mediacentre/news60/en/index.html> (Jan. 11, 2008).

United Nations Programme on HIV/AIDS. (2006b). *2006 Report on the global AIDS epidemic.* Joint United Nations Programme on HIV/AIDS. <http://www.who.int/hiv/mediacentre/news60/en/index.html> (Jan. 11, 2008).

UN-WATER/WWAP. (2006). *Executive summary of water: A shared responsibility. The United Nations world water development report 2.* UN-WATER World Water Assessment Programme, <http://unesdoc.unesco.org/images/0014/001444/144409E.pdf> (Jan. 11, 2008).

Wagner, E. G., and Lanoix, J. N. (1958). *Excreta disposal for rural areas and small communities.* WHO Monograph Series No. 39. World Health Organization, Geneva, Switzerland.

World Health Organization (WHO). (2002). *World health report 2002: Reducing risks, promoting healthy life.* WHO Press, Geneva, Switzerland.

World Health Organization (WHO). (2004). *Death and DALY estimates for 2002 by cause for WHO member states.* <http://www.who.int/healthinfo/bodestimates/en/index.html> (Jan. 11, 2008).

Community Approaches to Project Design and Management

3

Participatory Approaches and Community Management in Engineering Projects

Communities will eventually be left to manage, operate, and maintain projects on their own. Establishing the proper ownership, skills, and management capacity is as important as any aspect of the physical construction of a project. Of course, each community will have its own idiosyncrasies. Some communities will be ready for a project and familiar with effective management. Others will be disorganized, some to the point where organization seems impossible. These latter communities may benefit most from a well-run project. When effective participatory design and management is used, a project may be the catalyst for broader community changes that allow further development.

Manikutty (1998) studied the implementation of water and sanitation projects in rural communities in various settings in India. He found that the ability to maintain community participation throughout the project relied heavily on a strong commitment to it from the outset. Engineers may want to rush a project to get things underway and have something to show on the ground. The unfortunate result is that technically well-designed engineering projects fail because local citizens are not key members in the development plan to design, build, and manage the projects. Early participation in the project cycle is critical. Apportioning responsibility in the implementation stage of a project continues community participation and allows adjustments to the project to be made with the local community as a full partner.

The word *project* thus encompasses more than the physical structure that is designed and constructed. It includes the social setting where the project is located and the people who will operate, manage, and benefit from the project.

⟩⟩⟩ 3.1 The Project Cycle in a Participatory Setting

Development agencies frequently use some form of the project cycle (Fig. 3-1) to plan and implement rural community projects (Baum 1982; Kanshahu 2000). Large agencies with large projects have formal methods to identify, complete, and evaluate projects. Various phases of a project (shown in the outer circle of Fig. 3-1) can have specific activities and required reports before the next project stage of the cycle is initiated. In addition, the process

This chapter was written by Blair Orr and Jonathan E. Annis.

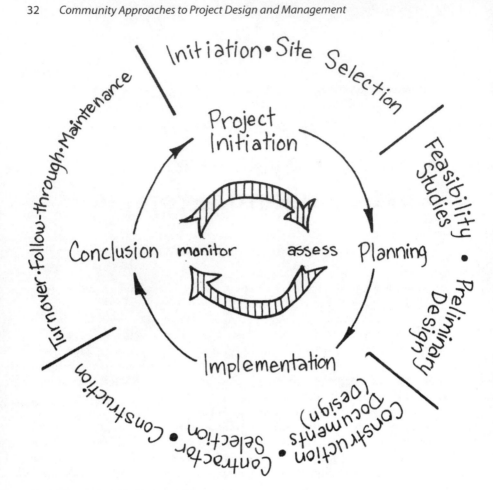

Figure 3-1. The Project Cycle.

Note: Project cycles can involve varying steps, but the overall stages of project initiation and investigation; planning; implementation; conclusion; and monitoring and assessment are found in most projects.

of monitoring and assessment (shown in the center of Fig. 3-1) may be iterated multiple times (Margoluis and Salafsky 1998). After implementation of one project phase, evaluation may indicate that another phase needs to be revisited. Monitoring and assessment are then often built into each phase of a project, allowing for corrections in the entire project cycle.

Thus, before construction (implementation) even begins, there are two stages that help to determine whether a project should be undertaken at all. Given this forethought, one would think that engineering projects undertaken at the community level would rarely fail. Unfortunately, many development projects are not successfully completed or break down after completion and are never repaired.

Annis (2006) found that more than 90% of water projects in Ikongo District, Madagascar, failed to develop or implement adequate financial management schemes to collect money from community members for routine maintenance and spare parts. Physical failure of the water systems was also common. Often, repair work would have been simple but was never attempted. In Santa Barbara and San Pedro Sula, Honduras, approximately 80% of the designated projects were either not completed or failed after completion (Reents 2003). Many water projects studied by Reents and Annis showed that the community was not involved in the early stages of project planning and had not been actively involved in developing the project or the management structure needed to operate and maintain the water supply system.

The importance of local community participation in project development and implementation has been recognized in the field of rural and agricultural development since the late 1950s and 1960s, the beginning of the modern development era. The early theoretical work of Paulo Freire in the late 1960s (translated to English in Freire 2000) and its practical applications included transformation of power through literacy and radical participation. Agricultural technicians developed participatory appraisal methods, participatory research models, and farmer-to-farmer programs. These methods all owe their origins to early concerns about the ineffectiveness of top-down development programs (Sumberg and Okali 1989; Holt-Giménez 2006). Many development workers see participation by local community members as important both for successful implementation of a specific development project and as a catalyst for social change. The latter topic is addressed briefly at the end of this section.

3.1.1 Project Initiation and Investigation

Assessing the Resources and Needs of the Community

In the first iteration of the project cycle, development practitioners must work with the community to determine community resources and community needs. Determining physical resources can be done using participatory rapid appraisal (PRA) methods. PRA has several underlying fundamental principles.

1. *Respect for the community*. Failing to respect the local culture and knowledge is likely to not develop the proper rapport needed to work effectively in the community. This respect and rapport can be developed by living and working with a community. The importance of respect is well documented (Tayong and Poubom 1999).

   This respect can be manifested in the exchange of power in a participatory process. Chambers (2002) calls this exchange "handing over the stick." At times, this may be a true physical stick, such as a piece of chalk or a marker, which is given to a community member who will facilitate a community meeting or workshop. At other times, it is the less tangible transfer of authority to a local organization, such as a community water committee. Such transfers of power can be made evident through signed formal documentation of the transfer or with an inaugural event that signals the start of a new community-driven stage in the development project.

2. *Triangulation*. Triangulation is the use of many different types of people, both outsiders and community members who represent the diversity of the community, who use a wide range of techniques to learn the resources and needs of a community. Together they determine and implement a development project.

Teams conducting a community analysis should be made up of individuals with different backgrounds and traits. An assessment team should have an appropriate mix of people. Without creating too large a group, think about including at least one male and one female; people who represent different disciplines: e.g., an engineer, a social scientist, and a health worker; an expatriate, an official from the national ministry, a local citizen, and a local extension agent. A team interviews different groups of people (women, younger households, smallholders (tenants), local merchants, and the village authorities) using a range of PRA methods. This variety of techniques and people is called triangulation, and it generates a range of information. The survey team can sort through the information and determine what is important to the community as a whole and what is important to only a segment of the community. Invalid information can also be eliminated through triangulation.

Assessment methods are subject to different types of bias, and PRA is no exception. Bias may occur when development practitioners only speak to one or two groups within the community, for example, ignoring women, small landholders, more impoverished households, communities not on the major transportation routes, or other critical groups (e.g., Datta 2006). Bias may also occur when a community member responds to a survey with an inaccurate response, hoping that it will generate a benefit for him or her. Seasonal bias may occur when responses to questions vary during different times of the year, for example, during the wet season versus the dry season. Successful PRA uses triangulation to reduce the problems introduced by bias.

3. *Optimal ignorance.* It is not possible to gather every piece of information about a community and the individuals within a community. A good participatory process stops when sufficient information is available to move on to the next stage in the project cycle. Gathering too much information wastes precious resources, including the patience of community members (Chambers 2002).

4. *Flexibility.* Engineers must be ready to listen to community members and incorporate their concerns and ideas in the earliest stages of a project. This method means that the engineer may have to discard preconceived notions and design a project that meets the needs and constraints of a community. Throughout a project, unexpected obstacles and opportunities may present themselves. The engineer and community must be ready to assess them and make appropriate changes to the project (Davis and Garvey 1993; Ingles et al. 1999).

Assessing the Physical Resources

A community's physical resources are often inventoried through transects, village walks, and maps (Eckman 1996; Selener et al. 1999). Unlike traditional mapping, participatory mapping involves walking with community members or diagramming maps in a community meeting and letting the community develop the maps. Figure 3-2 shows an example of a community water map. These maps include important local social and cultural dimensions of the physical resources present in the community.

The mapping process should be triangulated. The most common method is to work with different segments of the community and develop several different maps of the same

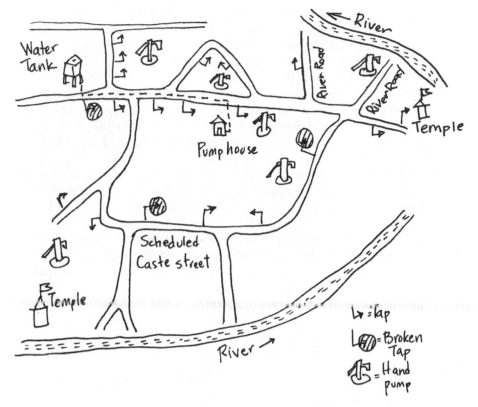

Figure 3-2. Community Water Map Developed by World Neighbors in India.

Note: The map includes local social and cultural dimensions of the physical resources present in the community.

Source: Johnson 1992.

area. Transects (straight lines) and walks (a purposeful meandering) are essentially maps in only one spatial dimension. Walks are useful in water development projects to determine proposed routes for pipelines. The walk may be used to determine ownership of the land where the pipeline may be located. Multiple walks can be critical if land ownership or boundaries are disputed. Different people may lay claim to the land, and only multiple walks will make these conflicts evident. Securing right-of-way is one of the first critical stages in water projects in developing countries (Reents 2003).

Walks can also be used to determine current water use. In some communities, women may travel several hours to collect water. They may use different water sources for different purposes or at different times of the day, as described in Chapter 9. For social reasons, women may collect water at locations that do not seem to be optimal in terms of water quality or minimizing effort. A walk with different women, especially a walk with a set of casual questions, can often reveal preferences women have. Those preferences can then be incorporated into the project design.

The current physical resources related to a project should also be assessed. For example, for water projects, the current water source should be analyzed. What kind of water source is used by the community, where is it, and what state of repair is it in? Are there agricultural or human waste disposal activities in the area that could contaminate the source? Is the area forested or cleared farmland?

Additional questions also require participatory techniques to elicit answers. How does the community feel about the water source? Do they see problems associated with the current source? Is it too far away, does it run dry during parts of the year? Are there illnesses that the community believes are related to the water source? What would they like to see improved? If a system is not working, one would like to know the duration of the problem and why the problem has not been corrected. Answers to these questions are necessary when developing appropriate solutions. Informal group discussions and conversations with key informants are typically used to explore and develop this type of information.

Assessing the Human Resources: Identifying Community Leaders

Projects need the support of the community to be successful. This step requires the support, or at least the approval, of the influential members of the community. Some influential community members may be obvious, such as a chief or mayor. Others may not be evident to someone new to the community. Lists or Venn diagrams are effective methods of gathering information about the internal dynamics of the community. The lists are developed with key informants, individual surveys, or group meetings.

As always, triangulation is important. First, develop a list of the organizations or development projects within the community. Ask people if the organizations or projects were successful. Do responses indicate that most people have the same opinion about the degree of success, or is there a range of opinions? Next, determine who the leaders of the organization or project were. The leadership may include both formal and informal leaders. While questioning people about the quality of the leaders can be a delicate endeavor, it is often worthwhile. Identifying poor or weak leaders or obstructionists may help later when a key committee must be formed. Other community members can be added to the lists in two groups: active and inactive members.

Instead of using lists, the same information can be displayed in Venn diagrams. Different color circles can represent organizations or projects with different levels of success, and people can be placed within the circles (Fig. 3-3). If a person is found in many of the circles, that individual may be critical to the success or failure of a project.

One word of caution: Some people may be overburdened with leadership roles in a community, so caution must be used to protect such people from overcommitment when introducing a new leadership role. Overburdened leaders may burn out and become ineffective in their roles. Organization of a management committee can begin with a core group selected from the group of critical positive people, while attempting to avoid the influential people who hinder projects. At times, it may not be possible to ignore the obstructionists, but the group can strive to work with them and limit their ability to derail a project.

If Venn diagrams or lists are used as a rapid appraisal method and little time is allotted to reflect on the information collected, the lists of projects and participants may be incomplete. Snowball sampling (Bernard 2002) provides a method to determine if the list of influential people is complete. First, determine the names of several influential people

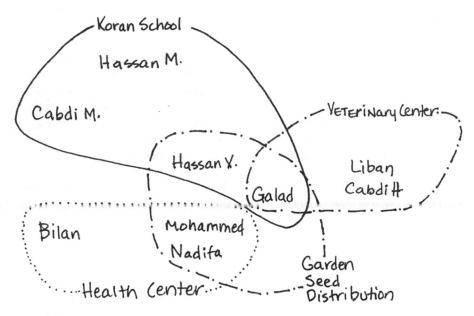

Figure 3-3. Venn Diagram Used to Determine Important Community Leaders and Institutions.

Note: According to this Venn diagram, Hassan X. and Galad are probably key leaders in the community, and involving them in a project will increase the chances that it will succeed.

and their association with particular organizations and projects. Each of these people is then asked who they believe is important or influential. The new people are briefly interviewed to determine their associations and whom they believe is influential. This process is repeated until no new names of organizations or people are added to the list.

A modified version of snowball sampling may be used with Venn diagrams and lists for individuals or many small groups. Instead of directly asking people for new names of influential people, one continues the survey work with groups and notes whether new people continue to be added to the Venn diagram or list. A word of caution: In divided communities, people on one side of the divide may intentionally not name people from the other side.

Assessing Community Needs

The purpose of community needs assessment is to ascertain what development interventions will be of greatest benefit to the community. Some agencies have their own framework for community needs assessment, and others rely on more generic participatory methods.

Two fundamental approaches exist when developing a needs assessment: (1) asking directly about the needs and (2) asking about the current problems in the community. Determining current problems, although apparently a less direct approach, can often be the most useful. Needs assessments ask people to determine how they envision their future. Most other aspects of participatory planning ask about the past or current conditions. This method leaves needs assessment more vulnerable to strategic bias among those who are involved with planning. For example, people may have a particular goal that is a desire rather than a need, or respondents may ask for a project that they really do not need or want, but which has indirect benefits, such as short-term employment, which they desire.

A dialogue about community problems is likely to reduce both strategic bias and some bias held by the people running the survey. A development worker may enter a community with a particular solution, task, or project in mind—even if there is no problem that matches the solution. Asking about problems can open a technology-free discussion of what the desired final state is. Only after the current problem and desired outcome are determined by a community should one look at potential technical solutions.

In the case of water projects, needs assessment can include a discussion of the prevalence of diarrheal and other water-borne diseases. PRA tools such as seasonal calendars show different seasons when water-borne disease is particularly problematic. Local hospital or health center records can be used to gauge community health needs. These statistics must be taken with caution, however, because many of the vulnerable poor in a community may not visit hospitals or health centers.

Assessing a Community's Capacity for Project Success

Over the long term, outside organizations will not remain in the community to provide assistance. The current capacity of the community to manage development projects should be considered in project design. What current and recent projects have been implemented in the village, and how successful have they been? Have the projects continued after the initial funding and support ended? Why have projects succeeded or failed? One can develop a community map and ask people to place the physical location of the projects on the map. Ask questions about the successes and failures and the reasons for each project as the map is developed. Again, triangulating to obtain the views of different groups within the community will give more reliable and robust information. Many development projects are targeted to reach certain segments within a community, and only by working with a cross section of the community will the development worker obtain a realistic picture of the community.

"Capacity" also includes the ability to pay for the ongoing operating costs of a development project. It is often not the initial capital costs that are the financial downfall of a project, but the ongoing costs of operation and maintenance. Capital costs are often subsidized by an outside agent, whereas operation and maintenance costs are not. What experience does the community have in collecting funds for other projects? How was it done? How much was collected? Do they have experience using a bank? These questions can be incorporated in the mapping exercise described in the previous paragraph.

It may also be important to determine income levels to gauge an upper limit to people's willingness or ability to pay for a service. Gathering income data can be a sensitive undertaking, and one must have the trust of the community for this work to be effective. Sometimes a survey of current expenditures of households will provide insight into the financial profile for the community. In some communities, cash savings can be minimal, with expenditures equal to income. Loans may be acquired during certain parts of the year. It is important to look at expenditures over the year and to find out about loan activities. Using the average income in a community when determining ability to pay is not appropriate because poorer community members will probably be excluded if use or maintenance fees are established based on average income. It is better to develop a range of income profiles for the community. Triangulation of different groups is critical in obtaining this type of information.

>>> *According to the World Health Organization, households should not have to spend more than 3.5% of household income for water projects or 1.5% of household income for sanitation projects.*

Prioritizing the Solutions

People may have their own solutions, based on indigenous knowledge or experiences of other communities that they have observed. When facilitating a discussion of problems, future states, and (especially) solutions, it is easy to fix on one idea. Facilitators should keep the discussion moving, although this form of communication is not easy or appropriate in some cultures.

Only after local people present solutions should a development worker consider proposing a solution. Often people generically present an idea, such as a well or a water tank. This presentation can be an introductory point for the engineer to describe the technical and cost details associated with the broader suggestion. Free-format and open group discussions of problems may reveal a long list of concerns and solutions.

Developing a clearer picture of each problem and problem solution can be time-consuming. In some cases, a second meeting will be necessary to develop reasonably accurate information about each solution. How much labor and money will be required? What are the capital costs and what are the ongoing maintenance and operation costs? How long will it take to implement the project? Who has to approve the project and work with the community as the project is implemented? Without sufficient detail, a community cannot realistically prioritize the problems to address first. Although the engineer may develop this information, the purpose is to help the community understand the details of a particular solution. Communities should be involved in project analysis of potential projects. They will be further invested in the project, have a deeper understanding of the solution, and be more likely to continue the project after the development group leaves the community.

Prioritizing can be done by asking participants to rank the lists of problems and solutions. With small groups, this procedure can be confusing. Time may be spent listening to individuals argue over the top priority on the list, and the group may never begin to discuss other items. Pair-wise comparisons can speed up the process, allow many people to voice their opinions, and act as a starting point for further discussion. In participatory pair-wise comparisons, community members are asked to select among preferred alternatives one pair at a time. They may have four possible locations for community tap stands: *a*, *b*, *c*, and *d*. Which one is preferred? People would be asked if they prefer *a* or *b*. Then they are asked if they prefer *a* or *c*. This line of questioning is continued until all pairs are exhausted. This is an effective tool when the choices are limited in number and the issue at hand is complicated (usually more complicated than the location of a tap stand). Participatory pair-wise comparisons have been used in natural resources (Mendoza and Prabhu 2000), irrigation (Tesfai and de Graaff 2000), and sanitation (Wiwe 2005).

Participatory methods are structured tools to gather specific information and to generate discussion within the community. Their value has been documented by Kersten (1996) in matrix ranking and pair-wise analysis. Participatory methods should ultimately bring a deeper understanding of the needs and resources of a community through open discussion. Facilitators should probe beyond the superficial by constantly asking "why" until the underlying causes are evident (Banez-Ockelford 1999; Paulete 2006). The results of the matrix ranking or pair-wise analysis can be a useful starting point for what might have been a rambling discussion.

A word of caution is necessary with participatory information gathering, especially in the early stages of the project or during a needs assessment. It is easy to raise the expectations of the community beyond what the project can provide (Selener et al. 1999). Even after a close look at the costs and commitment necessary for a particular project, a community may place it at the top of the list of preferred solutions. Community members may suggest activities in a needs assessment that are beyond the scope of the agency. People may suggest a particular project, which although within the broader mandate of the agency, is not appropriate for the community, either because of the cost, management capacity required, or technological suitability. In most cases, the needs of the community will be broader than the mandate of a particular development group, whether it is a ministry, agency, or nongovernmental organization (NGO). Communities may request a health clinic, an all-seasons road, improved food storage, and a new water system. An NGO may only have the authority and skills to work in one of these areas. It is important to let a community know what an NGO can and cannot provide to the community.

In less obvious cases, it may be possible to use another participatory technique to continue the dialogue with the community. Based on technical expertise or the results of earlier participatory investigations, it may seem that the community has made a poor choice from among the alternatives. For example, communities may choose a technology that has higher operating and maintenance costs than they can support.

If several people from outside the community have been involved in the dialogue, one can set up an insider–outsider poll (Jones 1996). Individuals are given beans; people from within the community get one color, and outsiders receive another. People vote for their choice of alternatives by placing beans in containers that represent the different alternatives. When the voting is complete, results are tallied by bean color. The results are a starting point for a discussion of why people allocated the beans the way they did. This method tends to open discussion, is nonconfrontational, and gives everyone an equal voice at the start of the process. Even when there is a fairly good early consensus among community members, this technique can be used to generate a clearer understanding of the community's choices.

Is this really participatory development if the development worker intervenes with a technique designed to change the opinion held by the community? In fact, there is a good chance that the outside "expert" may be wrong. If the costs are not prohibitive and the two alternatives are not mutually exclusive, it may be possible to initiate a feasibility evaluation and design of two of the proposed alternatives at the same time and monitor each for progress. If one project appears noticeably superior, the weaker alternative can be dropped. Before starting a dual-track process, everyone should help develop and be aware of the criteria that will be used to judge between the two projects and of what criteria will be used to determine when to stop working on a project. If there is a good consensus about which easily measured parameters will be used for judging projects and their progress, there will be fewer arguments once the implementation has started.

Many of the concerns about community versus engineer control, implementation, and management are linked to a broader question identified by Chambers (2002): "How participatory can and should [the process] be?" How much preparation does a community have for a specific project? How much scope and authority does the NGO have? What is the capacity of the community to manage change? Answers to each of these questions will help determine the specific techniques, the broader goals for social change, and the participatory nature of a project. A person facilitating a project should work toward greater community participation as the project progresses.

Communities may also learn about successful implementation of a similar project by visiting other communities that have already implemented a project. These community visits are now commonly used across the spectrum of development, although they developed from the early use of farmer-to-farmer exchange programs in agricultural extension work (Selener et al. 1997; Holt-Giménez 2006). If a community is considering a specific type of water system, they might visit a community that already has that type of system in place. When considering several different types of designs, it might be possible to visit more than one community, looking at how each of the designs was implemented. The community can learn what worked and what did not, the benefits of the system, and how difficult and expensive it is to operate a particular system. Learning from other communities is one of the most effective participatory tools (Box 3-1).

3.1.2 Planning

The Transition to Participatory Implementation.
Although diagrams of the project cycle (Fig. 3-1) make it appear as if there are distinct phases, the reality of project management is that each stage in the project cycle slowly

---

))) Box 3-1. The PHAST Approach to Water, Sanitation, and Hygiene Improvement within Communities

In 2000, the World Health Organization, in collaboration with the World Bank, introduced the Participatory Hygiene and Sanitation Transformation (PHAST) approach to working with communities. PHAST is an innovative participatory approach designed to promote hygiene behaviors, sanitation improvements, and community management of water and sanitation facilities using specific participatory techniques. For more information on PHAST, see their website: http://www.who.int/water_sanitation_health/hygiene/envsan/phastep/en/.

blends into the next, even when there may be end points built into a specific part of the cycle (Davis and Garvey 1993). At the community level, the transition from one stage of the project cycle to the next may be fairly gradual, rather than the abrupt transition often envisioned in theoretical models. A project may have a report due at the end of the assessment phase, and it may be possible to end or modify the project at this point. However, if the project is to continue, important groundwork will have been completed that moves the project forward.

Defining the Goals and Objectives of a Development Project

Through participatory planning, general goals may be enunciated by the community. However, a broad general consensus may not provide the objectives necessary to reach those goals. A working group or management committee is established (possibly not until the implementation stage of a project) and held responsible for the project outcome. This group is best suited to develop objectives necessary to reach the community's goals. The goals act as guidance, and the management committee should not stray from the direction provided by the community. As long as the management group remains transparent and engaged with the broader community, it can develop the specific objectives and strategy of the project. Goals and objectives, once the domain of large organizations, can be developed by small communities using participatory methods (Ely 2005). Box 3-2 lists characteristics of objectives that improve the likelihood of success for a community project.

One method often used to develop objectives that are consistent with community needs is to transform the problems and needs that have been identified by the community

---

⟫⟫ Box 3-2. Characteristics of Objectives That Result in Successful Community Projects

According to Margoluis and Salafsky (1998), each objective should be

1. impact-oriented—to move the community toward its goal;
2. measurable—such that the community can clearly know when the objective has been met;
3. time-limited—so that the community can have a reasonable expectation of when the objective will be completed;
4. specific—so that all community members have the same understanding of the objective, because fuzzy objectives lend themselves to dissent over whether something has been accomplished and, at times, to failure of the community to follow through on a commitment; and
5. practical—so that the community and the development group agree that the objective can be achieved and is appropriate for the community.

Note that "time-limited" may conflict with flexibility. Agencies may be under pressure to meet deadlines and skip or shorten critical steps in project development and implementation. Unfortunately, participatory processes, including triangulation, are often the first victims of time shortcuts (e.g., Datta 2006).

into objectives (Davies 1997). "Rats consume too much of the food stored by the community" can become "By the end of the next harvest season, 40% of the households in the village will have and use improved concrete-sealed silos."

3.1.3 Implementation and Conclusion

Community members who have actively participated in identification and planning of a development project are likely members of a community management committee or group that will implement and operate a project. Participation in the implementation stage does not use as many standard participatory methods as the early stages of the project cycle. Once a project has reached the implementation stage, it usually has unique factors determined by the design requirements, technology, and the specific needs and characteristics of the community. To improve success rates, the community requires similarly specific methods of participatory implementation. One common method of implementing a project effectively is a series of small steps or "playing catch with the community" (Orr 1985). "Playing catch with the community" implies that the NGO and the community have fairly separate tasks and that when one group completes a task within the project, the responsibility for continued implementation is "tossed back" to the other group. Creating a project delivery with multiple steps (called step design) is a broader concept because several consecutive tasks may be the responsibility of the same group or individual.

In step design, work plans are designed so that the project is broken into small, discrete, and manageable tasks. Each task is designed to be completed in a short time, with someone (or, at worst, a small subset of the community) responsible for its completion and an identifiable output. This method allows projects to move along sequentially, with no task started before the earlier tasks are completed. This system allows community members to view larger projects in small increments and to complete each step in a timely fashion. Clear responsibility is delegated. For example, if a trench for a waterline segment is not complete, then the pipe will not be laid.

Clear delineation of responsibility can help a community with projects where parts of the community, often geographic subsections or households, move at different rates. In a household latrine project where materials are supplied or subsidized by an NGO, some households may acquire the supplies to build a latrine, but not use them for the intended purpose. If the project were implemented in a stepwise fashion, the household would first be required to dig the pit for the latrine. After completing the task, some supplies, such as materials for flooring, might be provided. Next the household would have to complete floor construction before the remainder of the supplies would be provided. Some community members may work more quickly than others, and a step design allows each person or household to work at its own pace.

3.1.4 Monitoring and Assessment

Step design allows closer program monitoring for assessment and evaluation. Whereas monitoring, evaluation, and assessment can be viewed as parts of the final stage of a project, they can also be used effectively within a project. At any point in the project cycle, it is useful to know if progress is being made, and if not, why. Monitoring and assessment should be built into project design and is best developed with community participation. Triangulation should be used in the monitoring and assessment. In a community water project with a goal of reducing childhood illnesses, one could ask households to self-report illnesses, a health worker could visit a subset of households that are reporting, and local health clinic records both before and after the construction could be gathered. The three data sets should be consistent. If they are not, ask why.

Effective assessment is participatory. Community members are included when plans are developed so that they can develop expected project outcomes and ways to assess them. They can then feel comfortable with any adjustments that need to be made to a project (Roche 1999). The checkpoints and criteria for success are their own, so it becomes easier for people to also claim ownership of and participation in the project. People can identify when a project is not advancing as they had hoped or anticipated and can then make the changes necessary for the project to move forward again.

Assessment must also cover a range of scales and viewpoints to determine the success of a project. A common error in project management is to measure inputs or number of people served. Although one does not want to ignore these factors, they are not truly indicative of the progress or effect of a project. Measuring pipe purchased or installed does not mean people are receiving water.

Many water projects have multiple objectives: to supply water, improve health, and reduce the amount of time women spend obtaining water. Each of these objectives can be measured, but they may also have unexpected consequences. For example, it is assumed that if water supply is closer to the house, women will have more time available to engage in more productive work. If this is the case, then a monitoring and assessment plan should determine how women are using the time now available to them. If the additional water creates greater expectations of women, it may be that the new water supply has merely caused a reallocation of time from one type of drudgery to another. Monitoring superficial details would not uncover this new time allocation.

Brett (2006) provides an example of a case in which the wrong details were monitored to track progress of a development project. In a microfinance project, measuring money loaned or number of loan recipients may provide a good number to use early in the project implementation, but there are other indicators that are more telling. Many microfinance projects use the loan default rate as an indicator of the success of their loan programs. However, the objective of a microfinance project is to improve the quality of life of the recipient. If the loan is not used effectively, the quality of life can deteriorate. In at least one case, Bolivian women who received microfinance loans did not receive adequate counseling or instruction and soon found themselves losing money by obtaining loans and then investing in poorly designed household development projects. They were forced to obtain other loans or rearrange household spending patterns; and they ended up in a poorer financial position than before they had obtained the loans. From an institutional perspective, the program was a success, as loans were made to poor women and were repaid. From a household perspective, the program was often a failure (Brett 2006).

If a project or project objective meets the five criteria of Margoluis and Salafsky (1998) (described in Box 3-2), then monitoring the project or objective is usually straightforward. The specific, measurable unit can be observed at the expected time of completion to determine if the measured value meets or exceeds the target set for the objective.

In addition, because the objective is supposed to have an effect on the community (e.g., reduced incidence of intestinal disease among children), a clear measure of the final positive outcome can be developed and monitored.

Frequent monitoring allows the community to determine the next step and to make adjustments to a project. Early adjustments are less costly and easier to implement than late changes. Not only should the project have milestones and measurable indicators, but there should be a plan in place for changing course. Who makes the decision to continue or change the project? Is it a committee or a local community organization? Is there a vote? Local organizations that have been in place for some time may already have a procedure. New committees or work groups set up for the project should establish procedures before project implementation begins.

Project monitoring can be a positive experience. With each successful step, a community should gain confidence in its ability to design, build, and operate the project. The successes can be recognized through awards, ceremonies, and parties (Paulete 2006). For example, toilet festivals have been used to mark progress and educate the broader community (Burra and Patel 2002).

Each successful project has a conclusion and assessment. Some conclusions are natural: the water supply system has been constructed, or each family in the program has built its latrine. Assessment should include transition of operation and maintenance to the community. A well-planned and well-constructed engineering project will fail without the proper social, economic, and political support structure. Other projects may have less obvious conclusions, possibly determined more by an agency funding cycle than by rational perspective from the field. In these cases, the agency should have a phase-out and transition plan (Strahan 1996), complete with assessment. A well-designed project delivery will continue monitoring and assessment long after the development agency leaves the community.

))) 3.2 Community Management: Project Example

Community management means that communities themselves, rather than government institutions or the private sector, have primary control of a local project. This section discusses community management of rural water supply projects. Many components of community management are applicable to other projects, and the lessons can be adapted. The community management model for water supply evolved from the principle of maximizing community participation in water supply projects that gained universal acceptance during the 1980s International Drinking Water Supply and Sanitation Decade (IDWSSD). Community management is now a viable option for managing rural development projects because of an acceptance from three principal stakeholders with different agendas and priorities: governments, NGOs and private donors, and multinational lenders.

3.2.1 Defining Community Management

In practice, the term *community management* means different things depending on the level of decision-making authority given to the community. Narayan (1993) defines community management as a means to "optimize community participation in order to achieve sustainability through human development." In essence, communities create a set of statutes that guide the function and maintenance of communal infrastructure. These rules reflect the individuality of communities and create an equitable, cost-effective framework for managing rural water supply.

3.2.2 Community Management Model Components

In the context of this discussion, the main factors influencing the adoption of community management are presented in two categories: those that are internal and those that are external to the community. Factors external to the community are those over which the community itself has little control. These factors include, for example, institutional adjustments and policy decisions made by lawmakers concerning the rural water sector (see Box 3-3). Internal factors are those issues that are directly influenced by community members who implement and support rural water supply systems (Table 3-1). The remainder of this section will focus on those key factors within communities critical to actuating effective community management.

3.2.3 Establishment of Local Water Committees

Community management of water supply is often represented by the presence of a community water committee. Water committees are the local forums where decisions dealing with the running of the water system should be made.

))) Box 3-3. Important Factors External to Communities That Influence Community Management

Legal framework: National laws should exist that support the community management model. These laws officially recognize community management as the national approach to managing rural water infrastructure and acknowledge local communities' legal rights to own and make decisions about the management of their water supply systems.

Institutional support: Institutional support mechanisms (ISMs) are critical for the community management model to be successfully sustained. ISMs should include periodic site visits to communities after completion of the project, long-term assistance in technical problems beyond the scope of routine maintenance, and assistance with conflict mediation.

Technical design: A poorly designed or improperly installed system can spell doom for the most motivated communities' attempts to successfully actuate the community management model. In these cases, repair will be more frequent, causing maintenance costs to exceed collected funds. This design results in increased dependence of the community on outside support.

Table 3-1. Four Important Internal Factors Affecting the Achievement of Equitable Sustainable Community Management

Factor	Components
1. Community dynamics	Quality of leadership within the community; gender divisions and inequality within the community; social cohesion within the community; and management capacities within the community
2. Operation and maintenance	Properly trained technicians who are capable of making repairs and local supply of spare parts
3. Ownership: Rules, regulations, and enforcement	Clearly defined rules of operation; enforcement of rules; and a community sense of ownership
4. Recovery of costs	Ability and willingness to pay; transparency, or knowing where money is kept and how money is spent; and a method of funds collection

Source: Adapted from Schouten and Moriarty 2003.

In the early stages of participatory assessment and planning, the entire community may be involved, but some key community members may have a larger role than others and may take on a leadership role for the project. The informal formation of these leadership groups often precedes the formal or official recognition of a group. Some groups may function for years before they receive official government recognition. The official organization and recognition of the management committee are often critical steps. Care must be taken to see that the committee represents the interests of the entire community and is viewed as the legitimate authority for the project. Too often, top-down designation leads to a committee with little interest or capability for management (e.g., Duyne 1998).

In many instances, water committees are democratically elected and include a cross section of interests within the community. There can be a varying number of members with different roles within a village water committee. Usual tasks of the village water committees are summarized in Box 3-4.

⟩⟩⟩ Box 3-4. Typical Roles of a Community Water Committee

- Structure community decisions around system management.
- Organize contributions and control finances.
- Inform the community.
- Act as a liaison when dealing with water users.
- Ensure proper operation of the water system.
- Oversee technicians; coordinate maintenance and replacement of parts.

Source: Adapted from Bolt and Fonseca 2001.

Recognizing that communities are complex, nonhomogeneous compositions of stakeholders with differing levels of wealth and education is the first step to helping the community establish a water committee. As such, for village water committees to be effective, multiple community interests must work together for the common goal of maintaining the village water supply. The importance of gender equity in community management has been well documented. Women, because of their traditional role in most societies as water fetcher, cook, and household manager, play critical roles and must have significant input in decisions made by water committees. The inclusion of women might be difficult in some cultures but should be taken into consideration to the extent possible when negotiating the structure of the village water committee (Fig. 3-4).

Traditional management structures should be well understood before assisting a community to establish a water committee. It might be more effective to support an existing social structure than insist that a new council with a president, vice president, treasurer, etc., be elected in the community. In some cultures, this type of foreign management structure will be only a proxy assembly compared to the traditional structure

Figure 3-4. Education Can Take Place at Several Stages of the Project Cycle.

Note: The knowledge flow should occur in both directions. Activity-based presentations are typically more effective than lecture presentations.

that has managed local resources and conflict in the community for generations. However, when using an existing social structure, specific individual tasks and responsibilities should be well defined.

Similarly, the existence of an influential leader will often dictate how effective the water committee is in the community. These leaders can be a blessing for the committee by providing a local figure who legitimizes the committee. However, in some cases, community leaders who lack the dynamism or enthusiasm to support the project can be a burden to the water committee. The participatory approaches to community assessment outlined in this chapter should be performed to uncover the existing conditions before deciding how to establish and support the village water committee at the outset of the project.

In water supply projects, there is often a preference for individual tap stand committees or point source committees to be formed as part of a centralized water committee led by a president. These water point committees oversee the day-to-day operations of individual tap stands and collect tariffs or yearly contributions. Variations of the community management structure should be fully discussed with the community before making a consensus decision.

3.2.4 Operation and Maintenance

Operation and maintenance as related to community water projects is presented in Box 3-5. The task of maintenance of community water supply is often placed in the hands of trained local technicians. For all but the most major repairs, these technicians should be self-reliant and have access to tools and spare parts within or close to the community.

When establishing community management systems, it is important to acknowledge that in some exceptional cases, major repairs will be required that are beyond the capability of the community. These scenarios will be more common for technologies with

---

))) Box 3-5. Definition of Operation and Maintenance in the Context of Community-Managed Water Supply

Operations: Operation refers to the everyday running and handling of a water supply. This definition includes the proper handling of facilities by users to ensure long component life, e.g., the use of taps at tap stands.

Maintenance: Maintenance refers to the activities required to sustain the water supply in a proper working condition.

- *Preventive maintenance:* Regular inspection and servicing to preserve assets and minimize breakdowns.
- *Corrective maintenance:* Minor repair and replacement of broken and worn-out parts to sustain reliable facilities.
- *Crisis maintenance:* Unplanned responses to emergency breakdowns and user complaints to restore a failed supply.

Source: Lockwood 2004.

moving parts, such as deep boreholes with hand pumps. In these cases, the aforementioned institutional support mechanisms are critical to assist the community in finding outside support needed to fix the system. However, costly emergency scenarios can be avoided by the selection of appropriate technologies and adherence to the operations and maintenance agreements established after system construction.

3.2.5 Ownership: Rules, Regulations, and Enforcement

Successful community management requires that clear ownership of the water system be defined. Ownership issues lie at the heart of the paradox that communities with long histories of successful natural resource management in other domains are not always successful at managing water supply systems (Schouten and Moriarty 2003). Ceremonies common at the conclusion of construction of rural water supply systems, meant to "hand over" rights to rural community, should be accompanied by the supporting legal documentation to make this transfer binding. These documents should list rights and responsibilities of both the community and government relating to the infrastructure.

Enforcement of rules and regulations is usually the responsibility of the water committee, through the levying of some sort of social sanction or fine. In many cultures, it is difficult for such fines to be levied by nontraditional leaders of the community. In these cases, a blending of the traditional community social structures and the management committee will need to be decided on to ensure that fines are collected.

3.2.6 Cost Recovery and Financial Management

It is well documented that successful community management cannot be attained if a system of recurrent cost recovery is not properly created and followed by the community during the life of the system. The management of communal funds is an important litmus test for assessing progress of community-managed rural water systems because "cost recovery is still today one of the major obstacles to achieving sustainable drinking water supply in developing countries" (Brikké and Rojas 2001). Functioning mechanisms for the recovery of communal funds are evidence that a water system is demand-driven and has tangible economic value to rural communities. One overlooked item is that water committee members may be entrusted with the care of funds without having experience with formal banking. In this case, the engineer would work with the specific committee members to ensure that they are comfortable opening and using a bank account.

Many dynamics within the community need to be evaluated before a system of cost recovery can be agreed on. These dynamics include, but are not limited to, willingness and capacity to pay, financial management of collected funds, a method of fees collection, and enforcement of payment (Schouten and Moriarty 2003). Finding consensus on the type of water tariff that should be charged in rural communities is challenging because of the idea that water is a free commodity. Typically, tariffs are either set per volume of water consumed or standardized as a uniform price paid by all members of the community regardless of usage. Water tariffs are often based on average monthly maintenance costs and are equally divided among the water users. Collected funds are typically used to cover recurrent maintenance costs but not to recover initial investments or future expansion of the system.

Alternative systems of cost recovery, such as collecting agricultural products (e.g., rice or beans), are sometimes found in rural areas. Because of the seasonal fluctuation in the prices of commodities, crops can be collected during the harvest season and sold for higher profits later in the year. Systems such as these can be a valuable approach when the community experiences distinct hungry seasons throughout the year.

))) 3.3 Social Change

Development is only successful if it increases the ability of a community to manage its own resources. However, development of a community is more than establishing a greater level of technical expertise. Each community has its own history and therefore a set of traits that define its current capacity to organize and govern itself. Whatever the starting point may be for a particular community, each development project should help the community increase its confidence, capacity, and organizational skills.

Some development theorists speak of this in broad terms. Others believe development demands an explicit shift in the balance of power to the community and within the community, enabling the poorer or more disadvantaged portions of the community to control their own destiny and to obtain control of political, social, economic, and physical resources (Freire 2000; Selener 1997). Some projects may focus specifically on improving the status of women in the community.

Specific techniques may also be used to empower local communities. Participatory geographic information systems using improved technology can improve governance and transparency (McCall 2003). Even projects that may not have explicitly committed to empowerment, but allow people to control project development and management, can lead to further development. Bunch and Lopez (1995) evaluated the long-term effect of development projects in Central America. In many cases, communities had abandoned technologies that had been adopted with the support of the development agency. When the communities had initially adopted the technologies, they were successful and appropriate for the community. Over time, the political, social, and economic environments changed and, because communities had learned the process of implementing a new technology, they were able to adapt. The realization that a specific engineering project can also be a catalyst for change and the long-term development of a community should help engineers incorporate specific participatory tools and education within the design of the project (Davis and Garvey 1993; Regmi and Fawcett 1999; Ely 2005).

Women are often the primary source of labor in traditional water supply systems and are critically important in the design and management of any new system. This discussion of women is not placed here as an afterthought, but because earlier topics in this section are brought together when discussing women. Whether it is carrying water, washing clothing in streams, or walking treacherous roads to the market laden with goods, women often suffer more than anyone else because of inadequate facilities. In regard to sanitation and hygiene, women are typically the family health providers and are responsible for cooking and cleaning. Involving women is not only an exercise in equal rights; it is an excellent strategy to make community development work.

Flexibility is important for women. They have many other responsibilities and in some communities may only be in a position to commit small segments of time to a project (Fernando 1996). Project managers must be careful to arrange meetings when it is convenient for women; implementation schedules must also be flexible to accommodate women.

Successful projects often use community members as extension agents and health educators (Davis and Garvey 1993, Fernando 1996). Placing women in these roles can be especially beneficial, particularly in communities where health concerns and discussions may be gender sensitive. However, finding women with enough time and who are sufficiently educated can be a problem. In some cultures, it may be difficult to find women in the community who are permitted to take paid employment. Hiring women can meet fierce resistance in communities where it is assumed that well-paying jobs provided by a development agency rightfully belong to men. None of these obstacles are arguments against including women, but we provide a caution to development workers to be prepared and ready to work toward empowerment of women.

Project managers may meet with local resistance as they work to incorporate women into a project. Maintaining a professional respect for each party and their views can be difficult when conflicts arise. Agencies that make a long-term commitment to living and working in a community will find it easier to take the steps needed to elevate the status of women.

Without triangulation, it is easy to lose women's voices. For example, in many countries, household finances are split, with men and women managing different parts of the household budget. A woman may be the person in the household who pays for water from a new water system. In a case like this, total household income is not the important measure of ability to pay. The amount of money a woman has available will determine the feasibility of any revenue collection scheme.

Over the past decade, migration within countries and to other countries has increased. Men are more likely to migrate than women, especially in the early stages of migration from a community. As migration has left rural communities with fewer men, women have stepped in to fill the power vacuum. In some cases, the power is short-lived, and critical decisions must wait for men to return. In other cases, women have taken active roles, taken advantage of new political opportunities, and changed the focus of development within the community (e.g., Rudel et al. 2006).

Despite different philosophies and motivations, successful development projects require commitment to improving the management capacity of the community. This improvement can only be done with the direct involvement of the community in a participatory manner.

⟫⟫⟫ 3.4 Differences in Approaches for Rural, Periurban, and Urban Communities

There are many differences between urban, rural, and periurban locations. To conduct engineering projects within these distinct geographic units, engineers must understand the different political, socioeconomic, and organizational constraints that exist in each type of setting. Box 3-6 gives geographic definitions for each type of setting.

Because of high costs of living and lack of available housing, urbanization trends in the developing world have created "forced periurban" areas on the edge of many cities.

>>> **Box 3-6. Geographic Definitions for Urban, Rural, and Periurban Settings**

Urban and rural: Urban refers to territory, population, and housing units located within an urbanized area or an urban cluster. Urban area and urban cluster boundaries thus encompass densely settled territory. Rural refers to territory, population, and housing units located outside of urban areas and urban clusters.

Periurban: The border between urban and rural areas used to be considered a clear-cut line in the landscape, both in the physical and the organizational sense. However, rural and urban features tend to coexist increasingly within cities and beyond their limits. The interface between urban and rural domains, or the periurban area, as it is more often called, is under continuous change, driven by the expanding urban areas and the strong linkages with rural regions. Also known as the periurban interface, rural–urban fringe, or an area of rapid urbanization (Van Veenhuizen 2003).

Source: U.S. Census Bureau 2000.

These periurban areas appear as geographic entities unto themselves. The separate characteristics of these areas influence design, time lines, labor, and cost management and thus must be considered during project planning if long-lasting sustainable improvements are to be assured.

Ahrens (2005) and Ahrens and Mihelcic (2006) describe in detail differences in the project process for each setting for water and sanitation projects in Mali. The following paragraphs summarize important findings of their study.

The urban project process is primarily "top-down," whereas the rural and periurban project processes are principally "bottom-up." For example, in Malian urban settings, the policymakers determine both the need for and the process to implement the project, even though the project does not directly affect the policymakers themselves (i.e., they do not use the constructed wash area and soak pit).

In contrast, the beneficiaries in the rural and periurban cases often determine the need for and the process to implement a project that does, in fact, directly affect them. These rural and periurban beneficiaries are involved in the majority of the project process. However, in the urban case, the beneficiaries take on the role of a customer by paying the policymakers and the private sector to complete their facility needs.

Ahrens (2005) found that unique challenges exist in periurban areas, which typically exhibit a rural lifestyle coupled with the disadvantages of living near an urban center. In the periurban case study, a small neighborhood may indicate an interest in improving the areas within their living space. Because decision making occurs closer to the point of service or action (i.e., regional or local governmental offices), the affected individuals may be more aware of the issues affecting their communities and the possibilities that exist to resolve such problems. As a result, word of mouth is a powerful tool for promoting engineered solutions. For example, one community member may become aware that her greywater is a source of health problems through discussions with a local government official, and she might learn how to improve the situation by building a soak pit. This process could then start a chain reaction, with her neighbors becoming aware of sanitation problems and solutions. Additionally, being located near an urban center may mean that residents have increased knowledge of water, sanitation, and their influences on human health. It is common for some periurban residents to understand the need for a project to improve water, sanitation, and air quality because of the increased pollution concentration caused by high population density.

However, challenges in periurban areas may include residents (as renters or squatters) having little incentive to improve property themselves; residents not being able to afford the cost of materials and labor; limited space within the periurban neighborhood for construction; and large volumes of wastewater with high concentrations of pollution caused by the urban aspects of this location. Because population mobility is high in periurban areas, the knowledge and history of a particular project within the community may not be preserved.

))) References

Ahrens, B. T. (2005). *A comparison of wash area and soak pit construction: The changing nature of urban, rural, and peri-urban linkages in Sikasso, Mali.* <http://cee.eng.usf.edu/peacecorps/Resources.htm> (Jan. 17, 2009).

Ahrens, B. T., and Mihelcic, J. R. (2006). "Making wastewater construction projects sustainable in urban, rural, and peri-urban areas." *J. Engineering for Sustainable Development: Energy, Environment, Health,* 1(1), 13–32.

Annis, J. (2006). *Assessing progress of community managed gravity flow water supply systems using rapid rural appraisal in the Ikongo District, Madagascar.* <http://cee.eng.usf.edu/peacecorps/Resources.htm> (Jan. 17, 2009).

Banez-Ockelford, J. (1999). "Tips for trainers: 'But why?'" *PLA Notes,* 35, 97.

Baum, W. C. (1982). *The project cycle.* IBRD/The World Bank, Washington, D.C.

Bernard, H. R. (2002). *Research methods in anthropology, 3rd ed.* Altamira Press, Walnut Creek, Calif.

Bolt, E., and Fonseca, C. (2001). *Keep it working: A field manual to support community management of rural supplies.* IRC Technical Paper Series 36. IRC International Water and Sanitation Centre, Delft, Netherlands.

Brett, J. A. (2006). "'We sacrifice and eat less': The structural complexities of microfinance participation." *Human Organization,* 65(1), 8–19.

Brikké, F., and Rojas, J. (2001). *Key factors for sustainable cost recovery in the context of community-managed water supply.* Occasional Paper Series 32–E. IRC International Water and Sanitation Centre, Delft, Netherlands, Nov.

Bunch, R., and Lopez, G. (1995). *Soil recuperation in Central America: Sustaining innovation after innovation.* Sustainable Agriculture Programme of the International Institute for Environment and Development, Gatekeeper Series No. SA 55. International Institute for Environment and Development, London.

Burra, S., and Patel, S. (2002). "Community toilets in Pune and other Indian cities." *PLA Notes,* 44, 43–45.

Chambers, R. (2002). *Participatory workshops: A sourcebook of 21 sets of ideas and activities.* Earthscan, London.

Datta, D. (2006). "The role of local elites in development projects: An experience from Sudan." *Participatory Learning and Action,* 54, 128–134.

Davies, A. (1997). *Managing for a change: How to run community development projects.* ITDG Press, London.

Davis, J., and Garvey, G. (1993). *Developing and managing community water supplies.* Oxfam Development Guidelines No. 8. Oxfam, Oxford, U.K.

Duyne, J. E. (1998). "Local initiatives: People's water management practices in rural Bangladesh." *Development Policy Review,* 16(1998), 265–280.

Eckman, K. (1996). *Doing village assessments: A guide to action-oriented village research in developing countries.* Pact, Minneapolis–St. Paul, Minn.

Ely, M. D. (2005). *Analysis of renewable energy project implementation: Biogas and improved cook stoves in the village of Chhaling, Bhaktapur District, Nepal.* <http://peacecorps.mtu.edu/resources/theses.html> (Jan. 14, 2008).

Fernando, V. (1996). *Water supply.* Intermediate Technology Publications, Ltd., London.

Freire, Paulo. (2000). *Pedagogy of the oppressed.* Translated by Myra Bergman Ramos. (rev. ed.), Continuum International Publishing Group, New York.

Holt-Giménez, Eric. (2006). *Campesino a campesino: Voices from Latin America's farmer to farmer movement for sustainable agriculture.* Food First Press, Oakland, Calif.

Ingles, A. W., Musch, A., and Qwist-Hoffman, H. (1999). *The participatory process for supporting collaborative management of natural resources: An overview.* Food and Agriculture Organization of the United Nations, Rome.

Johnson, D. (1992). *Participatory evaluation: An attempt at using PRA techniques in evaluation.* MYRADA PRA-PALM Series–10. <http://www.myrada.org/series_pra.htm> (Jan. 14, 2008).

Jones, C. (1996). "Insider and outsider voting: Reflections from Scotland." *PI 4 Notes, 27, 81.*

Kunshahu, A. I. (2000). *Planning and implementing sustainable projects in developing countries: Theory, practice, and economics,* 2nd ed., AgBé Publishing, Hilvusum, Netherlands.

Kersten, S. (1996). "Matrix ranking: A means to discussion." *PLA Notes,* 26, 21–25.

Lockwood, H. (2004). *Scaling up community management of rural water supply.* (Thematic Overview Paper). IRC International Water and Sanitation Center, Delft, Netherlands, March.

Manikutty, S. (1998). "Community participation: Lessons from experiences in five water and sanitation projects in India." *Development Policy Review,* 16(4), 373–404.

Margoluis, R., and Salafsky, N. (1998). *Measures of success: Designing, managing and monitoring conservation and development projects.* Island Press, Washington, D.C.

McCall, M. K. (2003). "Seeking good governance in participatory-GIS: A review of processes and governance dimensions in applying GIS to participatory spatial planning." *Habitat International,* 27(4), 549–573.

Mendoza, G., and Prabhu, R. (2000). "Development of a methodology for selecting criteria and indicators of sustainable forest management: A case study on participatory assessment." *Envir. Mgmt.,* 26(6), 659–673.

Narayan, D. (1993). *Participatory evaluation: Tools for managing change in water and sanitation.* World Bank Technical Paper No. 207, World Bank, Washington, D.C.

Orr, B. (1985). "Refugee forestry in Somalia." *The Rural Sociologist,* 5(4), 267–272.

Paulete, F. E. (2006). *The Gambia all schools tree nursery competition: Promoting conservation in the Gambia through grassroots environmental education.* <http://peacecorps.mtu.edu/resources/theses.html> (Jan. 14, 2008).

Reents, N. (2003). *Design of potable water supply systems in rural Honduras.* <http://cee.eng.usf.edu/peacecorps/Resources.htm> (Jan. 17, 2009).

Regmi, S. C., and Fawcett, B. (1999). "Integrating gender needs into drinking water projects in Nepal." *Gender and Development,* 7(3), 62–72.

Roche, C. (1999). *Impact assessment for development agencies: Learning to value change.* Oxfam, Oxford, U.K.

Rudel, T., Bates, D., and Golbeck, S. (2006). "How poor remote rural places get child care centers: Patriarchy, out-migration and political opportunities in the Ecuadorian Amazon." *Human Organization,* 65(1), 1–7.

Schouten, T., and Moriarty, P. (2003). *Community water, community management: From system to service in rural areas.* ITDG Publishing and IRC International Water and Sanitation Center, London.

Selener, D. (1997). *Participatory action research and social change.* Cornell Participatory Action Research Network, Cornell University, Ithaca, N.Y.

Selener, D., Chenier, J., and Zelaya, R. (1997). *Farmer to farmer extension: Lessons from the field.* International Institute for Rural Reconstruction, New York.

Selener, D., Endara, N., and Carvajal, J. (1999). *Participatory rural appraisal and planning workbook.* International Institute for Rural Reconstruction, New York.

Strahan, P. (1996). "Handing over an operational project to community management in North Darfur, Sudan." *Development in Practice,* 6(3), 208–216.

Sumberg, J., and Okali, C. (1989). "Farmers, on-farm research, and new technology." *Farmer first: Farmer innovation and agricultural research,* R. Chambers, A. Pacey, and L. A. Thrupp, eds., Intermediate Technology Publications, London.

Tayong, A., and Poubom, C. (1999). "Convincing people to pay for water: Nkouondja in Cameroon." *PLA Notes,* 35, 52–55.

Tesfai, M., and de Graaff, J. (2000). "Participatory rural appraisal of spate irrigation systems in eastern Eritrea." *Agriculture and Human Values,* 17, 350–370.

U.S. Census Bureau. (2000). *Census 2000 urban and rural classification.* <http://www.census.gov/geo/www/ua/ua_2k.html> (Jan. 14, 2008).

Van Veenhuizen, R. (2003). "The rural urban interface." *Annotated bibliography on urban agriculture,* ETC Urban Agriculture Programme, Leusden, Netherlands.

Wiwe, S. (2005). *Participatory multi-criteria decision making in sanitation planning in Ecuador: Case study: Fanca (Bahia de Caráquez), Ecuador.* Unpublished master's thesis. Technical University of Denmark. Kgs. Lyngby.

⟩⟩⟩ Further Reading

McConville, J. R. (2006). *Applying life cycle thinking to international water and sanitation development projects: An assessment tool for project managers in sustainable development work,* <http://cee.eng.usf.edu/peacecorps/Resources.htm> (Jan 17, 2009).

McConville, J. R., and Mihelcic, J. R. (2007). "Adapting life-cycle thinking tools to evaluate project sustainability in international water and sanitation development work." *Environ. Eng. Sci.,* 24(7), 937–948.

Peace Corps. (1996). *PACA: Participatory analysis for community action.* Peace Corps Information and Exchange Manual M0053, Washington, D.C.

Schweitzer, R. N. (2009). *Community managed rural water supply systems in the Dominican Republic: Assessment of sustainability of systems built by the National Institute of Potable Water and Peace Corps, Dominican Republic,* <http://cee.eng.usf.edu/peacecorps/Resources.htm> (May 21, 2009).

4

Project Management

》》 4.1 Project Management in a Developing Country Context

The success of an infrastructure development project is measured by whether its benefits continue after project turnover. Therefore, project objectives should be not only to finish construction, but also to ensure that a system has been put in place that will allow the projects to continue functioning. Project success depends on a community's ability to operate and maintain the project. Development projects are many times collectively cared for, viewed as a public asset, and managed for the common good (Ratner and Gutiérrez 2004). When this collective ownership is not achieved in a community setting, infrastructure systems are not built to last and maintained properly. Box 4-1 lists project management goals for a successful project.

The World Health Organization (WHO) estimates that as many as 60% of the water systems in the developing world are not operational (Davis and Brikké 1995). This high percentage indicates that changes must be made in the way projects are managed. Implementing the strategies discussed in Chapter 3 can help navigate a project through many of the obstacles encountered. The outcome will be a well-built project that is possible for a community to maintain and operate.

》》 4.2 Laying the Groundwork: Project Organization

Infrastructure improvement has similar goals, regardless of whether it takes place in developing or developed countries. Project organization, however, can be quite different. In developing countries, lack of dependable transportation and communication, inadequate funding, and insufficient training present challenges that must be overcome through management.

Infrastructure projects in developed nations are typically owned by the government, a business, or a private entity, which provides the initial idea and the funding for a project and then sets it in motion. In contrast, in development work, the owner's role is often shared and executed by different participants. The funds may be provided by one or more entities, the project may be initiated and planned by others, and the beneficiaries

---

Kraig Lothe contributed to this chapter.

>>> **Box 4-1. Project Management Goals for a Successful Project**

Successful design and quality construction: A successful design focuses on community needs and future maintenance and facilitates quality construction. Quality construction ensures that a project lasts and functions with minimal maintenance.

Well-trained local personnel and follow-up: After project turnover, a community-based administrative committee (e.g., a water committee) is responsible for maintaining and operating the project. If local community members lack the experience and knowledge necessary for this responsibility, the development agency must train local staff and follow through with the community during their first months or years of maintenance.

Affordable maintenance: Infrastructure projects can be expensive, often beyond the means of local communities. Maintenance is also costly, but if considered during design, can be made affordable for the community. A community must establish a locally managed tariff system so that it has the means to pay for maintenance materials, labor, and operational expenses.

Local community ownership: Community members should act as project owners, not as mere recipients. The community administrative board should meet often to discuss issues relating to usage, maintenance, and tariffs.

are yet another group. These groups may share responsibilities. However, the participants may not always share the same motivation or goals for the project.

Project delivery commonly refers to the contractual relationships between the owner, designer, contractor, and management service used in a project. These relationships create the framework in which a project progresses from an idea to a completed facility (CSI 2005). Different project deliveries have been developed in industrialized nations to effectively address the unique demands that result from cost, scope, and time constraints. Project deliveries in development work address many of the demands present in developing countries but have not benefited from extensive research.

4.2.1 A Who's Who of Development Work: The Participants

This section gives a brief description of the roles that may be played by each participant. The general relationships between these participants are shown in Fig. 4-1.

- *Beneficiaries or community owners*—In development projects, beneficiaries are usually located in underdeveloped communities that the project is designed to serve. Often, the beneficiaries are responsible for the project initiation and should be involved until project turnover. Better terminology is needed to address communities as something more than mere users, beneficiaries, or clients. Terms such as "partners," "co-owners," or "community owners" deserve consideration in developing participatory models and community management (McCommon et al. 1990).

Figure 4-1. Responsibilities and Relationships of Participants in International Development Work.

- *Funding agency*—The funding agency provides financial resources for the project. This entity has a contract with the development agency, delegating to it the management of design and construction.
- *Development agency*—The development agency (e.g., a government office or aid organization) sets the project in motion. This agency performs a site investigation, provides or contracts a designer, awards the contract for construction, and manages construction. It also secures and handles money for the project. The development agency can also be the funding agency.
- *Designer or design engineer*—The design engineer's role varies in each international development project. The designer often works within the development agency but may also be independently contracted to do the design work and create a project proposal.
- *Builder and Contractor*—After the design phase of the project is complete, the development agency awards a construction contract to a contractor. In most projects, this is when contractor involvement begins. The contractor is usually a local builder with experience in construction techniques particular to the project and area. The contractor supervises the local labor and construction.

A project's long-term success depends on the participants involved, the implementation practices they use, and the checks and balances of each party. Projects require dedication and competence on behalf of all parties. In developed nations, the relationships between participants are defined through contractual agreements. In development work, contracts are often vague and responsibilities are unclearly defined. When participants work together and are involved from the beginning of the project, goals are united and the people involved are more likely to work as a team.

However, designers often work independently of the community, sending technicians to collect field data. As in all countries, contractors may cut corners to finish a project more quickly and cheaply. A development agency's efforts to involve the community in decision-making processes, training, and follow-up are time-consuming and rarely financially rewarded.

4.2.2 From Idea to the Finished Product: Project Phases

Project Initiation and Site Investigation

Project Initiation

It is common for projects to be initiated outside of communities. Development agencies or funding agencies may propose a project, provide funding, and take it to the community. This top-down approach may not involve the beneficiaries in early decision-making processes. Some projects even engage the beneficiaries as late as project turnover. Typically, projects that are initiated outside of the community and do not involve the community in early decisions are not as successful as community-solicited projects. By initiating a project, a community not only chooses an undertaking that meets their needs but also demonstrates that they are willing to engage in the long-term commitment required to maintain any new infrastructure.

Site Investigation

After a community solicits assistance to initiate a project, an engineer or technician from the development agency performs a site investigation. This investigation can be done at the same time as the community assessment process. Physical factors, including travel time and distance to the site, project size, and estimated cost, are assessed. It is also common to assess the interest of the local government because municipal support will help ensure project success.

Although profits do not drive development work, nongovernmental organizations and government agencies have a lot to lose, or gain, depending on the efficiency with which a project is completed. Future funding depends on a project's success. Development agencies want to invest their efforts in projects that can be completed on time and within budget. Difficult, labor-intensive projects in remote areas are risky and are often avoided by development agencies. Unfortunately, the most remote communities are often the most in need of assistance.

Feasibility Studies and Preliminary Design

After the initial site investigation, feasibility studies and preliminary designs are conducted. In this stage of a project, field data are recorded and compiled to create preliminary design documents, which include preliminary sketches, rough design calculations, and preliminary materials lists. This work is usually the responsibility of the development agency, which either sends its own engineer or technician or contracts an outside engineer. Preliminary design should be done in collaboration with the community to maximize the likelihood of success (see Chapter 3). These documents are then compiled in a proposal and used to secure funding. The feasibility study and preliminary design take an idea and turn it into workable plan.

An accurate preliminary design allows potential contractors and funding agencies to better understand the project and assess their possible involvement. The uncertainties of construction make it difficult to develop a perfect preliminary design, but by properly using preexisting designs, collecting sufficient and accurate field data, and involving expertise of construction personnel in this stage, the preliminary design can minimize discrepancies that surface during construction. Accuracy in the feasibility study and pre-liminary design will reduce the time and budget of a project.

Economics

Economics is an important factor in managing a project and making engineering deci-sions. For example, increasing a water tank's diameter by a meter might allow rural users to irrigate small vegetable plots, improving nutrition at a marginal cost. Material selection and sizing also play an important part in managing and delivering a successful project. For example, designing a pipe system with a limited number of pipe sizes will make con-struction move quickly because limited accessories and perhaps tools will be required. This design decision minimizes the potential for workers to install the wrong pipe size and allows the supervising contractor or engineer to spend less time at the construction site. Future repairs are also simpler for the community because they know what pipe size or accessory to purchase in case of failure. The initial material cost may be greater, but the intangible savings of less required supervision, ease of maintenance, and long-term project sustainability are likely to outweigh the upfront cost.

Preparing a Budget for the Project

Prepared budgets should always clearly include the entire community contribution to a project. The contribution may be cash, agricultural products that can be converted to cash (e.g., rice or beans), local materials, provision of meals for a work crew, use of an animal such as a mule, or labor. Pie charts are effective to display this information.

Calculating the community contribution is important not only for funding agen-cies, which often require a matching contribution from the community, but also for com-munities themselves. Being aware of how much they are contributing can enhance their sense of ownership. As one example, Table 4-1 provides a list of items to include in a bud-get estimate for a spring box project located in Cameroon (Fry 2004). In this project, the community was able to contribute more than 40% of the total project cost by providing labor, meals, and materials.

Securing Funding

Despite their substantial contributions, community owners often do not provide the entire funding for a project. A proposal, created from the data recorded during the preliminary design phase, is used by the development agency to solicit funds. It should be written in col-laboration with the community, ensuring that participants clearly understand their respon-sibilities and expected outcomes. It will include information about the community, acces-sibility, a summary of the design, material lists, drawings, and estimated costs. This proposal provides a funding agency with information to make a decision about their involvement. Securing funding is the pivotal point of a project. When funding is received, project partici-pants can be organized and plans can be made to execute the project. And after the funding agency finances a project, it will depend on the development agency to manage the funds.

Construction Documents (Design) and Contractor Selection

Construction document designs are based on the preliminary design documents. These final documents include detailed specifications and drawings, construction plans, and contracts.

Table 4-1. Typical List of Items to Include in a Spring Box Budget Estimate

Spring Box Construction Materials List		
Designation	Unit	Quantity
Small Project Assistance Request		
Cement	sack	50
Rebar, 8 mm	bar	25
Sikalite, a powdered waterproofing admixture	kg	50
PVC pipe, diameter 40 mm	pipe length	2
Saw blade	pack	1
Binding wire	reel	1
Nails, 70 mm	kg	10
Elbow	each	4
Faucet 20/27 (inside/outside diameter)	each	1
Stop valve 20/27 (inside/outside diameter)	each	1
Galvanized pipe 20/27 (inside/outside diameter)	each	1
Sleeve 20/27 (inside/outside diameter)	each	2
Tangit glue	each	1
Transport of materials		
Technician—mason		
Technician—plumber		
Contingency (10%)		
Community Contribution		
Clearing and grubbing and site excavation	days	2
Storing rocks and gravel	m^3	4
Stocking sand	m^3	5
Form lumber	boards	15
Nonspecialized labor	person-hours/day	80
Meals for workers	each	16
Technical supervision	day	20
Contingency (10%)		

Note: A community may be able to provide more than 40% of the total cost in locally available materials and labor.

Source: Fry 2004.

Additional data are gathered through site visits to create a detailed design. This design, which includes material lists and specifications, is used by contractors during the bidding process.

For larger projects, construction agreements and contracts are created to define the contractor's responsibilities and describe the scope of the work. These documents assign responsibilities to the contractor and the development agency. Agreements are signed to avoid disputes during construction. Once signed, the contract legally links the development agency and contractor, allocating risk to each party. Although the contracts are often standard forms that the development agency selects and adapts to particular projects, it is important to tailor these documents to each project so that there are no doubts about how responsibilities are assigned.

Although many projects are technologically simple, without qualified skilled labor, projects can suffer from poor workmanship and management. If a community lacks the specialized skills required, a contractor is hired by the development agency to assume responsibility for the construction. As in the developed world, the contractor rarely enters the project until immediately before the construction process.

The contractor is usually hired based on previous work experience or by lowest bid, but in many settings, noncompetitive relationships may hold sway over more effective and fairer selection methods. Contractor selection can be done in a single-stage or a two-stage process. In a single-stage process, contractors are given the opportunity to bid for the contract. They base their estimated cost on the design documentation that the development agency provides for them. The contract is awarded to the contractor with the lowest qualified bid. Local contractors may be accustomed to negotiating, not competing, for projects. When foreign participants and development agencies are involved, local contractors may try to inflate their price. The bid competition in single-stage contractor selection is an effective way to avoid this.

A two-stage process includes a prequalification stage before the bid evaluation and selection. The two-stage selection process aims to streamline the decision-making time after pricing and to improve work quality. In large construction projects, three to four bidders are often prequalified based on previous experience and management capabilities. These prequalified bidders are asked to submit proposals. After their proposals are submitted, their prices are weighed against their experience, qualifications of personnel, quality of service provided, quality of management, current workloads, strategic alliances, and resources (Love et al. 2004). This process is also common in international development work. The process can take more time, but it provides the designer and owner with the better contractor to match their needs. It can also provide an opportunity to involve potential contractors before the design phase is finished, allowing valuable construction knowledge to be incorporated into the design.

Construction

The construction phase includes actual construction and some additional planning. Strategies that take place in the planning stage can reduce the project schedule by 35% at no extra cost. Most prevalent are material procurement strategies, aimed to save time by ensuring that materials are always present (Songer et al. 2000). Materials, both local and nonlocal, can be acquired before construction begins if they can be stored securely. Materials can also be acquired during construction, but common problems in transportation and supply may cause delays.

A project's success depends on construction planning efforts. Though it is the responsibility of the contractor and the development agency, community committees and owners must also be included in decision making regarding material planning and organization of labor.

Before construction can begin, it is necessary to establish and begin training a local administrative committee. This work should be done before construction, while enthusiasm is high and community members are ready to make the long-term commitment required of any successful project. Through this administrative committee, labor is organized and a fee system is established to ensure that the community can afford future maintenance costs.

During project implementation, it is imperative that community members are trained for technical aspects such as concrete work, pipe fitting, and erosion control. The construction phase provides an excellent opportunity to learn these skills. The development agency and the contractor may facilitate this training through community meetings.

Project Turnover, Follow-Through, and Maintenance

At completion, the community becomes sole owner of the project. Although the goal is to turn a project over to a community that can handle all system operation and maintenance, the community may need some additional support to address challenges that arise. A good development agency will continue to work with the community to ensure that daily operations are carried out correctly and that problems are successfully resolved. Davis and Brikké (1995) wrote that water supply projects should not be regarded as an end but as "initiators of a range of benefits which continue long after projects have been handed over." With this type of thinking, it is easy to see why a project that lasts for 20 years is 10 times as effective as a project that lasts for 2 years.

4.2.3 Driving a Project to Completion: Project Delivery

Project delivery—the contractual teaming relationships between the owner, designer, contractor, and management service—creates the framework by which a project is implemented. Projects can be initiated by development agencies or funding agencies. This top-down style of project delivery is common in the developed world. However, the top-down approach does not typically involve the community until the end of the project. As these projects are set up, no effort is required from the community and there is no obligation to the community on behalf of the development agency. Top-down approaches to infrastructure development have been practiced for many years. Centralized governments still decide where resources may be targeted and what projects will be executed.

Another trend has also begun in which smaller agencies and local governments involve communities in the decision-making process and respond to their needs. This grassroots approach involves the community from the beginning of the project to project completion. Community-initiated and community-driven projects are better focused on the community's needs and its ability to maintain the project beyond turnover. Grassroots approaches to decision making are gaining acceptance in the international development community. They have proven to be an excellent way to address the needs of the community and to make sure that they are prepared to take on the commitment a project requires.

⟩⟩⟩ 4.3 Putting the Community First: Increasing Community Involvement

Lack of community involvement in decision making often results in communities without the capacity or desire to maintain a project. Without a sense of ownership, a community will not maintain a project. Lacking a sense of ownership may also be a sign that a project is not right for a community. Ownership is gained through involvement in the decision-making process throughout the life of a project.

Because communities often lack financial resources to commit to projects, manual labor is an important community contribution to a project's budget. Not only is this labor contribution imperative to execute a project, it also demonstrates the community's commitment and desire to have the project.

Labor organization is vital to a project's efficiency. The contractor may hire skilled labor to complete the project more quickly. During construction, there should be enough unskilled labor available to minimize the contractor's time in the field. When the contractor arrives, the community should be organized and ready to work. This organization is a shared responsibility of the development agency and the community.

Organizing labor can be difficult, especially in a community that has not previously organized a large construction project. In this case, the initial organizational effort must be facilitated, and this situation provides an opportunity to train community members in the organization process. This process involves calling a public meeting, setting meeting agendas, and facilitating constructive discussions from which issues can be resolved.

Some families may choose not to work, but this cannot disrupt the motivation of the other families. Many successful projects have "buy-in." If someone from one household works a certain number of days, they have earned an equal share of the project. If another household wishes to have a share of the project but cannot, or will not, work, they can "buy in" to the project. The community can then put this money toward the operation of the project.

When women take ownership of a project, the odds of success are much greater. However, labor forces in construction may be predominantly male. In some cultures, though, women will be engaged in strenuous tasks such as carrying materials to the project site. Organizational and administrative roles are also excellent ways to involve women in development projects. Women provide a valuable and often overlooked perspective. Some countries may even require female involvement, for example, by law, in Honduras there must be one woman on every water committee.

))) 4.4 Tricks of the Trade: Effective Strategies in Managing a Project that Incorporate a Community

When the sole goal is completion of construction, many social factors are ignored that should be incorporated into project goals. There are strategies that focus on social factors as well as the completion of a quality project. A general project management checklist to ensure a project is sustainable and incorporates social factors is provided in Fig. 4-2.

4.4.1 Include the Community in Project Planning and Development

As discussed previously (and in Chapter 3), communities should initiate projects and be involved in the decision-making process from the start. Not only does this foster a feeling

---

Project & Community Management Strategies Checklist

Project Initiation and Investigation

Conduct a local needs assessment.
Identify stakeholders and community leaders.
Analyze the organization of leadership within the community.
Assess the community's willingness and ability to contribute.
Analyze climate and environmental constraints.
Investigate site.

Feasibility Studies and Preliminary Design

Define clear project goals with all participants.
Evaluate community's ability and willingness to operate and maintain design.
Explore options to integrate local knowledge and resources into designs.
Consult pre-existing designs of similar projects.
Estimate project costs.
Analyze seasonal variation in water supply and demand.
Develop funding proposal and secure funding.

Construction Documents (Detailed Design)

Solicit community input regarding design.
Incorporate local knowledge and techniques into design.
Involve community members in project planning.
Establish an equitable system for community labor and resource contribution.
Consider seasonal labor constraints.
Analyze the traditional structure of community projects.
Define the roles and responsibilities of participants in a detailed action plan.
Consult with potential contractors regarding design.
Pre-qualify contractors using a two-stage contractor selection process.
Verify the costs and availability of resources.
Contractor identification and selection.

Construction Phase

Establish a locally-elected administrative board to manage the project.
Work with the local administrative board to establish a tariff system.
Work closely with the contractor and administrative board to organize labor.
Set a realistic work schedule and allow for the unexpected in planning.
Train community in techniques and tools related to construction.
Invite local agencies to assist in training and education.
Implement local knowledge and techniques daring construction.
Increase designer involvement throughout construction.
Monitor spending throughout construction.
Discuss project benefits, operation, and maintenance in public assemblies.
Collaborate with local agencies when planning for operation and maintenance.

Turn-Over, Follow-Through, and Maintenance

Ensure community members are able to perform operation and maintenance.
Facilitate community leadership by the locally-elected board.
Ensure that project costs and benefits are equitably distributed.
Encourage discussions about constraints to system use and maintenance.
Coordinate operation and maintenance of the system with local agencies.
Work with the local administrative board to implement a tariff system.

Figure 4-2. Checklist Detailing Strategies That May Be Implemented in Development Work to Improve Project Sustainability.

Source: Adapted from McConville 2006 and McConville and Mihelcic 2007.

of community ownership, it also allows the development agency to understand the financial, scheduling, and logistical constraints the community faces.

4.4.2 Incorporate Local Construction Techniques and Knowledge into Design

Construction knowledge should be introduced in early project phases to address the issue of constructability. Unfortunately, there is often a large separation of knowledge and responsibility between the design and construction phases, causing problems in both. For example, the designers do not always have practical construction experience. When local knowledge and construction techniques are not taken into account, unreasonable expectations and inadequate specifications result. To minimize this situation, development agencies can either staff a contractor or involve potential contractors in the design phase through a two-stage contractor selection process. This assistance provides continuity of construction personnel from design to construction.

Bringing contractors into the project earlier allows them to express their needs and learn the needs of the other participants. Studies have confirmed that integrating construction knowledge into the design process helps achieve a better quality project, completed in a safe manner, on schedule, and for the least cost (Arditi et al. 2002). These constructability practices have a high potential in development work because local innovation can positively influence the construction process.

Design can also be improved through better construction documentation. The specifications and drawings used in development work are often standardized. This practice saves time, but because these documents are rarely tailored to specific projects, they may not be appropriate given site conditions.

Engineers must revise standard documents, drawings, and specifications and review them with the contractor and local community to ensure that they are consistent with the community needs and demands of the particular site and project. This review can facilitate a discussion about the feasibility of the plan and the challenges that may arise. Having worked with similar standardized documents, contractors can explain how they have been inadequate in the past. This construction knowledge and local expertise should be incorporated into the design documentation.

Construction quality depends not only on the competence of contractors but also on the development agency. Development agencies and designers should increase the amount of time they spend on site. This presence will help maintain a higher quality of work and also provide an opportunity for designers to learn more about construction and local knowledge that can be used during the design phase of future projects. There is a wealth of knowledge in the field, but often designers are divorced from construction.

4.4.3 Build the Community's Capacity to Administer, Operate, and Maintain the Project

Through the creation of a local administrative committee, the framework of a project's organization is started. This educational aspect of development work may not have the obvious, immediate results that construction has, but its success multiplies the effectiveness of any project. Different cultures have unique organizational structures. It is important to work within these structures and follow models that have already succeeded in similar settings.

4.4.4 Increase Project Participation and Coordination Among Participants

Projects will suffer if there is a lack of participation by any party involved. If project participation and coordination are implemented, they can

- eliminate ambiguous and incomplete designs,
- prevent misunderstandings during construction,
- incorporate valuable construction knowledge into design,
- unite participants' goals,
- foster project ownership among community members,
- make use of available resources and local expertise,
- keep construction and maintenance manageable for the community, and
- improve the quality of design.

))) 4.5 Navigating the Stumbling Blocks during the Construction Process

The involvement and cooperation of participants is required to overcome the challenges of clashing personalities, financial limitations, time and scheduling problems, technical barriers, and various surprises of nature. Material delays, misunderstandings, and discrepancies in design are a few common problems that prevent projects from being completed successfully and on time. The following are some of the most common tasks where problems are encountered during the construction process in development work.

4.5.1 Scheduling

For effective project management, projects require scheduling and updating from initial planning to conclusion of construction, training, and project turnover. A schedule is a tool that is useful in communicating among the development agency, the community, the designer, and the contractor. Effective schedules include all project phases, not just construction. The construction portion of the schedule may be more detailed and may require more frequent updating and communication than the other project phases.

The construction process consists of a series of interdependent steps. For example, the spring box project budgeted in Table 4-1 involved five distinct construction steps, and the construction lasted approximately three weeks, as shown in Figs. 4-3 and 4-4 (Fry 2004). Figure 4-3 demonstrates that the actual construction phase is short compared with the overall length of the entire project cycle. Each step requires specific technical knowledge and the cooperation of participants.

Because of unreliable transportation, harsh weather, illness, a seasonal work and holiday calendar, material procurement difficulties, and numerous other constraints, successfully scheduling construction can be a monumental task. However, it is the responsibility of the

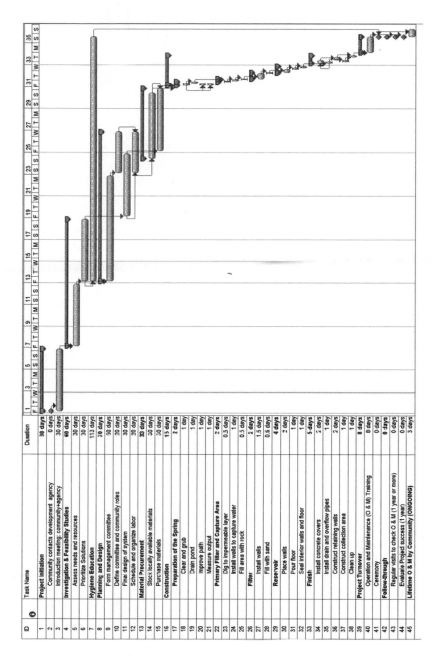

ID	⊙	Task Name	Duration
1		**Project Initiation**	30 days
2		Community contacts development agency	0 days
3		Introduction meeting: community+agency	30 days
4		**Investigation & Feasibility Studies**	60 days
5		Assess needs and resources	30 days
6		Prioritize Solutions	30 days
7		**Hygiene Education**	113 days
8		**Planning and Design**	70 days
9		Form management committee	50 days
10		Define committee and community roles	20 days
11		Fine design of system	30 days
12		Schedule and organize labor	20 days
13		**Material Procurement**	33 days
14		Stock locally available materials	20 days
15		Purchase materials	30 days
16		**Construction**	15 days
17		**Preparation of the Spring**	2 days
18		Clear and grub	1 day
19		Drain pond	1 day
20		Improve path	1 day
21		Measure output	1 day
22		**Primary Filter and Capture Area**	2 days
23		Dig to impermeable layer	0.5 days
24		Install walls to capture water	1 day
25		Fill area with rock	0.5 days
26		**Filter**	2 days
27		Install walls	1.5 days
28		Fill with sand	0.5 days
29		**Reservoir**	4 days
30		Place walls	2 days
31		Pour floor	1 day
32		Seal interior walls and floor	1 day
33		**Finish**	5 days
34		Install concrete covers	2 days
35		Install drain and overflow pipes	1 day
36		Construct retaining walls	2 days
37		Construct collection area	1 day
38		Clean up	1 day
39		**Project Turnover**	8 days
40		Operation and Maintenance (O & M) Training	8 days
41		Ceremony	0 days
42		**Follow-through**	0 days
43		Regular visits to check O & M (1 year or more)	0 days
44		Evaluate Project success (1 year)	0 days
45		Lifetime O & M by Community (ONGOING)	3 days

Figure 4-3. Example Spring Box Project Schedule Presented in Format Used by the Software Tool MSProject.

Source: Adapted from Fry 2004.

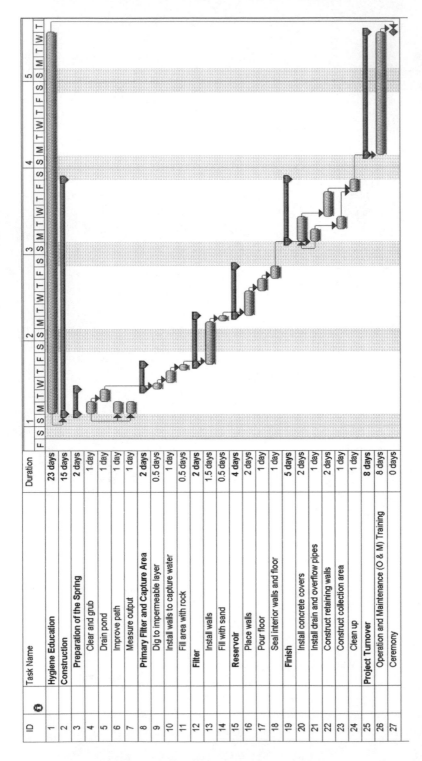

ID	⏱	Task Name	Duration
1		**Hygiene Education**	**23 days**
2		**Construction**	**15 days**
3		**Preparation of the Spring**	**2 days**
4		Clear and grub	1 day
5		Drain pond	1 day
6		Improve path	1 day
7		Measure output	1 day
8		**Primary Filter and Capture Area**	**2 days**
9		Dig to impermeable layer	0.5 days
10		Install walls to capture water	1 day
11		Fill area with rock	0.5 days
12		**Filter**	**2 days**
13		Install walls	1.5 days
14		Fill with sand	0.5 days
15		**Reservoir**	**4 days**
16		Place walls	2 days
17		Pour floor	1 day
18		Seal interior walls and floor	1 day
19		**Finish**	**5 days**
20		Install concrete covers	2 days
21		Install drain and overflow pipes	1 day
22		Construct retaining walls	2 days
23		Construct collection area	1 day
24		Clean up	1 day
25		**Project Turnover**	**8 days**
26		Operation and Maintenance (O & M) Training	8 days
27		Ceremony	0 days

Figure 4-4. Example Spring Box Project Schedule for Construction Phase Presented in the Format Used by the Software Tool MSProject.

Source: Adapted from Fry 2004.

development agency, in collaboration with the administrative committee, to manage risks and prevent delays when possible. This management is accomplished by using a detailed schedule as a tool that is continually updated and actually used to manage the activities. Although it is impossible to avoid or anticipate all disruptions, if both the community and the contractor can depend on the planning and scheduling efforts of the development agency, participants can work in a coordinated effort to keep a project on schedule.

4.5.2 Labor

Many labor problems are also unique to infrastructure development in developing countries. High levels of unemployment and low levels of education lead to a surplus of willing but poorly skilled contractors and laborers looking for income opportunities (Danert et al. 2003). On the other hand, a community may have educated members who ask for reading material or a website to inform themselves and then approach you with questions. The labor force of community members is the cornerstone of the construction effort.

4.5.3 Obtaining Materials

Material procurement requires significant planning because of road conditions, scarcity of transportation, and inconsistency of suppliers. Many sites may be remote and accessible only by small four-wheel-drive vehicles, horse or donkey, or by foot. Roads are often washed out, preventing any vehicles from passing. Seasonal rainfall may make access to gravel and sand impossible in areas where these materials are obtained from local riverbeds.

Nonlocal materials, such as piping, cement, and accessories, should also be purchased early and securely stored at the job site. Material costs can fluctuate because of price hikes caused by shortages, gas rates, or inflation. For this reason, materials may be ordered weeks in advance and stored in the community.

Material storage is also important. For example, during the rainy season, moisture-sensitive materials such as cement may be ruined if not properly stored. Cement, wood, rebar, and accessories often need to be carried to the site, which for water projects may span many kilometers. Items to consider when selecting a storage site include security, weather-proof construction, and proximity to the work site. Proximity to the work site should not be overlooked because additional materials may be required not only when materials are forgotten or miscalculated but also when design changes are made or repairs are required.

In developing countries, material costs will be high in comparison to labor, and materials are generally of more variable quality. Local contractors may also be accustomed to stretching materials to save money. Design calculations cannot assume the expected 3,000 lb/in.^{2} concrete design strength because it may not be attained in practice. In addition, inferior quality of materials may not be noticed until they arrive at site and must be either reordered or installed as-is, compromising work quality.

4.5.4 Safety

Safety should always be a concern. If machinery is not present, every task may be performed by hand, often on unforgiving terrain, using sharp tools and heavy objects. And because injuries may occur, the effect may be magnified because job sites can be located in remote areas where medical attention is not available. It is thus a good practice to train communities in the basics of first aid. Perhaps most important, communities should create an emergency plan so that they can act in a coordinated effort when needed. The involvement of local health posts and health workers in community training is an excellent way to tap into nearby resources.

4.5.5 Financing

Projects may be under financial constraints for several reasons in addition to everyday economic stresses. Construction is susceptible to large budget overruns caused by many unpredictable variables. Because projects deal with relatively large sums of money, there is often top-down bureaucratic oversight to monitor the use of appropriated funds. These bureaucratic systems may make the process of receiving funds appear difficult. And once a budget is set, obtaining additional funding to cover incidentals and budget overruns can hold up a project for long periods. It thus may be wise to have contingency funds ready to deal with issues that will ultimately arise over even the most well-planned project.

4.5.6 Seasonal Calendar

Weather affects not only roads, material procurement, and working conditions, but also the routines of community life. Many beneficiaries of development work rely on agriculture as an important source of income. Agriculture follows a strict schedule, so construction projects must be flexible enough to accommodate this. Most subtropical climates have a distinct rainy season that dictates when crops must be planted and harvested. At these times, local laborers have no choice but to work in the fields to provide for their families. Other events, such as religious or cultural activities, must also be taken into account when scheduling a project. Projects should therefore be planned, estimated, scheduled, and communicated well in advance so that the construction phase takes place when communities have time to work and the weather will not hinder the construction effort.

))) References

Arditi, D., Elhassan, A., and Cengiz Toklu, Y. (2002). "Constructability analysis in the design firm." *J. Constr. Engrg. and Mgmt.*, 128(2), 117–126.

The Construction Specifications Institute (CSI). (2005). *Project resource manual—CSI manual of practice*, 5th ed., McGraw-Hill, New York.

Danert, K., Carter, R. C., Rwamwanja, R., Ssebalu, J., Carr, G., and Kane, D. (2003). "The private sector in rural water and sanitation services in Uganda: Understanding the context and developing support strategies." *J. International Development*, 15, 1099–1114.

Davis, J., and Brikké, F. (1995). *Making your water supply work: Operation and maintenance of small water supply systems.* IRC International Water and Sanitation Centre. The Hague, Netherlands.

Fry, L. (2004). *Spring improvement as a tool for prevention of water-related illnesses in four villages of the Center Province of Cameroon.* <http://cee.eng.usf.edu/peacecorps/Resources.htm> (Jan. 17, 2009).

Love, P. E. D., Irani, Z., and Edwards, D. J. (2004). "A rework reduction model for construction projects." *IEEE Transactions on Engineering Management*, 51(4), 426–440.

McCommon, C., Warner, D., and Yohalem, D. (1990). *Community management of rural water supply and sanitation services*. UNDP/World Bank, New York. WASH Technical Report No. 67. Water Handbook UNICEF, 20.

McConville, J. R. (2006). *Applying life cycle thinking to international water and sanitation development projects: An assessment tool for project managers in sustainable development work*. <http://cee.eng.usf.edu/peacecorps/Resources.htm> (Jan. 17, 2009).

McConville, J. R., and Mihelcic, J. R. (2007). "Adapting life-cycle thinking tools to evaluate project sustainability in international water and sanitation development work." *Environmental Engineering Science*, 24(7), 937–948.

Ratner, B. D., Gutiérrez, A. R. (2004). "Reasserting community: The social challenge of wastewater management in Panajachel, Guatemala." *Human Organizations*, 63(1), 47–56.

Songer, A. D., Diekmann, J., Hendrickson, W., and Flushing, D. (2000). "Situational reengineering: Case study analysis." *J. Constr. Engrg. and Mgmt.*, 126(3), 185–190.

))) Further Reading

Dorsey, R. W. (1997). *Project delivery systems for building construction*, Associated General Contractors of America, Washington, D.C.

Haltenhoff, C. E. (1999). *The CM contracting system: Fundamentals and practice*. Prentice Hall, Upper Saddle River, N.J.

Jaselskis, E. J., and Talukhaba, A. (1998). "Bidding considerations in developing countries." *J. Constr. Engrg. and Mgmt.*, 124(3), 185–193.

Lothe, K. (2006). *An analysis of constructability strategies in project delivery: Making infrastructure construction sustainable in Copán Ruinas, Honduras*. <http://cee.eng.usf.edu/peacecorps/Resources.htm> (Jan. 17, 2009).

Midgley, A. R. (1999). "Working overseas: A fascinating professional challenge." *Engineering Management J.*, 9(5), 235–240.

Stern, P., and Longland, F. (1983). *Field engineering: An introduction to development work and construction in rural areas*. Intermediate Technology Publications, Ltd., London.

Basics of Construction

5

Topographical Surveying Using Abney Levels and Global Positioning Systems

))) 5.1 Topographical Surveying

A topographical survey provides the designer with data about the elevations, distances, and directions required for a potential infrastructure project (such as a water distribution system). A survey requires a few tools, a well-motivated workforce, and a lot of patience. A lack of experienced professionals combined with rugged terrain can make surveying an adventure as much as it is a study. This chapter applies surveying to water supply systems, which are discussed in Chapter 11.

An organized workforce can cut the duration of a survey in half and can also improve the accuracy of the data. In the beginning, take time to explain each person's role, the basics of surveying, and the importance of accuracy and consistency. It is best to keep certain workers doing the same job to gain a level of expertise. A workforce that understands the principles of surveying will work together more efficiently to reach a goal quickly, safely, and accurately.

5.1.1 Which Tool to Use?

A number of tools can be used to complete a survey. In developed-world projects, professional surveyors use modern survey equipment, including computers and satellite-based telecommunication equipment to identify coordinates, elevations, and distances. These may not be available for developing-world projects because organizations do not have adequate funds or technical knowledge. Occasionally, secondhand equipment is donated to a nongovernmental organization (NGO), but rough terrain, dense vegetation, and a lack of knowledge can add to the difficulty of surveying. In flat, open areas, however, this equipment may be useful.

In rural field work, Abney levels are teamed with measuring tapes and compasses, and the global positioning system (GPS). An Abney level (Figure 5-1) is a handheld sighting tube with an attached level, used to measure angles of inclination or declination of a line from the observer to a targeted point. The Abney level is relatively inexpensive and

Kraig Lothe contributed to this chapter.

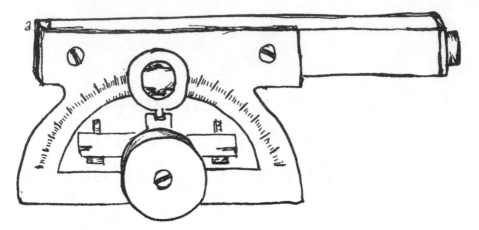

Figure 5-1. An Abney Level.

accurate for the basic technology it uses. The accuracy of GPS units is questionable in current models, but as technology improves and GPS units become more affordable, their place in field engineering may replace that of the Abney level. Each survey tool has its distinct advantages. Abney levels are best for short distances to minimize error accumulation. Each measurement may be slightly inaccurate, and for studies of 3 km or more, the accumulation of error may affect design if there are critical points in the line.

A *critical point* is a high point in the pipeline where the pressure in the pipe may become unacceptably low to deliver water past these points. This absence of water occurs if the point is less than 10 m below the source elevation or the last point where the pressure is broken, such as a break-pressure tank. If the high point is more than 2 km from the last pressure break and its elevation is within 20 m of the source or break-pressure tank, consider it critical. Low points are not as critical because one can be conservative. If there is any concern that the pressure is too high, use a stronger piping material.

GPS units are significantly more convenient and much faster to use, but each point may have a substantial error. The redeeming quality of GPS units is that the error will not accumulate. If there are no critical points in the pipeline, a GPS unit may be the obvious and easy choice.

5.1.2 The Preliminary Survey

Regardless of the survey tool, conduct a preliminary survey. This survey is a chance to check the water source, usually a spring or small stream. Document the quality of the water, the use and health of the watershed, and the flow rate of the source. After the source is checked, walk the proposed route of the water pipeline, noting distances, each high point in the line, and the tank site. For this preliminary survey, it is helpful to use a GPS. If possible, visit each house to take a point with the GPS. This step assists in determining the route for distribution lines. If time doesn't permit, measure the altitude at the highest house and the lowest house in the village.

These steps help to determine the feasibility of the project. Many projects and sources are abandoned at this point. It is difficult to say no to a proposed project, but if plans need to change, this is the best time to do it. If the project appears feasible, one or two leaders in the community should organize work teams for assistance with the topographical survey.

5.1.3 Information Required for Design

There are sets of information needed to complete the design:
1. measurements and photos of the source (to design a spring box or dam);
2. locations of possible concrete anchors, sediment clean-out valves, and air valves;
3. notes of stream crossings, road crossings, or terrain that cannot be excavated;
4. flow rate at the source, in L/min (from the preliminary survey);
5. the number of households and community buildings included in the survey; and
6. measurements: angles, cardinal direction, distances, and detailed notes.

Google Earth can also be used before the site visit and later on to confirm the selected pipeline path.

))) 5.2 Use of the Abney Level

5.2.1 Materials and Preparation

The Abney level requires little technology but a larger workforce. It requires the following items:
1. an Abney level;
2. a magnifying glass;
3. a pocket level;
4. a screwdriver or pocketknife;
5. a measuring tape, 30 or 50 m long;
6. a compass;
7. a notebook and pens;
8. three staffs of equal length, about 1.5 m;
9. one permanent marker;
10. many stakes, about 50 cm long, and a tool to pound stakes into the ground; and
11. a number of machetes (and people who know how to use them).

Cut staffs before the survey starts, but you can also cut the stakes while the survey is being performed. At a minimum, a team of six people, including the engineer, is needed to complete the survey.

The most important resources are the community members. They will be instrumental in selecting the best route and providing directions. The path should stay lower than the source, should avoid crossing unsuitable terrain, and should be the shortest route to the destination, not the convenient walking path. Make sure that the whole team clearly understands what should be accomplished in choosing the path.

Although it is possible for one person to both read the Abney level and record the results, it is helpful to have a second person assist with recording the data. This system not only makes it easier to organize the work, it will greatly reduce the required time. Draw a chart in the notebook that allows the data recorder to keep the data organized (Table 5-1). Too many details are never a problem.

5.2.2 Calibrating the Abney Level

Calibrate the Abney level. This step should be done before each survey, and perhaps twice for multiday surveys. It is done by following these four steps:

1. Find a level surface, such as a table or the top of a wall, using a pocket level until you find a section that is level.
2. Using a magnifying glass to ensure accuracy, align the zero (0) on the top scale with the zero (0) on the bottom scale and then tighten the knob to lock the sliding scale in place.
3. Place the Abney level on top of the level. Using a screwdriver (for most models), adjust the level vial on the Abney level until the bubble rests directly between the two lines.
4. Check the calibration. This step requires two people. Simply repeat a series of front and back shots as described below. If the readings are consistently within 40 min of each other, but opposite, the level is well calibrated.

5.2.3 The Process

Stations are points along the route at which measurements are taken and recorded. To ensure accuracy, there should be no more than 30 m between consecutive stations when using an Abney level. Stations should be located at each local high and low point, where air valves or sediment cleanouts may be needed. Stations should also mark locations of pipe anchors or required changes in the piping material (e.g., between PVC and galvanized iron). Also record detail related to changes in the terrain and direction, changes in land ownership, changes in piping material, and possible location of valves, anchors, stream crossings, fences, and other features along the line. When crossing water, take note of low- and high-water marks along both sides of the bank.

Table 5-1. Example of Data Recorded During a Topographic Survey Using an Abney Level, Tape, and Compass

Beginning Station	End Station	Inclined Dist. (m)	Forward Angle	Back Angle	Compass	Observations
1	2	10.00	2°10′	−2°30′	154°	At source
2	3	17.80	15°30′	−15°10′	150°	Rock, use GI pipe
3	4	8.45	13°40′	−14°00′	120°	Easy to excavate
4	5	8.85	18°40′	−18°40′	134°	Down to anchor
5	6	11.35	−5°50′	5°30′	123°	Cross stream to anchor, GI pipe
6	7	13.20	12°10′	−12°50′	120°	Leave forest

Note: GI pipe is galvanized iron pipe.

Because terrain near the source can be covered in dense vegetation, it is best to do a practice survey with the work team before reaching the source. Once the team arrives at the first station (the source), one team member (the surveyor) places a staff there and keeps it vertical. Another team member is placed in charge of another staff and takes it down the line, holding the end of the measuring tape. Once this person selects the next station, she places the staff upright as well and holds the end of the measuring tape firmly on top of the staff. Stretching the tape tightly so that it does not touch vegetation, and holding it to the top of each staff, the surveyor reads tape and records the distance between the two points. The staffs must stay vertical and not be moved until measurements are complete.

Once the distance has been recorded, the surveyor uses the compass to record the direction. He records this direction in degrees under the column "Compass" (Table 5-1).

Next, the surveyor uses the Abney level to measure the vertical angle from the first to the second point (Figure 5-2). This measurement is called the *forward shot* (recorded under the column "Forward Angle" in Table 5-1). A line of sight is needed for this measurement, which may require the path to be cleared with machetes. Once the line of sight is established, the person holding the staff opposite the surveyor places a hand on top of the staff perpendicular to the staff.

The surveyor then places the Abney level atop her own staff, sights the top of the opposite staff (Figure 5-3), which is visible with the help of the other team member's hand, and takes a measurement. Leaving the first staff with another team member, the surveyor proceeds to the second staff and measures the angle in the opposite direction. This measurement is called the *back shot* (recorded as "Back Angle" in Table 5-1). The angle measured here should be opposite the angle measured in the forward shot. An acceptable error between the forward and backward shot is 30 to 40 min; depending on the accuracy demanded by the circumstances (1 degree is divided into 60 min). If the measurements are consistently different, the Abney level may require recalibration.

Once the back shot has been taken and the angles are acceptable, complete all the measurements for the first station. At this point, remove the staff and replace it with a stake.

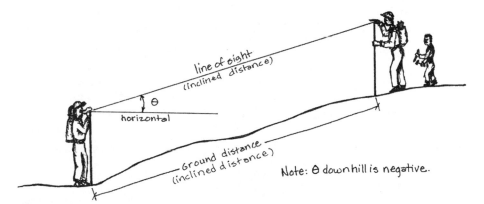

Figure 5-2. Using an Abney Level with Staffs to Take Elevation Readings, Measuring Vertical Angles.

Source: Redrawn with permission from Jordan (1980).

Figure 5-3. Measuring the Angle of Inclination with an Abney Level.

Each stake should be about 50 cm long. Sharpen one end, shave the bark off the other end, and write the number of the station in permanent marker. Stakes must be cut and carved during the entire survey so that they are always available. These stakes are important. They must be numbered correctly and should not be moved until the construction begins, which may be a year or two later. Inevitably, a handful of stakes will be missing, but with accurate data and field notes, the points can be found again or resurveyed.

5.2.4 Measuring Vertical Angles

Measuring angles with the Abney level requires a steady hand and a good eye. Using a staff as a support makes measurements consistent and accurate. Looking through the Abney level, the surveyor sees a cross hair. Hold the horizontal cross hair steady and align it with the top of the opposite staff (Figure 5-4a). While the cross hair is lined up with the top of the staff, adjust the level so that the bubble aligns itself vertically with the cross hair as well (Figure 5-4b). Then read the measurement. The marks are small on the Abney level; use a magnifying glass to avoid errors.

5.2.5 Picking up the Pace

When the team becomes comfortable with the different jobs, it is possible to speed up the process by adding another worker. This addition allows the team to leapfrog and work on more than one station at a time. After each back shot is taken and the back staff is no longer needed, the team member in charge of the back staff carries the staff past the next

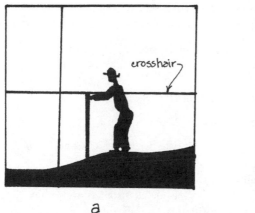 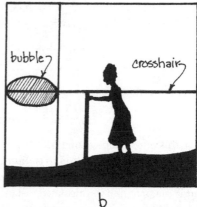

a   b

Figure 5-4. The View Seen Through the Abney Level.

Note: (a) First line up the cross hair with the top of the staff; (b) then line up the bubble with the cross hair and the top of the staff after the level is adjusted.

two staffs to the next station to be measured. As they pass the second staff, they leave the end of the tape measure to measure the next distance. In this way, the surveyor measures the angles while the tape is stretched to the third staff. The surveyor can read the tape and record the distance as soon as he or she reaches the next staff. This system can cut work time in half, but team members need to be proficient before increasing the pace.

5.2.6 Using the Data

After taking the measurements, use the data to find the vertical distance, horizontal distance, and compass direction of each station. This work can be done manually, as shown in Example 5-1, but the work is made easier with the aid of a computerized spreadsheet (Figure 5-5).

)))) Example 5-1. Sample Calculation Used to Find the Horizontal and Vertical Distances between Stations Using Data Recorded from an Abney Level Survey

This example demonstrates how to calculate horizontal and vertical distances using the field data recorded below from station 1 to station 2 (Table 5-1).

Inclined Dist. (m)	Forward Angle	Back Angle	Compass
10.00	2°10′	–2°30′	154°

Using trigonometry, calculate the horizontal and vertical distances between station 1 and 2. First, average the angles to minimize error. Because the forward angle is the angle used in calculations, make both angles positive.

The average of 2 deg, 10 min (written as 2°10′) and 2°30′ is 2°20′, or 2.33°. Now calculate the vertical and horizontal distances as follows:

$$\text{Vertical distance} = \text{inclined distance} \times \sin \theta$$

$$\text{Vertical distance} = 10 \ m \times \sin 2.33° = 0.41 \ m$$

$$\text{Horizontal distance} = \text{inclined distance} \times \cos \theta$$

$$\text{Horizontal distance} = 10 \ m \times \cos 2.33° = 9.99 \ m$$

Use this information, along with the compass direction, to derive the X, Y, and Z coordinates of station 2 in relation to station 1.

The degree of accuracy in determining angles of inclination and inclined distance limit the precision in calculated distances. In reality, calculated distances are within 0.3–0.5 m, so there is no need to report calculations with many significant figures to the right of the decimal point.   ⟫⟫⟫

⟫⟫⟫ 5.3 Global Positioning Systems

On a survey covering less than 3 km, the Abney level is generally more accurate. However, if the survey is going to be over a long distance or over a short distance with no critical points, a GPS unit can save time. One advantage of the GPS unit is that the distance between stations can be increased to 50 m.

5.3.1 *Materials and Preparation*

Fewer tools and people are needed for a GPS survey. The GPS unit replaces the Abney level, the compass, and the measuring tape. The items necessary for GPS surveying are the following:

1. a GPS unit;
2. the required combination of batteries needed per day, depending on battery quality;
3. a notebook and pens;
4. one permanent marker;
5. many stakes, about 50 cm long, and a tool to pound stakes into ground; and
6. 2 or 3 machetes.

A GPS unit also eliminates the need for a large work team (which can have a negative effect when attempting to involve more community members in the project). It is still important to work with a community leader who knows the route and at least two other community members who will clear the way with machetes.

The GPS unit referred to in this section is the eTrex Summit (Garmin, Olathe, Kan., www.garmin.com). This unit uses a barometric altimeter. In simpler models, which use an altimeter based only on satellites, it is possible for the altitude reading to vary significantly while the unit switches between satellites for communication. With the eTrex Summit and

Hydraulic Design of the Conduction Line
Project: Agua Buena Segunda, Copán Ruinas, Copán
Date of Design: September, 2004
Design by: Kraig Lothe, Cuerpo de Paz
Design Flow: 7.44 GPM

Station		Inclined Distance (m)	Forward Angle (deg)	Back Angle deg.	Compass	Pipe Material (GI or PVC)	Forward Angle (deg)	Back Angle (deg)	Average Angle (deg)	Vertical Dist. (m)	Horiz. Dist. (m)	Coordinates X (m)	Coordinates Y (m)	Coordinates Z (m)
Variable or equation		a			b				c	d=a*sin(c)	e=a*cos(c)	previous value + e*sin(b)	previous value + e*cos(b)	previous value + d
	dam											10000	10000	1000
dam	1	7.20	-4°30'	4°50'	28°	GI	-4.50	4.83	-4.67	-0.59	7.18	10003	10006	999
1	2	29.25	-8°00'	8°00'	53°	PVC	-8.00	8.00	-8.00	-4.07	28.97	10027	10021	995
2	3	5.35	5°50'	-5°50'	85°	GI	5.83	-5.83	5.83	0.54	5.32	10032	10027	996
3	4	25.25	1°30'	-1°30'	45°	PVC	1.50	-1.50	1.50	0.66	25.24	10050	10050	997
4	5	13.66	-21°00'	21°00'	65°	PVC	-21.00	21.00	-21.00	-4.90	12.75	10061	10055	992

Figure 5-5. Example Spreadsheet Used for Data Entry in a Topographical Survey. This Spreadsheet Presents Both Field Data and Calculated Values.

similar models, the satellite altimeter can be turned off after calibration and no error will result from switching satellites. Because of changing atmospheric pressures, there will be a margin of error. Ways to reduce this error are discussed later.

A chart must be drawn to record all data (Table 5-2). This chart is simpler than the one used for the Abney level because most of the data are stored in the GPS unit memory. The GPS unit will automatically number the waypoints. If data have previously been stored, the survey may not start from Waypoint No. 1. Record both the stake number and the waypoint number in the notebook as it appears on the GPS unit.

5.3.2 The Process

Not unlike an Abney level survey, using a GPS unit requires the surveyor to select stations (now referred to as *waypoints*) and record data. Locate waypoints at every point of interest along the line. As with the use of an Abney level, waypoints will include changes in the terrain, changes in direction, changes in piping material, high and low points, possible valve locations, anchors, stream crossings, fences, and other features along the line. Follow the exact line where the pipes will be laid for the survey. Leave the path well trodden and clear the underbrush where necessary. The more waypoints you take, the more accurate the survey will be. Even on a straight line, it is best not to space waypoints more than 50 m apart.

To mark a waypoint, the GPS unit must communicate properly with the satellites. Hold the unit with the screen facing upward. Wait until the unit is navigating to within 10 m of accuracy. Under a dense forest canopy or in a steep valley, it may be impossible to receive a signal. The altitude reading may fluctuate, so wait until it levels out. When ready to mark the waypoint, navigate to the menu screen (Figure 5-6a). Select the mark option. The mark waypoint screen will appear (Figure 5-6b). On this screen, the waypoint name and number will appear on the flag. Record this number in the notebook under the Waypoint column. Next, press ENTER to mark the waypoint, and you will be sent back to the menu screen. On this screen, the time is shown. Write the time in the corresponding column in the notebook. The GPS unit does not keep track of the time for waypoints. The time will be used later to adjust

Table 5-2. Example of Data Recorded in Notebook During a Topographic Survey Using a GPS

Stake No.	Waypoint	Time (hour:min:s)	Observations
1	2	8:30:13	From source, use GI
2	3	8:33:24	Begin descent to stream, use GI pipe
3	4	8:35:54	Across stream, between anchors, use GI
4	5	8:38:21	To top of small valley, soil can be excavated
5	6	8:40:43	Out of forest to coffee field
6	7	8:41:23	Coffee field

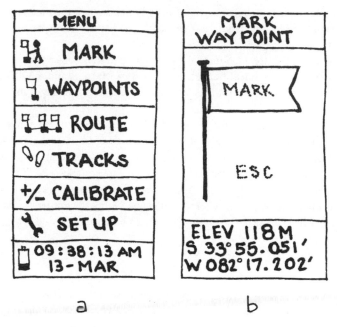

Figure 5-6. An Example of a GPS Unit Screen.

Note: (a) The menu page of the eTrex Summit GPS unit; and (b) the waypoint page shows the waypoint name, elevation, and location.

the data and reduce the margin of error. After noting the waypoint, number a stake and pound it firmly into the ground.

During a survey, the GPS unit will also take track points. *Track points* are like breadcrumbs that are dropped continuously to show the route even better. The track points can be helpful in graphing the route as they record the X, Y, and Z coordinates. Keep the unit screen facing upward at all times so that these points can be recorded.

5.3.3 Changes in Atmospheric Pressure

GPS altimeters measure altitude based on the atmospheric pressure, which is lower at higher altitudes. Atmospheric pressure depends on temperature and changes throughout the day. Because air is less dense at higher temperatures, any rise in temperature will be recorded as an increase in elevation. The barometric pressure is usually lowest early in the morning and reaches a maximum in the late afternoon. Depending on local weather, the apparent altitude can vary from 5 to 50 m daily.

A simple experiment can reveal more about pressure changes in a particular region. Place the GPS unit, facing upward, in a location where it can remain for a 24-hour period without risk of theft or getting wet. Make sure to have charged batteries on hand and be prepared to change them once or twice during the experiment. Turn the unit on and let it record track points in the same location for 24 hours. Record the data and graph it to better understand the results. Figure 5-7 shows the results of this experiment in Copán Ruinas, Honduras.

The results were scattered, but the data show a trend that ranges over approximately 25 m and is approximately linear between the hours of 9:00 A.M. and 4:00 P.M. With this knowledge, it is possible to manipulate data to compensate for changes in atmospheric pressure.

To do this, mark checkpoints during the course of the survey. A *checkpoint* is a waypoint that the surveyor returns to at least once during the day. It is possible to use more than one set of checkpoints. Examining checkpoint data taken at different times will indicate changes in atmospheric pressure over the time period. Because the trend during the workday appears linear (Figure 5-7), each waypoint can be adjusted proportionally, depending on when the waypoint was taken. This method will compensate for the pressure changes that have occurred since the first checkpoint was recorded. This calculation is shown in Example 5-2.

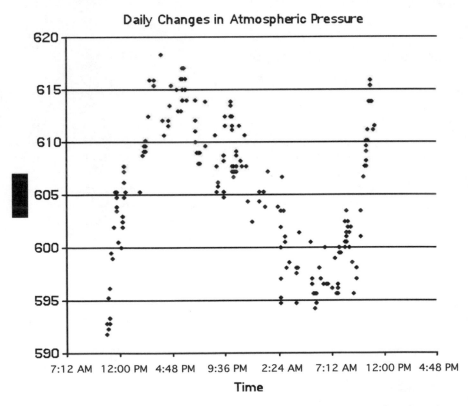

Figure 5-7. An Example of Daily Changes in Instrument Readings for Altitude Caused by Atmospheric Pressure Changes Recorded by a GPS Unit in Copán Ruinas, Honduras (*y* Axis Is Altitude in Meters).

》》》 Example 5-2. Sample Calculation Used to Adjust GPS Data to Account for Changes in Atmospheric Pressure

Assume that during a 3-hour period, the pressure increases 15 m. The data can be changed to compensate for the atmospheric pressure change as follows.

Assume the following data are recorded:

Stake No.	Waypoint	Time (hour:min:s)	Observations (Elevation)
Checkpoint A	1	9:00:00	Checkpoint at school (615 m)
1	2	10:00:00	Rocky, above stream, use GI pipe (548 m)
2	3	11:00:00	At stream, need anchor, use GI pipe (530 m)
Checkpoint A	4	12:00:00	Checkpoint at school (630 m)

Note: It is not necessary to record the altitude in your chart because it is stored by your GPS unit; however, a paper record serves as a backup.

Referring to Checkpoint A, the apparent altitude changed 15 m in the 3-hour period from 9:00 A.M. until noon. Assuming that the pressure changed linearly (Figure 5-7), adjust the measurements as follows:

Waypoint 2 was recorded 1 hour into the 3-hour period or one-third of the way through the time period. Subtract one-third of the 15-m change from waypoint 2.

$$548\,\text{m} - \left(15\ \text{m} \times \frac{1\,\text{h}}{3\,\text{h}}\right) = 548\,\text{m} - 5\,\text{m} = 543\,\text{m}$$

The adjusted elevation at waypoint 2 is 543 m.

Similarly, subtract two-thirds of 15 m from waypoint 3, as it was recorded two-thirds of the way through the period.

$$530\,\text{m} - \left(15\ \text{m} \times \frac{2\,\text{h}}{3\,\text{h}}\right) = 530\,\text{m} - 10\,\text{m} = 520\,\text{m}$$

The adjusted elevation at waypoint 2 is 520 m.

Now each waypoint has been reduced to represent the elevation as if the atmospheric pressure had not changed after 9:00 A.M. Perform this same calculation for each waypoint. Using a spreadsheet, this takes only a short time (Figure 5-8).   **》》》**

5.3.4 Processing the Data

The software, *GPS Utility*, is used to transfer the data recorded by the GPS unit to a computer. Each waypoint will have a northing value, an easting value, and the altitude. These values can be readily copied, pasted into a spreadsheet (Figure 5-8), and used in a GPS

Elevation Adjustment Box 1								
Absolute Apparent Altitude Change Since Last Calibration Group=0.0 m								
check 1	750164	1664779	1009.0	1	1009	12/9/2004	10:36:40	Checkpoint at car
check 2	270717	1633612	1072.0	1	1009	12/9/2004	10:50:22	Checkpoint at car
Est.	*Easting, m-UTM*	*Northing, m-UTM*	*Elevation, (m)*	*Elevation Adjustment Box*	*Adjusted Elevation, (m)*	*Date, (day/m/yr)*	*Time*	*Field Notes*
check 1	750164	1664779	1009.0	1	1009.0	12/9/2004	10:36:40	Checkpoint at vehicle
check 2	270717	1633612	1072.0	1	1009.0	12/9/2004	10:50:22	Checkpoint at vehicle
dam	271028	1633154	1149.4	1	996.1	12/9/2004	11:10:00	Source
1	271030	1633150	1149.1	1	988.9	12/9/2004	11:11:30	Before stream anchor
2	271032	1633165	1145.5	1	977.7	12/9/2004	11:13:10	Anchor across stream
3	270995	1633201	1133.5	1	945.0	12/9/2004	11:17:40	Top of hill
4	270958	1633202	1131.3	1	934.3	12/9/2004	11:19:30	At fence, rocky, GI pipe

Figure 5-8. An Example Spreadsheet Used for Data Entry with a GPS Survey. Data Are Either Obtained from the Field or Calculated.

design template. If *GPS Utility* is not available, the computer data input can be done manually. As shown in Figure 5-8, the calculation in Example 5-1 can be done automatically within the spreadsheet.

))) Reference

Jordan, T. D., Jr. (1980). *A handbook of gravity-flow water systems*, Intermediate Technology Publications, London.

))) Further Reading

Brinker, R. C., and Wolf, P. R. (1977). *Elementary surveying*, sixth ed., Harper & Row, London.

6

Engineering Materials

This chapter discusses earthwork, stone, bricks, block, cement, concrete and mortar, steel reinforcement, and wood. Chapter 7 provides the specific construction techniques that use these materials.

))) 6.1 Earthwork

Earthwork involves excavating, moving, and placing earth. This work may be as simple as using a hand shovel to dig a hole or as complex as using heavy equipment to quickly excavate and transport large quantities of earth. The types of earth involved include solid rock, gravel, sand, clays, silts, and organic material (topsoil). Soils are generally characterized by particle size and plasticity (the behavior of soil under wet conditions), using one of several soil classification systems. (See Chapter 8 for information on soil properties and identification and foundations.)

Examples of earthwork include burying a water distribution pipe below grade to protect it from agricultural activities, removing undesirable (from the perspective of construction) organic soils and clays to reach sound soil to support a foundation, leveling a particular area, and creating a slope so that waste will flow by gravity into a septic tank.

In a narrow excavation such as a trench, hole, or well, care must be taken to ensure that walls do not collapse and bury workers within the excavation. At a minimum, cave-in precautions should include sloping and benching the excavation walls greater than 1.5 m to prevent slope failure or using shoring or trench boxes to support the earth, and keeping heavy equipment and materials away from the excavation edge.

Earthwork is also used (because it is a source of local materials) as fill for an embankment or excavation to manufacture building materials, such as bricks or aggregates. In the simplest case, excavated mud can be combined with straw and sticks to construct walls. Cement may be added to soil and then compacted to form rammed earth walls. This technique requires appropriate and consistent proportions of clay, sand (also referred to as fine aggregate), and gravel (also referred to as coarse aggregate) in the soil (organic soil should not be used). Sand, gravel, stiff clay, cemented soil, and rock are ideal for foundations, whereas organic soil and plastic clay are problematic because of settlement over time. When used for fill, earth disturbed during the construction process or relocated from elsewhere must be properly compacted if it is to support a structure or be used as backfill (see Chapter 7).

Christopher G. Gilbertson, Brooke Tyndell Ahrens, Daniel M. Hurtado, and Thomas Van Dam contributed to this chapter.

))) 6.2 Stone

Stone is one of the oldest building materials. Simply gathering and stacking fieldstone can create a wall. When used with *mortar* (a mixture of sand, water, and cement), stones are held together to create a durable structure (often seen in fences, walls, and chimneys). An alternative to fieldstone is *rubble masonry,* which is created when rock is blasted. Approximately 1.25–1.5 m^3 of loose material is required to create 1 m^3 of a finished rubble masonry wall (Stern and Longland 1983).

Irregularly shaped stones can be used to create walls (Fig. 6-1). Cornerstones, or *quoins,* provide structural rigidity and may be stone, reinforced concrete, or brick. *Bond-stones* project into the wall and are used to anchor smaller stones. Twenty-five percent of the wall face should consist of bondstones (Stern and Longland 1983). *Through stones* serve a similar purpose but are sized for the entire wall thickness. Smaller stones are used to minimize voids, which decrease the integrity of the wall. Stones can be arranged in various patterns. For example, Fig. 6-2 contrasts stones arranged in a *random rubble wall* and the more layered arrangement of a *coursed rubble wall.* Large stone walls may consist of two mortared walls with rubble fill placed between them. Through stones are used to tie the two double wall faces together.

Cut stone is versatile because it is shaped to fit together tightly without mortar. The stone must be durable and workable; however, not all stone has these two attributes.

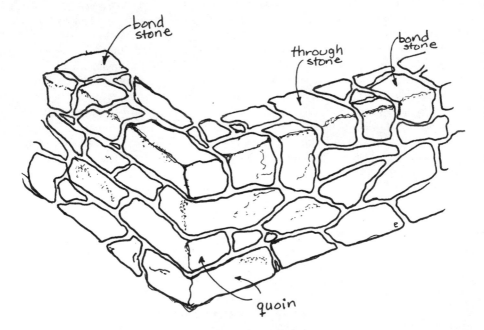

Figure 6-1. Walls Can Be Constructed of Rubble Masonry.

Note: Cornerstones (or quoins) provide structural rigidity and may be constructed of stone, reinforced concrete, or brick. Bondstones project into the wall and are used to anchor smaller stones. A through stone is a type of bondstone that occupies the complete width of the wall.

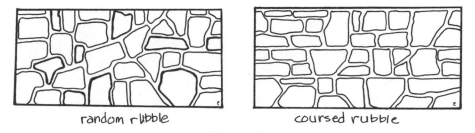

random rubble   coursed rubble

Figure 6-2. Stone Walls Can Be Constructed in Various Patterns.

Note: Here, the same source of stones is used. Stones in the coursed rubble wall are arranged in a linear, layered fashion, which contrasts with the random rubble wall.

Source: Redrawn with permission from Allen and Iano 2003.

Durability is a measure of the capability of a material to withstand the elements, such as moisture and thermal cycles. *Workability* is a measure of a material's ability to be cut, shaped, and formed into useful shapes. In general, *sedimentary* rocks are easier to work but less durable, whereas *igneous* rocks are quite strong and durable but difficult to work. Cut stone must be broken free, or *quarried*, from natural rock formations using fires and rapid cooling, wooden wedges that swell when moistened, or precise blasting. In some locales, modern stonecutting facilities allow cut stone to be used similarly to brick and block.

))) 6.3 Bricks

Different types of bricks (Fig. 6-3) can be manufactured from clay and loam that is mixed with water and compacted in a mold (Fig. 6-4). Rice husk, sawdust, or ash is added to clay to make the bricks less dense and also to minimize cracking during firing. Brick mixtures can be tested visually by placing a sample in the shade to dry and harden, and then observing whether the sample brick has cracked. Termite hill material is ideal for brickmaking because of the uniformity of the clay-rich soils. Termites build their nests with clay-rich soils that they collect below ground and cement together with saliva. The saliva contains a bonding agent. When selecting soil for brick production, avoid areas where salts are present because the salt will dissolve to yield porous, poor-quality bricks.

Bricks are molded and can be dried or fired in a kiln or clamp to increase durability (Figs. 6-5 and 6-6). Both kilns and clamps work by distributing heat from a fire (wood,

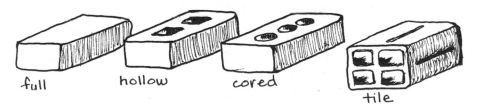

full   hollow   cored   tile

Figure 6-3. Brick Styles.

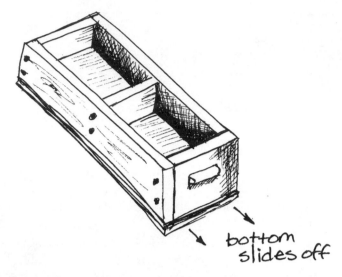

Figure 6-4. A Mold for Making Two Bricks at a Time.

Note: The mold should be wetted before each use, either by washing or by sprinkling with water.

coal, gas, or other combustibles) to the clay bricks. A *kiln* is a permanent structure and features an insulated enclosure that produces a uniform heat distribution across all bricks. A kiln produces a greater percentage of properly fired bricks with greater fuel efficiency than a clamp. A *clamp* is a temporary arrangement of unfired bricks, stacked to create holes where fires are built. A clamp can be built anywhere bricks are being

Figure 6-5. Dried Bricks Stacked in a Clamp and Ready for Firing.

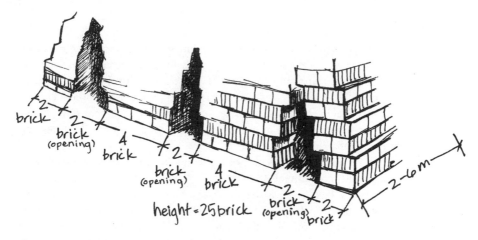

Figure 6-6. A Clamp Used for Firing Dried Bricks.

Note: Clamps differ from kilns because they are not a permanent structure. After the bricks are stacked, the exterior sides are covered with clay. Wood fires are built inside the openings. The top is covered with several centimeters of earth after the first flames make their way through the top. The fires are kept burning for 3–6 days, after which the stack is allowed to cool.

Source: Based on information from Stern and Longland 1983.

formed. The disadvantage is that uneven heat distribution can cause some bricks to be improperly fired and therefore wasted. However, it can be made more fuel efficient with a lower percentage of waste bricks by providing a means of insulation around the outside. Insulation can take many forms, including a plaster made of clay and materials such as rice husk or ash.

Alternatively, sun-dried earth bricks (adobe) are effective. *Adobe* consists of a mixture of 75–80% sand and 20–25% clay. Stabilizers such as cement can be added to increase strength and longevity, and the use of natural fibers is common. For example, mixtures of 50% sand, 35% clay, and 15% straw can be used. Walls (0.25–0.8 m thick) are constructed with adobe blocks (or bricks) joined together with mud mortar. Block and wall size depend on the geographical location, being thinner in coastal areas and wider in cool climates or hot desert climates. Construction is usually limited to one-story structures, although in space-limited mountainous areas, construction up to three stories is observed.

Common architectural features of adobe structures include a rectangular floor plan, a single door, and small windows. The foundation usually consists of larger stones joined with mud or coarse mortar. The roof's wood joists rest directly on either the walls or indentations made on the inside at the top of the walls. Severe cracking, wall disintegration, and separation of the structure at corners and where the roof meets the wall are common during seismic events. Reinforcement (using concrete or timber) can be added to improve resistance to earthquakes. This technique requires a strong and continuous bond beam, which ties the walls together (Blondet and Villa Garcia 2008).

⟫⟫ 6.4 Block

In general, block is larger than brick. It is made by binding soil or aggregate together using cement, not through firing or drying. The lowest quality block is referred to as *landcrete* (a word combining laterite, land, and concrete; ITP 1995). Lateritic soils, a mixture of fine and coarse sand mixed with clay, are common in the tropics. The addition of cement stabilizes the material, making the blocks stronger and more resistant to rain. Once made, the blocks need to be wet cured for a minimum of 5 days (longer is better). Curing can happen by submerging the blocks in water.

Concrete block or *sand–cement block* is of higher quality and greater expense and is structurally stronger and more durable than landcrete. The blocks can be solid or hollow cored. Hollow blocks are lighter and easier to transport and use. Concrete mix is typically placed dry into a block-making machine that uses pressure or vibration to compact the block (Fig. 6-7). A typical mix is 1:6 cement to sand by volume. The sand can be

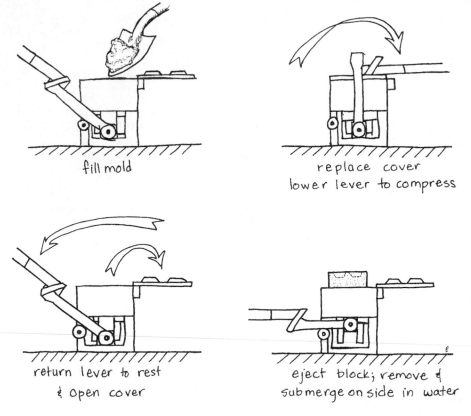

fill mold

replace cover
lower lever to compress

return lever to rest
& open cover

eject block; remove &
submerge on side in water

Figure 6-7. Mechanical Block Maker.

Note: Vibration is sometimes provided by an engine.

Source: Adapted with permission from Davis and Lambert (2002).

well-graded fine to coarse aggregate, depending on the block shape. Alternatively, a mix of $1:2:4$, $1:2.5:5$, or even $1:5:8$ cement to sand to coarse aggregate can be used if the aggregate passes through a 13-mm screen. After molding, blocks must be carefully handled and placed on a level surface for curing. Curing should last 7 days, at which point the blocks can be stacked for additional curing of 21 days before use (Stern and Longland 1983; Davis and Lambert 2002).

))) 6.5 Cement

Concrete is a mixture of Portland cement (and possibly supplementary cementitious materials such as fly ash), fine and coarse aggregate, and water. When blended together, the materials form a viscous mass that can be moved, placed, and consolidated, taking the shape of any form. Through a series of chemical reactions called hydration, the wet concrete solidifies into a hardened, rocklike mass. In developed countries, concrete often contains one or more chemical admixtures that improve its wet or hardened properties. Exposure to cement causes skin discomfort, and the dry powder should not be inhaled. Thus, proper protective clothing and polyvinyl chloride (PVC) gloves are recommended while working with both dry and wet cement.

Portland cement, the binding agent in concrete, is produced from *clinkering* (heating to high temperatures) sources of calcium (e.g., limestone or seashells) and silica (e.g., clay or shale) in a kiln. The resulting nodule or clinker is then ground with a source of soluble sulfate (gypsum) to form a grey powder. Portland cement is primarily calcium silicate cement, which produces calcium–silica–hydrate (C–S–H) and calcium hydroxide (CH) when it reacts with water. The reaction is exothermic (heat producing) and temperature sensitive, doubling in rate with every increase of 10 °C.

The production of Portland cement is a major contributor to worldwide CO_2 production. On average, manufacturing 1 kg of cement produces 1 kg of CO_2. The 1.45 billion metric tons of global cement production account for 2% of global primary energy use and 5% of anthropogenic CO_2 emissions.

Pozzolans are a supplementary cementitious material that provide a source of reactive silica (and sometimes alumina). The pozzolan is not cementitious; however, when combined with Portland cement, the silica in the pozzolan reacts with the calcium hydroxide produced from the hydration of the Portland cement to form calcium–silica–hydrate. Because calcium hydroxide is a less desirable hydration product than calcium–silica–hydrate, this reaction has a beneficial effect. Typically, 25% of cement can be replaced with a pozzolan. This replacement slows the rate of hydration, requiring a longer curing period before loading the structure with a force. It also reduces the heat of hydration, which can be an advantage when placing large amounts of concrete. In this case, strength will be increased and permeability reduced, producing a more durable concrete. Natural pozzolans (such as unaltered and weathered volcanic ash, pumice, diatomaceous earth, opaline cherts, and shales) have been in use for thousands of years to create durable, long-lasting concrete structures.

Cement is commonly sold in sacks, the weight of which varies by region. A sack weighs 94 lb in the United States, 49.5 kg in Panama, and usually 50 kg in Africa. It is important to keep cement dry because it hardens by absorbing moisture in the air. In coastal areas, materials are transported by boats, so each sack must be sealed in plastic to prevent the cement from getting wet and hardening. Covers can also help prevent tears in the sack and contain the cement if a sack bursts. In areas of high humidity, sacks enclosed in plastic bags can keep several months without hardening. It is advisable to purchase recently manufactured cement because cement that is stored 1.5 years may only reach half the final strength of recently manufactured cement.

⟫⟫ 6.6 Concrete and Mortar

Figure 6-8 shows a typical composition of concrete. Cement is about 10% of the total composition by mass. The cement and water form a paste that binds the other constituents together. Cement is usually the most costly component and should be used as efficiently as possible while taking into account its importance to ensure strength.

Other components of concrete are air, water, and coarse and fine aggregate. Aggregate is usually obtained locally. Its primary purpose is as an inexpensive and inert filler. Air is entrapped through the mixing process. Excessive air creates unwanted porosity, reducing strength and increasing permeability. Therefore, concrete is consolidated after placement by rodding or vibration. In the case of entrained air, a chemical additive is purposely incorporated to minimize adverse impacts of freezing and thawing.

Mortar is a mix of cement, water, and sand. It is used for laying masonry (brick, block, or rock) and *parging* (applying a thin layer of mortar). The sand is sifted through a fine screen as shown in Fig. 6-9 to remove larger particles. The ratio of cement to sand is typically 1 : 3 to 1 : 4 for masonry work.

Properly sized aggregate reduces the need for additional cement. This fact can be explained by imagining a container filled with equal-sized large spherical aggregate (Fig. 6-10). Here there are large voids between the contact points of each aggregate. In concrete,

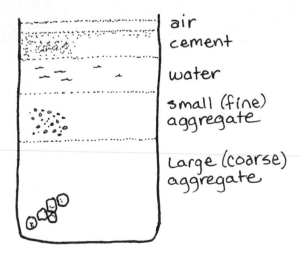

Figure 6-8. A Typical Composition of Concrete (by Mass) May Be 5% Air, 10% Cement, 15% Water, 25% Fine Aggregate, and 45% Large Coarse Aggregate.

Figure 6-9. Sifting Sand to Be Used for Mortar or Concrete.

this space is filled by cement paste and entrained air. In the case of poorly consolidated concrete, there is also entrapped air, which greatly reduces the strength. Adding smaller and differently sized aggregate increases the points of contact between individual pieces of aggregate and decreases the void space. This mixture requires less paste (a savings in cement) while decreasing the likelihood of entrapped air.

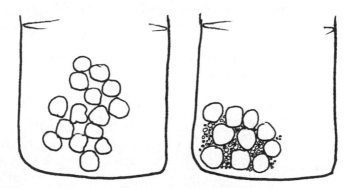

Figure 6-10. Using a Mixture of Different Aggregate Sizes (as Shown on the Right) Increases the Number of Points of Contact Between Individual Aggregates and Also Decreases the Amount of Void Space. This Method Results in a Savings in Cement and Also Lessens the Likelihood of Entrapped Air.

Common aggregate includes crushed rock, gravel, and sand. Broken pottery and bricks may also be used. Aggregate is typically referred to as either *coarse aggregate* (larger particles, greater than 4.75 mm) or *fine aggregate* (smaller, sand-sized particles). It is common on the job site to see community members breaking large rocks by hand into smaller aggregates.

Aggregates should be clean (free of organic matter, clay, wood, and other deleterious materials), sound (strong enough to prevent disintegration, avoiding chalk, sandstone, and chert) and round or cubic (not flat or elongated). Aggregate quality is also important to concrete strength. Sand containing clay, silt, salt, mica, or organic material weakens concrete. Aggregate is often obtained locally and may be of low quality. Local masons and workers who have routinely used these materials may be unaware of future problems that this use can cause and of how proper aggregate selection will improve the overall quality of the finished product.

In the developing world, concrete is typically proportioned by volume instead of by weight. Material volumes are likely to be measured in buckets, wheelbarrows, donkey carts, or shovels, rather than in cubic meters. Table 6-1 provides typical guidelines for the different volumes of material used in various concrete mixes. Communities are often accustomed to a specific way in which concrete is mixed, which may include making batches using only one sack of cement at a time, or using the trained eye as the only measuring tool.

Workability is an important attribute of concrete, reflecting its ease of placement. A good concrete mix design is useless if it cannot be used in the desired application. Aggregate shape plays a key role in this. Large aggregates composed of crushed rock with angular faces tend to make a less workable mix that is difficult to consolidate and finish. The consolidation process, typically done by manually rodding the concrete, helps to pack the aggregate into a tight configuration. The benefit of angular aggregate is that there is greater contact area for each rock to bond to the cement paste. However, rocks do not slide easily over each other, making it difficult to place. The tendency, however, is to use more water in an angular aggregate concrete to increase workability, which in turn reduces concrete strength. Some communities prefer to blend coarse aggregates (50:50 round and angular), yielding a more workable concrete while preserving the benefits of angular aggregate.

Aggregate that is too large can prevent a thin slab of concrete from being adequately placed and finished smoothly, or stones may not fit within the formwork and steel rein-

Table 6-1. Concrete Mix Designs (Cement to Sand to Coarse Aggregate)

Mix (volume)	Cement (50-kg sack)	Sand (m³) (U.S. gal)	Coarse Aggregate (m³) (U.S. gal)
1:2:4	1	0.07 (18.5)	0.14 (37.0)
1:3:6	1	0.10 (26.4)	0.20 (52.8)
1:4:8	1	0.14 (37.0)	0.28 (74)

Note: The strongest mix design is 1:2:4.

Source: Reproduced with permission from Stern and Longland (1983).

forcement. A general rule of thumb is that the diameter of the largest pieces of aggregate should be smaller than ⅓ the thickness of a slab or ¾ the open space between rebar or wires and formwork to allow for proper consolidation of concrete around the reinforcement.

Sometimes a blend of fine and coarse aggregate is obtained from a riverbed with the sand-to-rock ratio ignored. This mixture usually contains much less coarse aggregate than desired. Riverbed-derived aggregates may also contain higher amounts of organic matter, which decreases concrete strength. Use of beach sand and stone is acceptable, but they may contain salt residue, which decreases the concrete strength and can cause corrosion of embedded steel. Accordingly, it may be necessary to wash the aggregate and sand with freshwater before mixing concrete to keep the mix strength high.

A general rule for water used in concrete is that if you can drink it, you can use it for concrete. Avoid water that has a high organic, sulfate, or alkali content.

Many factors contribute to the ultimate strength of concrete, including aggregate strength, water-to-cement ratio, amount of cement, and curing conditions. The water-to-cement ratio is important because it directly affects the workability, strength, permeability, stress–strain relationship, durability, and resistance to chemical attack. Achieving an ideal water-to-cement ratio is critical to produce long-lasting, durable, and strong concrete. Concrete needs water to hydrate. However, too little water makes concrete difficult to place, and too much water makes it porous and weak.

It is common for masons to add more water than is desired to make placement easier. Water should be added to provide adequate workability, but no more should be added. This factor is especially important for structures that will hold water (such as water storage and septic tanks) or structures requiring structural integrity. Too much water results in concrete with low strength and greater permeability that also undergoes greater strain at less stress. A high water-to-cement ratio also decreases the amount of entrained air, which results in a less durable concrete that is more susceptible to thermal effects. The *ball test* is an indicator of water content in a mix. A handful of concrete that is formed into a ball and squeezed should stay intact. If the ball does not stay intact, the mix is too wet (Box 6-1).

Conversely, concrete with too low a water-to-cement ratio will be difficult to place and finish, may not reach a high strength, and may contain an abundance of nonhydrated cement particles. Nonhydrated cement does not serve any purpose in the concrete other than being an expensive filler (like fine aggregate).

Within 2–4 hours of adding water to cement, it will have set and can no longer be worked. The chemical hardening process of cement is most rapid for the first few days. Table 6-2 shows how time influences concrete's compressive strength. Even after the structure has begun to be used, the cement will continue to cure for at least a year.

Moisture is essential in the curing process. Concrete should be covered soon after placement and kept wet for as long as possible. Dehydration of concrete results in extensive loss of potential strength, as well as substantial shrinkage, which results in cracks. Moisture during curing is especially important in dry and hot environments. Concrete that is curing should be covered with wet burlap, wet cement bags, wet straw, or banana leaves and kept moist during the entire curing period. Seven days of moist curing is desirable. You can even cure concrete underwater. For example, a concrete-encased water tap can be cured underwater in a mold that prevents washout of the fines.

⫸ Box 6-1. Estimating Material Quantities

For each project, the volume of materials must be estimated for purchase and installation. This example shows how to estimate the volume of cement, sand, and coarse aggregate required to construct a 2 m × 2 m wash area and associated soak pit (Chapters 7 and 21). Other estimation methods can be used. The example assumes the following materials requirements (on a volume basis):

- concrete for wash area surface and underlying wall supports (1:3:5 cement to sand to coarse aggregate mixture);
- mortar for the wash area's rock masonry wall, finishing, and pipe placement (1:4 cement to sand mixture); and
- mortar (parge) for finishing the wash area floor (1:2 cement to sand mixture).

The wash area surface consists of a 5-cm-thick layer of concrete that is 2 m × 2 m. The volume of the wash area surface is

$$200 \text{ cm} \times 200 \text{ cm} \times 5 \text{ cm} = 200{,}000 \text{ cm}^3$$

Assuming that 33% additional volume of concrete is required to account for the concrete mixture that serves as the underlying wall supports, the total volume of concrete (V_t) required is

$$V_t = (1.33) \times 200{,}000 \text{ cm}^3 = 266{,}000 \text{ cm}^3$$

A 1:3:5 mix (by volume) of cement to sand to coarse aggregate is used. The volume of cement in the concrete mix can be found by assuming 50% compaction (which explains the use of 1.5 in the equation below) of the concrete after it is placed and a 10% loss during mixing and placement (1.1 in the equation below). Also, the unit volume of cement required will be one-ninth of the total volume (remember that the mix is 1 volume cement to 8 volumes of sand and coarse aggregate). The unit volume of cement is calculated as

$$\frac{V_t \times 1.5 \times 1.1}{9} = 49{,}000 \text{ cm}^3$$

Thus, 49,000 cm³ is the volume of cement required for the wash area surface and the wall support. The volume of sand required for this part of the wash area can be found by multiplying this value by 3 and the volume of coarse aggregate required for this part of the wash area can be found by multiplying this value by 5 (these results are not shown).

Cement is purchased on a mass (weight) basis, so the mass of dry cement that needs to be purchased (assuming a density of 1,300 kg/m³) is

$$49{,}000 \text{ cm}^3 \times \frac{1 \text{ m}^3}{(100 \text{ cm})^3} \times \frac{1{,}300 \text{ kg}}{\text{m}^3} = 63.7 \text{ kg}$$

Similar calculations can now be made for the 1 : 4 mortar mix used in constructing (a) the four 2-m-long rock masonry walls of the wash area (assuming that they are 7 cm wide × 7 cm high and that ⅕ of the total volume of the masonry wall is occupied by cement) and (b) the 1 : 2 parge mix that is used to provide a 2-cm-thick finish on the wash floor area. This assumption results in 16.8 kg of cement required for the 1 : 4 mortar mix and 57.2 kg of cement required for the 1 : 2 mortar mix to finish the wash area floor. The total cement requirements for the project are

$$63.7 + 16.8 + 57.2 = 138 \text{ kg of cement}$$

The total amount of cement required for the wash area is three 50-kg sacks. The total volume of sand and gravel will also need to be determined, as will the amount of rock that goes into the wash area walls and the floor base. For the soak pit, approximately 1.5 m of PVC pipe would be purchased and perhaps a 2 m × 2 m sheet of plastic to serve as a cover while the wash area cures if natural plant materials (e.g., banana leaves) are not available. Materials may be ordered based on the size of the local rice sack or the space occupied by a donkey cart.

))) 6.7 Steel Reinforcement

Reinforcing steel is commonly referred to as rebar. Rebar should not be smooth. It should be ribbed (also referred to as deformed) to facilitate bonding to the concrete. It is available in many sizes. Table 6-3 provides a comparison of rebar size for English and metric designations.

Reinforce concrete in any structural application and where cracking cannot be tolerated. This requirement includes in slabs and roofs (e.g., floors of latrines or water storage tanks), concrete under hydrostatic or lateral earth loads (e.g., walls of water storage tanks and retaining walls), or for any structural member, such as a beam, column, or elevated slab.

Concrete is much stronger in compression than it is in tension. This fact is why vertical concrete columns are commonly used in building design to support loads. The

Table 6-2. Percent of Ultimate Compressive Strength of Portland Cement Concrete Reached over Time

Time	Percent Ultimate Compressive Strength Reached
3 days	20
7 days	45
28 days	60
3 months	85
6 months	95
1 year	100

Table 6-3. English and Metric Designations for Rebar

English Designation (Diameter of Corresponding Plain Round Bar in in.)	Metric Designation (Diameter of Corresponding Plain Round Bar in mm)
3 (0.375)	10
4 (0.500)	13
5 (0.625)	16
6 (0.750)	19
7 (0.875)	22
8 (1.00)	25
9 (1.128)	29
10 (1.270)	32
11 (1.410)	36
14 (1.693)	43
18 (2.257)	57

Note: In English designations, the size corresponds to the eighths of an inch in the diameter of a plain round bar. In metric designations, the size corresponds to the nearest mm in diameter of the plain round bar.

top drawing in Fig. 6-11 shows a beam that has deflected after a load is placed on it. In this situation, the top of the beam is undergoing compression and the bottom of the beam is in tension. This figure shows the same phenomenon that occurs in the floor, roof, and sidewalls of a water storage tank or spring box.

Steel is added to reinforce concrete in locations of tension. In Fig. 6-11, the two drawings on the bottom are cross sections through the beam (note that the centerline corresponds to the centerline of the deflected beam). The drawing on the lower right shows that compression is at a maximum near the top of the beam and tension is at a maximum at the bottom of the beam. Rebar should be placed where the tension is the greatest while still providing a minimum cover to allow for proper reinforcement and concrete interaction (note the placement of three pieces of steel reinforcement in the drawing on the lower left). Rebar placed at the neutral axis (the centerline location, where stress converges from compression to tension) of a structural member will not engage (take any load) until the beam has cracked and the neutral axis shifts toward the compression zone. For this reason, on water storage tanks, Jordan (1980) recommends placing the rebar 3 cm from the top of a floor slab and 3 cm from the bottom of a roof slab.

Cover is defined as the distance between the rebar surface or bar end to the nearest face of concrete (Fig. 6-12). Cover helps protect the rebar from corrosion and also allows it to properly interact with the concrete. The cover required over rebar depends on many factors, including aggregate size, location of the member, and environmental considerations, such as exposure to saltwater. Rebar or wire smaller than 16 mm must have at least 4 cm of concrete covering it in all directions if it is exposed to earth or the elements. A minimum of 2 cm should be provided if the concrete is used indoors. If the concrete is to be cast directly against earth without formwork, the cover should be 8 cm (ACI 2005).

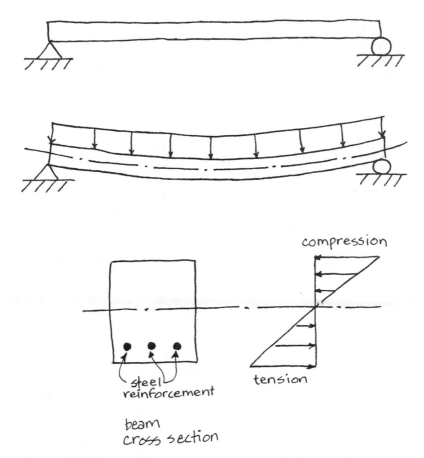

Figure 6-11. Compression and Tension in a Simply Supported Concrete Beam.

Note: The top drawing shows a beam with a load placed on it. The bottom two drawings are cross sections through the beam (note the centerline). In the lower right drawing, compression is greatest on the top of the beam, and tension is greatest on the bottom of the beam. Steel reinforcement is placed near where the tension is greatest. In the lower left drawing, three pieces of rebar have been placed near the bottom of the beam, where the tension is the greatest, while still providing a minimum cover to allow for interaction between the steel and concrete.

Rebar is placed in a grid pattern and spaced according to the total amount of steel required per cross-sectional unit (for example, 13 cm^2 per 960 cm^2 of concrete, which equals 2 in.^{2} per ft^2). It is often best to know steel requirement in terms of steel area per concrete cross section because your location may determine the available rebar sizes. The rebar spacing can then be calculated based on the size of rebar that is available.

Jordan (1980) indicates that for the floor of a water storage tank, the rebar cross-sectional area must be greater than 0.225% of the total cross-sectional area. For the roof of a water storage tank, the area of rebar must be greater than 0.30%. Table 6-4 provides guidance on spacing of rebar for slabs of water storage tanks.

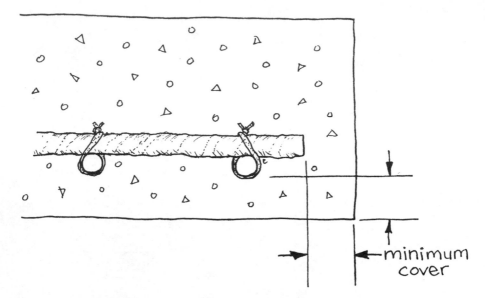

Figure 6-12. A Minimum Cover of Concrete Protects the Steel Reinforcement.

The steel reinforcement is also placed on *chairs* so that the steel remains in the correct place within the slab cross section and does not move to the bottom of the slab when concrete is placed over it. Chairs are typically made of broken block or brick or of flat, stable stones. The correct size of chairs allows for adequate concrete cover around the rebar.

Structural sections like beams and columns require steel around all sides of the cross-sectional perimeter. Rebar that has been bent (typically as a rectangle or a circle) to fully encompass the cross section is referred to as *stirrups* (see Fig. 7-5 in Chapter 7 for an example of stirrups). Stirrups may take the appearance of hoops or may be spiraled (like a

Table 6-4. Spacing of Rebar in Concrete Slabs Incorporated in Water Storage Tanks

Type of Slab	Thickness of Slab (cm)	Rebar Spacing (in cm) for 6-mm Rebar	Rebar Spacing (in cm) for 8-mm Rebar	Rebar Spacing (in cm) for 10-mm Rebar
Floor	8	15	30	40
Roof	8–9	12	21	33
Roof	9–11	10	17	27
Roof	11–13	8	14	22
Roof	13–15	7	12	19
Roof	15–17	6	11	17

Source: Reproduced with permission from Jordan (1980).

spring) and should meet the minimum cover requirements. Stirrups should be spaced at a distance no greater than the width of the column. Size, spacing, and location of stirrups should be determined by a qualified engineer. The purpose of this reinforcing depends on the particular application. Typically, the purpose for columns is to confine the concrete in the center of the structural member. Axial compression (from a loaded column or beam) causes radial tension (an outward force) through the cross section. If not properly reinforced, the concrete can crack and flake away from the structure. When used in beams, the stirrups provide reinforcement against shear loads.

As discussed in Chapter 7, rebar is tightly tied together with wire. Bars must be tied securely to prevent slipping and misalignment while concrete is placed and finished. In some instances, such as for large projects or where only small lengths of rebar are available, it may be necessary to splice several sections of rebar together. When this is done, it is important that the rebar is placed such that one piece overlaps the other to allow for a full transfer of the load being carried by the rebar. The amount of overlap is referred to as the *development length*. The rebar is overlapped at least 30–39 cm (12–15 in.) and is tightly tied several times. The development length varies by each application and depends on the size of the rebar, the load being applied, concrete strength, and other factors.

Steel mesh or welded wire mesh (WWM) is reinforcement prewelded into large fence-type mats. It may not be readily available in rural areas. Steel mesh may be used as reinforcement against temperature and shrinkage cracks in concrete slabs and as reinforcement in ferrocement tanks. Steel mesh is not considered a structural component of a concrete system. It is used to deter cracking caused by shrinkage caused by water loss during the curing process and to resist cracking caused by expansion and contraction during thermal cycles. Steel mesh should not be used in place of rebar in situations that rely on the concrete to serve as a structural component of the project, including but not limited to columns, beams, supported floors, roofs, foundations, and retaining walls. A fine or overlapping mesh that does not allow the concrete components such as aggregates to easily pass through causes discontinuity within the concrete mass, increasing the likelihood of spalling, or flaking. In addition, it will not allow proper steel and concrete interaction, lessening its strength and its ability to transfer loads.

The determination of the proper amount of steel reinforcement for a particular application is a complex process and involves many considerations (MacGregor and Wight 2004). Steel reinforcement should be properly designed for any application that involves human occupancy. In these cases, its use should be double-checked against design recommendations and principles and never based solely on generalizations or rules of thumb.

))) 6.8 Wood

Wood is often economical and is easily manufactured and handled, making it widely used for light construction. However, wood has several disadvantages in construction, particularly flammability and decay. Wood exhibits a high degree of variation in material properties. Although they may be generalized for similar species, even within one species, properties vary depending on location and environmental factors during growth. Additionally, flaws and imperfections (e.g., knots) yield variability in the material properties of each board cut from a single tree.

There are two basic types of trees: deciduous (leaf bearers) and coniferous (commonly called pines). Deciduous trees have broad leaves, grow slowly, and are referred to as hardwoods. Coniferous trees have needlelike leaves, grow faster, are typically used for commercial lumber, and are referred to as softwoods. Bamboo is native to every continent except Europe and Antarctica. A few dozen species that grow widely in the tropics can be used for roofing, walls, floors, scaffolding, and supports. It is considered lighter than steel and stronger than concrete. Use of bolted mortar at the joints has eliminated the need to lash joints with vines or ropes. It can also be used to reinforce roads and construct check dams.

A tree has three specific dimensions: radial (at a 90° angle to the growth rings), tangential (tangent to the growth rings), and longitudinal (the length of the tree). One of the most significant structural considerations when designing with wood is the fact that wood is an *anisotropic* material: it has different material properties on all three faces, because of the tubular geometry of wood cells. The cellular structure of wood may be likened to a bundle of straws or thin-walled tubes glued together.

Moisture content is another important concept to consider when designing with wood. Moisture content is defined as the wet weight of the wood minus the oven-dried weight, divided by the wet weight. Water is identified based on its location within the wood structure; bound water is held within the cell walls, and free water is held within the cell cavities. The fiber saturation point is the point at which the bound water is at 100% of its capacity and there is no free water, typically 21–32% moisture content. Adding or removing water above or below this point greatly affects the physical and mechanical properties of wood. Wood will shrink as a result of moisture loss, and this shrinkage is greatest in the tangential direction. Shrinkage in the radial direction is approximately half of that in the tangential direction. Shrinkage in the longitudinal direction is negligible.

Wood is considered dry when it is below 19% moisture content, and in the United States, below 15% if it is kiln-dried. Wood will continue to dry until it comes into equilibrium with ambient conditions (5–12% in the northern United States) and will vary with ambient conditions thereafter. Wood that has been dried and allowed to reach ambient moisture conditions is more dimensionally stable and has less tendency to warp or twist after construction than wood that has not.

The strength of wood depends on species and growing conditions. However, a few generalizations can be made. The tensile strength of wood loaded parallel to the grain is two to four times greater than the compressive strength under the same load conditions. Both tensile and compressive strength of wood loaded perpendicular to the grain, however, is substantially lower. Design guidelines and material properties are available for various species of commercial trees, for example, those found in North America (AF&PA 1996). Manuals can also be found for other parts of the world.

The species of wood should be carefully considered if the wood is to be placed in an environment that makes it susceptible to moisture or insects. Some woods are naturally resistant to insect attack and wet conditions, such as African ekki wood.

Local carpenters should be familiar with the material behaviors of the species of wood available to them and should have a feel for the appropriate size and spacing of structural components. Things to consider when working with wood on a project are exposure to the elements; whether the wood is green or has had a chance to dry; and any significant defects, such as large knots, dry rot, or bark.

⟫⟫ References

Allen, E., and Iano, J. (2003). *Fundamentals of building construction: Materials and methods*, 4th ed. John Wiley & Sons, New York.

American Concrete Institute (ACI). (2005). *Building code requirements for structural concrete (ACI 318-05)*, ACI Committee 318, American Concrete Institute, Farmington Hills, Mich.

American Forest and Paper Association (AF&PA) and American Wood Council. (1996). *Manual for engineered wood construction*, American Forest and Paper Association and American Wood Council, Washington, D.C.

Blondet, M., and Villa Garcia, G. (2008). *Adobe construction*, Catholic University of Peru, Lima. <http://mitigation.eeri.org/files/adobe.pdf> (Jan. 25, 2008).

Davis, J., and Lambert, R. (2002). *Engineering in emergencies: A practical guide for relief workers*. ITDG Publishing, Rugby, Warwickshire, U.K.

Intermediate Technology Publications (ITP). (1995). *Rural building course*, Volumes 1–4, Intermediate Technology Publications, London.

Jordan, T. D., Jr. (1980). *A handbook of gravity-flow water systems*, Intermediate Technology Publications, London.

MacGregor, J., and Wight, J. (2004). *Reinforced concrete: Mechanics and design*, fourth ed., Prentice Hall, Upper Saddle River, N.J.

Stern, P., and Longland, F. (1983). *Field engineering: An introduction to development work and construction in rural areas*. Intermediate Technology Publications, London.

⟫⟫ Further Reading

Ashby, M. F., and Jones, D. R. H. (1980). *Engineering materials: An introduction to their properties and applications*. Pergamon Press, New York.

Houben, H., and Guillaud, H. (1994). *Earth construction: A comprehensive guide*, Practical Action, London.

Mamlouk, M., and Zaniewski, J. (2006). *Materials for civil and construction engineers*, second ed., Pearson Prentice Hall, Upper Saddle River, N.J.

Portland Cement Association (PCA). (2004). *Design and control of concrete mixtures*, 13th ed., Portland Cement Association, Skokie, Ill.

7

Construction Techniques

))) 7.1 Construction in Developing Countries

This chapter provides an overview of various construction techniques used in developing countries. It complements the material, design, and project management information provided in Chapters 4, 6, and 8.

Construction methods used in developing countries vary by region and with availability of local materials, ability and knowledge of local labor, cost, and tradition. This chapter provides general guidelines and advice for construction alternatives. Always investigate local conditions and compare them to the information on materials, design, and construction techniques provided in this book. Sometimes local methods are preferred, whereas in other situations, recommendations should be made to modify traditions.

No definitive literature proves that one type of structure is better than another because of complicated factors, such as topography, transport, and material availability. The challenge is to assess the local situation, review design guidelines, and select an appropriate and workable construction technique to produce a successful project. For example, a design may recommend a masonry block wall that is not available in the local area, whereas a poured concrete or rock wall could be used with equivalent results. This chapter presents construction techniques, however refer to Chapters 3–6 and 8 in addition to the other chapters in this book when designing and implementing the construction project.

Tools and techniques used in the developing world are generally simpler and slower than industrialized world projects. Along with this simplicity comes some inherent safety. However, an attitude toward maintaining the safety of the workers, community, and animals is important. Simply tripping over a line stretched for site layout can be dangerous. Wear head and eye protection when needed. Cement will burn the skin, so when working with concrete, protect hands and exposed skin. If skin is exposed to cement or concrete, wash with clean water. Tools should only be used for what they are intended.

The general construction project sequence is to
1. identify the project site,
2. clear and grub the area,
3. lay out the construction,
4. excavate,
5. construct footings and foundations, and
6. construct walls, floor slabs, and roof as required by the project.
The organization of this chapter also follows this progression.

Marilyn M. Phillips, Dennis J. Magolan, and Daniel Nover contributed to this chapter.

))) 7.2. Site Preparation

7.2.1 Site Survey, Layout, and Clearing

Before actual staking and construction, general clearing and grubbing of the location may be needed to survey and lay out the site. In addition, it is important to plan the construction site layout for storing and securing, transporting, and using construction materials. Planning can minimize the amount of work that has to be redone, such as stacking and restacking brick, while surveying a few extra stakes can avoid relayout of the site. Safety of workers and interested community members (especially children) should also be considered when planning adequate work space. Another point to plan and design is site drainage and erosion for the construction period, as well as the final project. Clearing too much area can be as detrimental as not clearing enough.

Final clearing and grubbing remove trees, brush, roots, other organic material, and rocks for an open and safe construction site. These materials should not be buried as a method of disposal. Strip topsoil and pile it for later reuse in final construction.

After selecting and surveying potential project sites (see Chapter 5), final layout and staking of the construction project will follow. Survey and lay out the project from a single established survey point to minimize possibilities of mistakes. If the original survey stakes remain, they can be used as reference. If not, stake and measure two new benchmarks. Stake construction corners and offsets in case the corner stakes are knocked down. Use string and wooden or metal stakes. Secure important corners using batter boards and nails. As shown in Fig. 7-1, offset batter boards to avoid losing corner stakes during

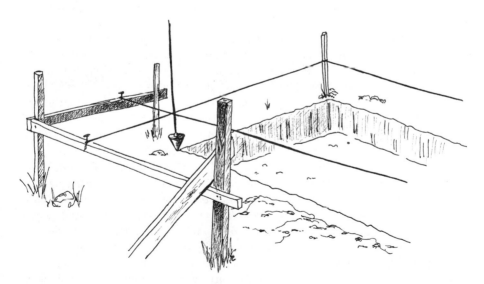

Figure 7-1. Batter Boards That Are Offset from the Actual Corner Point Are Shown in the Left Foreground. A Corner Stake Without an Offset Is Shown in the Right Background.

excavation. To make sure the layout is square, check diagonal measurements to ensure that they are equal. Use a tape measure or string, paying attention to the tension or stretch.

7.2.2 Establishing a Level

Establishing accurate and equal elevations or levels is necessary when laying out the construction project, erecting formwork, constructing masonry walls, and in establishing level or sloping floors. Use survey equipment, a carpenter's level, or a clear hose filled with water to establish points that are equal in vertical elevation. The advantage of a clear water hose is that it is readily available and can establish elevations at points that are substantially further apart than the carpenter's level. It can also double as a hose when necessary (later, Fig. 7-A in Box 7-2 shows this leveling technique).

If a slab is to have uniform slope or various slopes to a common drain, establishing the slope(s) can be easily accomplished by stretching string between the various points (and elevations) and building forms, screeds, or rock or brick fill to these level lines (Fig. 7-2). Excavation for the slab and gravel subbase should be a uniform distance below the string.

7.2.3 Excavation and Backfill

Excavate as needed for the staked construction, providing enough room in the excavation for worker access and safety. Build major construction on bedrock or stable native (in situ) materials, not on backfill, if possible. Chapters 6 and 8 provide advice on acceptable subgrades. Use shovels and pickaxes for manual excavation. If the soil is very hard, first soften it by pouring several buckets of water over the area and allowing it to soak for several hours. Be careful not to disturb foundation subgrade soils. Any soils softened by water should be removed.

If construction must be located within an area of fill, ensure that the backfill has been properly placed and compacted, and use proper fill materials. On small projects, backfilling and compaction are often done by hand; therefore carefully select fill materials. Use granular soils containing less than 15% fines (refer to Chapter 8 for soil classification) because cohesive soils (i.e., clay or silt) require specialized motorized equipment (e.g., pad foot rollers or sheep foot rollers) to provide adequate compaction. Tamping granular soil backfill by dropping tree trunks, large rocks, concrete, or timbers is a common method to achieve compaction by hand (Fig. 7-3). Place fill for hand compaction in layers or lifts no more than 5–10 cm (2–4 in.) and compact until the fill is stable under foot traffic and compaction equipment no longer leaves indentations. On larger projects using granular fill soils, lifts of 15 cm (6 in.) are typically used with special compaction machinery, such as vibratory compactors, pad foot rollers, or smooth drum rollers. Without this equipment, a loaded dump truck can be rolled over the fill area several times to achieve compaction.

Fill that is not well or equally compacted will produce differential settlement and cracking in walls and slabs that could lead to leaking or other types of failure. Regardless of the amount of fill that is present, construct structure foundations on natural undisturbed subgrade soils. Building footings should extend through any fill to bear on natural, undisturbed soils. Backfill for pipelines must be free of rocks that could damage pipe. Do not use heavy equipment to compact a pipe trench backfill.

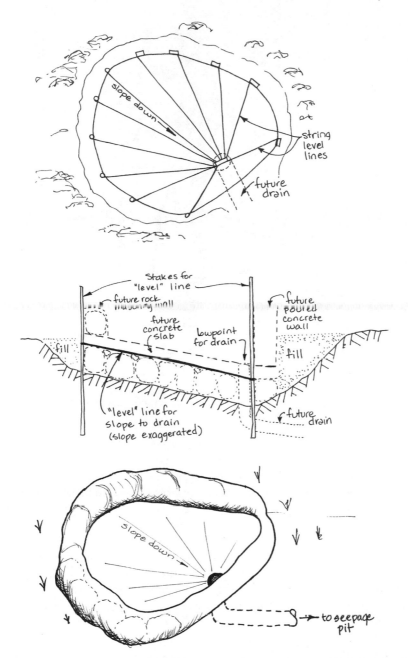

Figure 7-2. In the Top Figure, the Level Line for a Wash Area Is Used to Establish the Slope for the Initial Rock Subslab, as Well as the Final Concrete Slab. Stakes and String Provide a Guide for Installing Stone to a Proper Slope. Two Alternate Designs for Future Rock Masonry or Poured Concrete Walls Are Shown on the Cross Section (Middle). The Lower Figure Depicts the Finished Wash Area.

Figure 7-3. Hand Compaction of Loose Soil or Backfill Can Be Accomplished by Dropping a Tree Trunk, Concrete Weight, or a Timber.

Protect excavations deeper than 1.5 m (less in cases of unstable soil) for the safety of workers or excavate to the soil's angle of repose. Use excavation support systems, shoring, or temporary bracing made from reused lumber, trees, corrugated metal, or sheet piling to prevent cave-ins and retain excavated walls safely. Figure 7-4 shows how shoring and whaler support can be used to hold back deep excavation walls. Reusing these materials is efficient and provides a safe worksite. Install barriers to keep people and animals, particularly curious children, from excavations, especially deep ones. In rural areas where farm animals roam freely, fill in excavated pipeline trenches by the end of the day's work. Excavation support systems should be designed by someone familiar with support system design and versed in geotechnical as well as structural disciplines. An improperly designed or installed system can be more dangerous than not using an excavation support system.

7.2.4 Formwork

Formwork, also referred to as *forms* or *shuttering*, are supports used to contain wet concrete construction (Fig. 7-5). Typically, forms are reused lumber set in place and anchored with stakes. Concrete blocks, timber, or a firm soil wall can also serve as forms. The quality of the lumber is not as important as for permanent wood construction. However, forms should

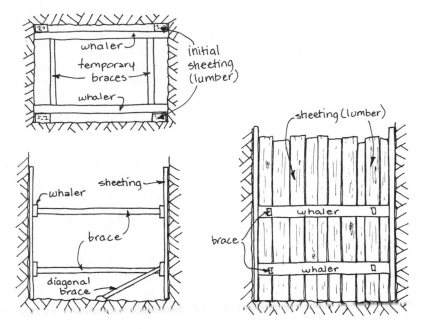

Figure 7-4. Shoring and Whaler Support Retains Deep Excavation Walls.

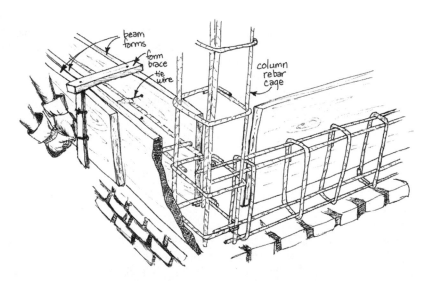

Figure 7-5. Grade Beam Formwork and Reinforcing Is Shown as It Ties into the Column.

Note: Stirrups encircle the longitudinal rebar, forming the rebar cage that is suspended in the forms. Wire is tied to hold the forms together and can suspend the rebar cage. A wooden form brace is shown as an alternative to the wire form ties.

be able to be installed straight, level, and true. They are usually installed along a string line that has been established as level to the finished elevation. Use pieces of wet, discarded concrete sacks to plug small holes in the forms. Other form material, such as plastic or fiberglass, can also be used, depending on availability, if it is properly supported.

Waste nothing on the construction site. Because lumber may be expensive and a scarce commodity, disassemble and reuse forms, stakes, and *scabs* (small pieces of wood), or borrow the lumber from locals who use it for other applications. Accordingly, design formwork to avoid or minimize cutting. In addition, remove, straighten, and reuse nails to save cost. Lumber is usually the actual dimension in other countries, rather than the nominal dimensions used in the United States. Different wood species may be heavier and more difficult to cut and nail. Greasing nails, even by rolling them on the forehead to pick up body oil and sweat, will make it easier to drive and remove them from hardwood.

After about 3 days of wet-curing (4–5 days for walls, roofs, and bridges), remove concrete forms. Before placing concrete, apply a form release agent to form surfaces that will touch the concrete to make removal of forms easier; it also prevents the forms from drawing water away from the curing concrete. Used motor oil is a common release agent. However, plastic, banana leaves, or anything that facilitates form removal without affecting the final product is acceptable. Do not oil or grease the forms if the surface is to be painted. Usually oil and grease are not used if the surface will be parged; however, the forms can be oiled if the concrete surface is scored using a rock hammer or similar instrument after form removal, before parging.

Provide adequate exterior form support. When forms are filled, the pressure and weight of wet concrete may cause the forms to bow, tilt, or burst. This problem can be difficult to repair without costly loss of concrete. For walls, beams, or columns, where the two forms are parallel and close together, often opposite sides are tied together using rebar tie wire that passes through holes drilled in both forms. The wire is cast into the concrete. After curing, the wire is cut to remove the forms and then trimmed to the face of the cured concrete. Figure 7-5 shows steel reinforcement that is encased in formwork before the placement of concrete. The hands shown in the left of the figure are tying two parallel (beam) forms together with tie wire. Optionally, wooden form ties can be used.

7.2.5 Screeds

A *screed board* is a wooden strip nailed to stakes that are driven into the subbase. Screeds serve as a guide for making a level surface on a concrete slab (Fig. 7-6). For small slabs or pours, the forms act as screed boards. In large flat concrete pours, screed boards are installed to enable laborers to construct and finish the entire slab to the proper elevation. A screed is also the board or metal strip dragged in a sawing motion across the top of forms or screed boards while placing concrete to give it its proper level, removing the excess concrete.

For large concrete slabs on grade, interior screed boards divide the large slab into sections manageable for screeding. Interior screeds are installed after outside forms are set in place. They are cast into the concrete slab and remain in place until the concrete has set enough to remove them without causing the surrounding concrete to subside. (See concrete finishing later in this chapter.) The void left by the screed can then be filled with fresh concrete and finished.

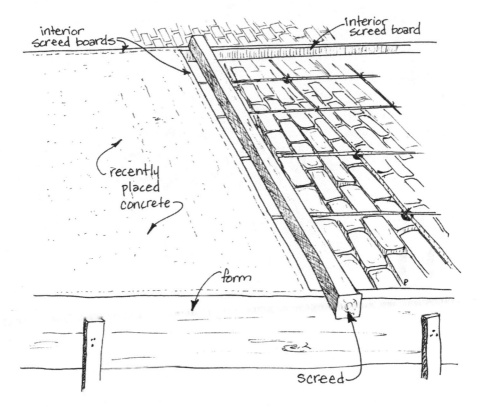

Figure 7-6. Screed Boards and Forms Establish the Finished Elevation and Support the Recently Placed Concrete (Left) Poured on Top of a Brick Layer (Right). The Concrete Has Been Screeded, with the Screed Resting on the Forms and Screeds.

7.2.6 Reinforcing Steel

Chapter 6 discussed the use of reinforcing steel for concrete that is used in structural applications or where cracking cannot be tolerated. Cut, bend, tie, and place correctly sized and designed rebar. Rebar can be cut with metal handsaws. Bend it using a short pipe as a lever in a jig, as shown in Fig. 7-7. Beams, columns, footings, and other structural components require stirrups or ties, which encircle the longitudinal rebar. Bend stirrups sharply and cleanly. Overlap the tie at a corner by at least 2 in. in each direction. Space stirrups at a distance no greater than the width of the column. When the longitudinal rebar and stirrups are assembled, it is called a "rebar cage," which is depicted in Fig. 7-5. Splicing (extending) rebar is discussed in Chapter 6.

Tie rebar together tightly with tie wire (16 gauge) at each intersection of two bars. Figure 7-8 shows a tying method using a nail to tighten the wire. Fold the appropriate length of tie wire (approximately 30–39 cm (12–15 in.)) in half. Place the center of the folded wire length across the top of the rebar cross. Then pull the tie wire under both

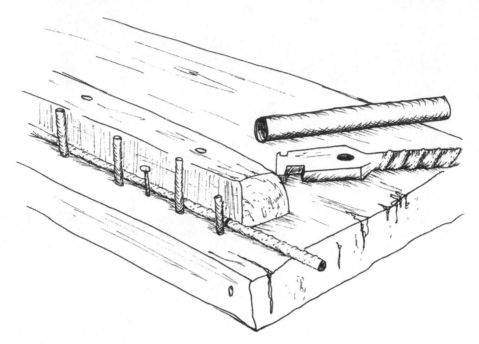

Figure 7-7. Rebar, Which Has Been Marked for the Bend, Is Placed in the Jig Between the Wood and Metal Rods and Is Bent with the Hand Tool or the Pipe.

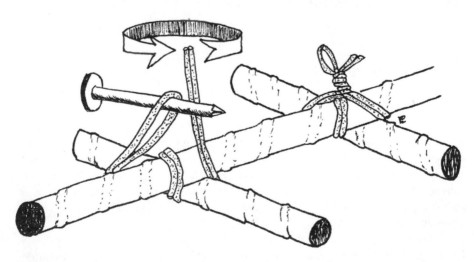

Figure 7-8. Tying Rebar Can Be Done with a Nail. Tighten It Until There Is No Movement.

rebars, cross it, pull it up in the opposite corner of the rebar intersection, and then tighten it by inserting a nail or similar tool in the loop and bending the loop around the loose wire. Tighten it completely until there is no movement possible between the bars to prevent slipping and misalignment while the concrete is being placed and finished.

The placement of rebar within the forms (and the concrete member itself) is critical because the rebar requires a certain thickness of concrete cover. *Cover* is defined as the distance between the rebar surface or bar end to the nearest face of concrete form or subgrade (see Chapter 6 for concrete cover dimension requirements). Place rebar on *chairs* or bolsters to keep it above the slab bottom and to achieve the design location and required concrete cover within the slab cross section. Chairs prevent rebar from moving to the bottom of the slab when the concrete is placed or if workers accidently step on it. They can be specially made metal or plastic holders, broken block or brick, or flat, stable stones. Chairs should be the correct size to provide adequate concrete cover below the rebar. The placement of chairs is shown later, in Fig. 7-11.

))) 7.3 Construction

This section discusses mixing, placing, and finishing concrete, as well as methods used to construct foundations, slabs, beams, and walls.

7.3.1 Measuring and Mixing Concrete

Mixing concrete at the construction site requires organization of the site, materials, material flow, and workforce. Because mixing and placing concrete is a continuous operation, avoid delays due to material and labor shortages or poor planning. Clear the location where aggregate will be stored and where concrete will be mixed. Remove rocks, roots, organic material, and topsoil. Maintain aggregate piles neatly by shoveling aggregate from the bottom to the top of the pile and into a bucket located on or adjacent to the pile to avoid waste.

As discussed in Chapter 6, aggregate size, selection, and mix design are key to concrete and mortar performance, as well as workability. In developing countries, concrete is often proportioned by volume, which is likely to be measured in buckets, wheelbarrows, donkey carts, or shovels, rather than in m^3 or ft^3. In some locations, it is customary to only use entire sacks of cement at a time.

Use 5-gal (20-L) buckets (or other readily available containers) for measuring aggregate. Measurement using sacks of cement is not recommended because of the subjective nature of determining the volume of a partial sack. Wheelbarrows are often used as a measuring device, and experience suggests that they work, although they vary in size. It is important to be consistent on a given job, so use the same measuring container throughout.

Generally, in any mixing, first add aggregates, followed by cement, and finally water. Mix concrete thoroughly to yield the strongest product. Mix concrete by hand, by small hand- or motor-powered mixers, or purchase it ready-mixed in trucks. Local concrete workers using powered mixers or ready mix should have experience, and the attention of the development worker will be focused more on their mix design than the mixing itself.

When mixing by machine, put about 10% of the water in the drum, and then add dry material and water, uniformly leaving 10% of the water to be added after the dry materials are in the drum. Allow 5–6 min for mixing after all materials are in the drum.

Mixing concrete by hand takes skill. Mix it on a clean, watertight surface such as a brick or concrete slab, a wood or metal mixing trough, or an anchored tarp to avoid contaminating the concrete with foreign matter. Mix concrete on the ground only if there is no other option. In this case, clear the ground of debris and organic matter and use the same location for sequential mixes.

Figure 7-9 shows the common *volcano method* of hand mixing concrete or mortar. Proper mixing takes time and is critical to concrete performance. In any hand mix method, thoroughly mix the dry aggregate and cement before adding water. Shovel a single pile of the measured dry materials from one pile to another until a uniform mix is obtained. Repeat this process a minimum of three times. Alternatively, two or three workers can shovel the outside edge of the pile to the top, advancing around the pile in a clockwise or counterclockwise motion, shoveling until the dry mix is uniform. Once the dry materials are thoroughly mixed using either method, shape the dry mixture to form a mound with a crater in the top. Add water into the crater. Then slowly and carefully shovel the dry mix from the "volcano" sides up to the crater edge so that a little falls into the water. Progress carefully around the volcano, one shovel at a time. As the concrete absorbs the water, open up the crater carefully and add more water.

Add water slowly, in small amounts. Do not fill the crater to the top with water because the crater may collapse and material may be lost. When most of the mix has absorbed water, turn over the pile with shovels several times to ensure a uniform consistency. Do not rush this process. Enough water has been added when the concrete can hold itself up in large clumps when it is placed on a level surface. Most masons and local workers prefer to add too much water to ease transportation and placing of the concrete. However, added water weakens the concrete.

Figure 7-9. Hand Mixing Concrete Using the Volcano Method.

If concrete or mortar freezes, the curing process stops and does not resume, so no further strength can be gained. Concrete generates heat as it cures. In hot weather, the curing process speeds up. The rate of curing doubles for every 10 °C increase in temperature, similar to biological processes (refer back to Chapter 6 for curing times).

7.3.2 Placing and Finishing Concrete

Before placing concrete, dampen the forms and subgrade so that the water within the concrete mix is fully available for the hydration process and is not absorbed into the dry forms or subgrade. Mixing and placing concrete is a continuous operation, and mixed concrete must be placed within 40 min because the setting process begins immediately. Thus, carefully plan the pour and lunch or other breaks.

Within 2–4 h, concrete sets and can no longer be worked. *Cold joints* or pauses between pours (as short as 1–2 h) where concrete is allowed to partially or fully cure need to be strategically placed because this will be a location of concrete cracking. These joints must be vertical within a slab pour and should not occur within a beam, column, or structural member. For a concrete slab, a cold joint should be vertical through the depth of the slab, not horizontal (i.e., for a 10-cm-thick slab, do not place 5 cm, let it partially cure, and then add the top 5 cm).

Avoid *segregation* (the larger aggregate separating from the cement and water mix) by placing the concrete instead of pouring it. Segregation results in enriched paste or aggregate areas that weaken the concrete. Use shovels to place concrete where it is needed. Do not drop, throw, or pour concrete mix into the forms. Segregation can also occur in a wheelbarrow moving from the mixing location to the placement site. Therefore, the mixing site should be as close as possible to the concrete placement site, and concrete should be mixed with shovels or trowels as it comes out of the wheelbarrow.

The concrete is typically *consolidated* into the formwork by rodding (stabbing the wet concrete) with a piece of rebar, a trowel or a concrete vibrator to minimize voids or *honeycombing*. Pay particular attention to corners, around reinforcement, and in hard-to-reach areas. However, too much rodding also causes segregation.

During concrete placement, ensure that reinforcement stays in the designed position on the chairs or within the forms with appropriate concrete cover. Laborers may need to use a hook to pull rebar up from the bottom of a slab or readjust beam or column rebar.

Trowels and bull floats are used to finish concrete surfaces. *Bull floats* are wooden or metal panels attached to long handles. They are pushed and pulled across the wet concrete surface after screeding to further smooth large surfaces. *Trowels* come in various sizes and shapes and are used to finish concrete (and apply mortar when constructing masonry walls). Rough concrete finishing with trowels can occur soon after bull floating. However, final finishing must wait until the concrete has hardened somewhat. When finishing, do not apply excessive pressure or work the surface too much, causing the cream to rise to the surface. The *cream* is water and cement that has been drawn out of the mix. This separation results in loss of uniformity and also spalling, or flaking, of the finished surface.

When applying the final trowel finish, masons usually kneel on a board, long and wide enough to distribute their weight, over the fresh concrete so that they don't imprint the final surface. Interior screed boards that have been cast into the slab are removed when the slab has cured for a few hours and workers are applying the final trowel finish. The void left by the screed board is filled with fresh concrete.

Consider the final concrete finish. If the surface is to be watertight and easily cleaned, a hard steel trowel finish may be required, or the surface may be parged later. We have observed in some instances that in addition to the hard trowel finish, the surface may also be "rubbed" with a hard stone a day or two after the pour to develop a sheen. For concrete surfaces that are to receive a parge, or where there are safety (slippage) concerns such as in a public wash area, a rougher final finish can be done with a trowel or a broom.

Moisture is essential in the curing process. Cover concrete soon after placement and keep it wet. Dehydration of concrete results in loss of strength, as well as shrinkage, which results in cracking. Cover curing concrete with burlap, cement bags, straw, sand, or banana leaves, and add water as often as needed to keep it moist the entire curing period. Seven days of moist curing is recommended. At a minimum, cover freshly placed concrete with plastic to avoid evaporation, which leads to cracking. It is especially important to wet-cure concrete in dry and hot environments. Protection from sun and wind is also recommended.

7.3.3 Footings and Foundations

Footings and foundations are generally at least twice the width of the wall to be placed on the footing and are used to provide support to the structure. Build the footings on suitable native subgrade soil or a gravel subbase that has been properly designed and prepared or that is acceptable for the planned loads. Construct footings with local materials of timber, rock, poured concrete, or masonry with poured concrete. Timber will degrade over time and is not a permanent choice. Brick or rock masonry with a poured concrete top is a lower cost alternative to poured concrete because the amount of cement is reduced and steel reinforcement is reduced or eliminated. However, the strength and ability to bridge differential settlement in the subgrade is not as great as with poured, reinforced concrete.

One method of constructing footings underwater is by using rice bags filled with dry concrete mix (aggregate and cement) placed on bedrock or firm subgrade (Fig. 7-10). The dry concrete mix hydrates and cures in the water. This method is advantageous for applications such as spring box foundations or for anchoring pipe in stream crossings.

As shown in Fig. 7-11, during footing construction, L-shaped vertical rebar is tied into the footing rebar at every column location. Install vertical rebar at each block cavity location in masonry block walls and every 21–31 cm (8–12 in.) for poured concrete walls. This rebar does not need to protrude more than 31–39 cm (12–15 in.) above the footing because additional vertical rebar will be spliced to the L-shaped rebar embedded in the footing.

7.3.4 Slab on Grade

Slab on grade construction is used in installing slabs such as floors, tank bottoms, or open wash areas. Slabs are typically 10–15 cm (4–6 in.) thick, depending on loads, materials used, and concrete quality. They are usually designed with foundations under the perimeter (see above) for edge support, especially if a wall is to be built on the slab.

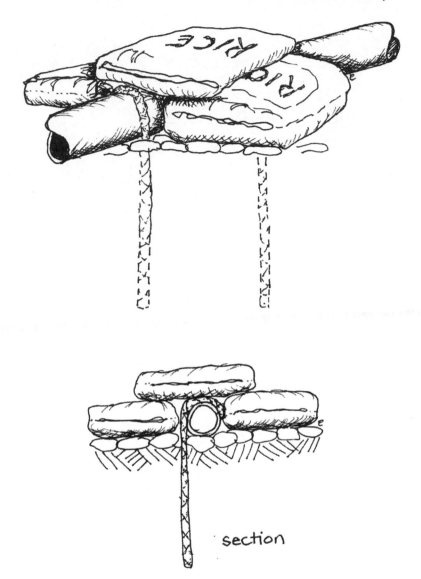

Figure 7-10. Dry Concrete Mix in Rice Bags Used for Pipe Anchors in a Stream Crossing. They Can Also Be Used as Foundations Underwater (e.g., for Spring Boxes).

Footings and slab concrete may be cast separately or integrally. The walls can either be built first on footings and then the slab on grade installed inside the walls or the slab on grade with internal footings can be built first with the wall then installed on top of the slab. For larger construction, the walls are usually built first on foundations, but for valve boxes and similar small projects, it may be easier to construct the floor first (without a

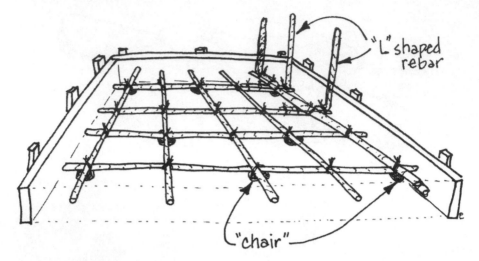

Figure 7-11. Rebar Tied in Slab on Grade Forms with Partially Installed L-Shaped Rebar to Tie into Future Walls. "Chairs" Sit Under the Rebar to Provide Cover Between the Bottom of the Slab and the Rebar.

foundation, depending on box size). For slab construction where walls are to be built on the slab, L-shaped rebar must be embedded into the slab at each block cavity location for masonry block walls and for poured concrete walls. Dimensions are as provided above for footings.

Slab on Grade: Poured Concrete

Slabs that are placed on grade are constructed by following the sequence listed in Box 7-1.

))) Box 7-1. General Sequence of Activities for Constructing a Slab on Grade

- Excavate and prepare the subgrade, removing rock and loose material.
- Plan the quantity of concrete to be used, taking into account cold and expansion joint placement.
- Install oil forms and intermediate screed boards, as necessary.
- Cut, bend, tie, and install reinforcing steel.
- Wet forms and subgrade.
- Mix and place concrete, rodding to eliminate voids.
- Screed, bull float, and trowel to finish, removing screeds as necessary.
- Remove forms after 3 days of wet-curing.
- Wet-cure for a total of 7 days.

Slab on Grade: Brick or Rock Covered with Concrete

A less costly and lower strength option for slab on grade construction is to install brick (one or two layers) or rock on the subgrade, which is then covered with approximately 5 cm (2 in.) of poured concrete, as shown in Fig. 7-6. This method requires less concrete than a poured concrete slab, and it has been used successfully for roads, wash areas, tank bottoms, building floors, and ball courts. A firm subgrade is important because steel reinforcement is either absent or substantially decreased, compared to traditional poured concrete slab on grade designs. Sometimes a 5 cm (2 in.) sand layer or a composite layer of sand, sheet plastic, and more sand (totaling 5 cm) is used as a base, as shown in Fig. 7-12. Sand aids drainage, and plastic protects the floor from high subsurface water. The added cost of the plastic and second sand layer must be weighed against the benefits.

Composite slab construction follows the sequence outlined in Box 7-1, with some modifications. The subgrade must be excavated and prepared to receive the rock or brick (and sand subgrade if required). The subgrade must be leveled (or sloped) to the correct elevation, accounting for brick or rock and concrete topping. If forms are used, they are

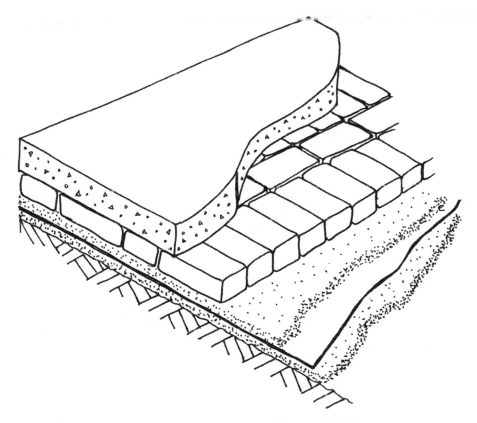

Figure 7-12. Composite Slab on Grade: Sand, Sheet Plastic (Optional), More Sand, Brick, and Poured Concrete Topping.

installed to enclose the poured concrete and to establish the finished concrete elevation. If the slab is large, intermediate screed boards are installed.

A line is then strung between stakes to the proper intermediate elevation for the top of brick or rock. The brick or rock is then installed to the level line. If two layers of brick are used, the adjacent layers should be staggered so that joints are not directly above each other. Rocks should be uniform and approximately the size of the fist of an adult hand.

Loose or unstable brick or rock can lead to slab cracking and failure, so it must be firmly grounded in the subgrade. The first two rows of brick or rock are installed to the correct elevation or slope along the length of two string lines. Then rock or brick is placed in the inner area between the two installed rows. Using a straight edge or board laid across the two lines of installed brick or rock allows verification that level is being maintained between the established string lines. Figure 7-13 shows a level line for a composite slab on grade that consists of rock and poured concrete topping. Figure 7-14 shows the use of a straightedge or screed board to verify the level of brick in a composite slab on grade that consists of brick and poured concrete.

When the field of brick or stone has been installed and concrete formwork and screeds have been rechecked for alignment to provide adequate concrete cover, place the

Figure 7-13. Composite Slab on Grade: Rock and Poured Concrete Topping, Showing the Use of Level Lines to Establish Elevation.

Figure 7-14. Composite Slab on Grade (Brick and Poured Concrete Topping) Construction, Showing Use of Straightedge or Screed to Verify Level of Brick Being Installed Before Concrete Placement.

concrete. Too little cover over rocks will lead to the concrete cracking. Generally, a slightly wetter concrete mix with smaller aggregate is used because the slab is thinner and to ensure that voids between the rocks are filled (Fig. 7-15). Follow the general concrete mixing and placing guidelines.

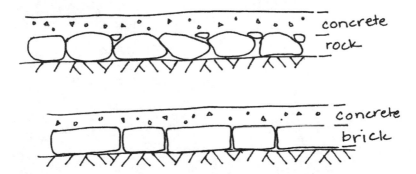

Figure 7-15. Composite Slab on Grade Construction Can Use a Base of Rock or Brick.

Note: When installing rock, smaller stones can be used to fill in spaces between the larger rocks, as long as they don't exceed the elevation of the string line and are properly seated (not loose).

7.3.5 Poured Concrete Beams and Columns

Beams are structural members that span and transfer load. Beams around the perimeter of the construction at grade or foundation level are called *grade beams*. Perimeter beams tying columns together are called *bond beams* and are in essence elevated grade beams. Masonry (brick, rock, or block) walls are typically constructed between the grade beams or footings, bond beams, and columns.

Columns support vertical loads. Construct concrete columns before or after construction of a non-load bearing masonry in-fill wall. Beams or columns may contain protruding rebar to tie into future wall construction, or columns may be drilled later to insert rebar to tie into future adjacent walls. Figure 7-16 shows foundation and column rebar

Figure 7-16. Rebar Being Readied for a Water Storage Tank That Will Be Situated Partially Below Grade.

cages readied for forms, and Fig. 7-17 shows the column formwork installation around a rebar cage. Figure 7-18 shows a bond beam and half-poured column for a concrete water storage tank.

Rebar design and placement is critical to beam and column performance and safety. To center the rebar in the form, a common technique is to hang the cage from the forms. Drive nails into the tops of the forms and use tie wire to suspend the rebar in the center

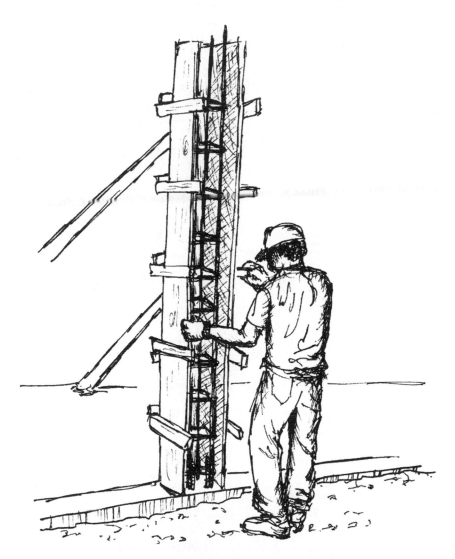

Figure 7-17. Column Formwork Installation Around a Rebar Cage.

Note: The column is poured before the masonry wall construction. Angled wood braces keep column forms vertical and plumb. The column rebar cage is tied at the bottom to rebar that was cast into the poured concrete foundation.

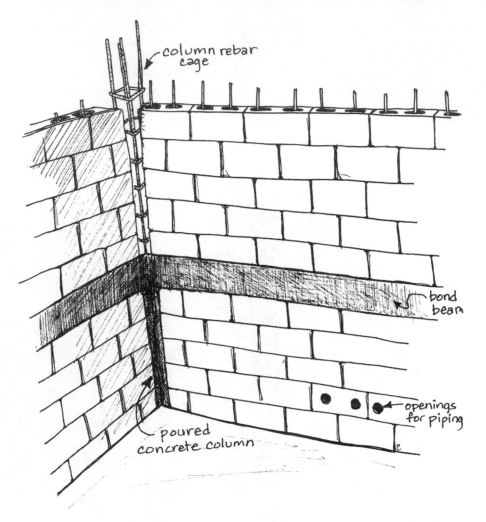

Figure 7-18. Poured Concrete Bond Beam and Column (Both Are Shaded) for Concrete Block Water Storage Tank. The Column Is Being Poured After Masonry Wall Construction.

of the form. Once the concrete has cured and forms are removed, cut off wires at the concrete face (see Fig. 7-5).

Place concrete in beams and columns according to previously outlined guidelines. Pay particular attention to mix design and aggregate size with respect to rebar cages and forms. Aggregate that is too large will not pass between the rebar and forms, and segregation and weakness will occur (see Chapter 6 recommendations). When column forms are filled with concrete, properly place and rod or vibrate to avoid segregation and honeycombing. Light pounding on the exterior of the column or beam form can aid in consolidation and vibration, but take care not to knock the form out of alignment.

7.3.6 Elevated Slabs

Poured in Place Elevated Concrete Slabs

Poured in place elevated concrete slabs are typically roof slabs for tanks or boxes, floor slabs for composting latrines, or second floors for buildings. These slabs are poured in place rather than precast and moved into place. At this point in tank construction, the floor must have already been completed because there will not be easy access to the tank interior once the roof is poured in place. Also, if the bond beam at the top of the masonry wall has not been poured separately, it can be cast at the same time as the elevated slab (i.e., the roof).

Pouring elevated slabs in place requires special falsework or formwork. In some cases, special precast beams and formwork remain in place after construction, but typically formwork is removed. Use plywood or locally available lumber for formwork; or use metal panels or multiple boards in place of plywood for short spans. Well-supported 5 cm (2 in.) thick rigid extruded foam board (not polystyrene or bead board) has been successfully used with proper support beneath the panel on short spans.

To pour the elevated slab in place, install wooden formwork. This work is similar to building a table within the tank or space. Adequate support of the "table top" is critical because of the weight of the heavy wet concrete and rebar. Formwork failure during a pour could be catastrophic. Figure 7-19 shows an example of columns and bracing

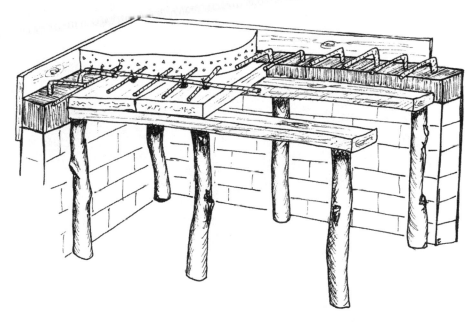

Figure 7-19. Falsework or Forms for an Elevated Slab for the Roof of a Water Storage Tank or the Top Slab of a Composting Latrine.

Note: The bond beam is shown darkened. Vertical bond beam rebar is bent to tie into the elevated slab, just as L-shaped rebar is cast into a foundation or slab for the future wall. For added stability, bracing can be added to vertical posts.

required for elevated slab formwork. Design each piece of plywood or lumber formwork to be removed through the access opening once the concrete has cured.

If holes are required in the finished elevated slab, form them with wood; a plastic or rice bag filled with small stones, sand, or rice; or bundles of bamboo or other stalks, depending on the opening shape desired. Openings are also sometimes a prefabricated metal access hatch frame. Doors typically provide access to the inside of a tank for inspection and cleaning. If a prefabricated metal access door is being used, tie the doorframe into the roof slab rebar and set the top lip to the elevation of the finished concrete.

After constructing the interior "table" formwork and installing steel reinforcement and hatch frames, place forms around the outside walls at the top edge of the wall. Nail the wooden forms in place and support them with braces. Figure 7-20 shows a cut-away view of a bond beam and falsework forms for an integral bond beam and elevated slab pour. The vertical wall rebar is bent to tie into the elevated slab just as the (L-shaped) rebar that was cast into the foundation was tied into the wall. After securing the formwork, mix and place the concrete as previously discussed, being sure to wet-cure the concrete and protect it from the sun and drying for a week. Figure 7-21 depicts the partially completed slab pour.

Movable Precast Elevated Slabs

Movable precast elevated slabs are lids for valve boxes, tank access hatches, movable tank roof sections, roof panels for storage tanks or spring boxes, and latrine floor slabs. Manufacture these precast reinforced concrete panels or hatches on site or buy them premanufactured elsewhere. On site they can be cast in forms on the ground, as shown in Fig. 7-22.

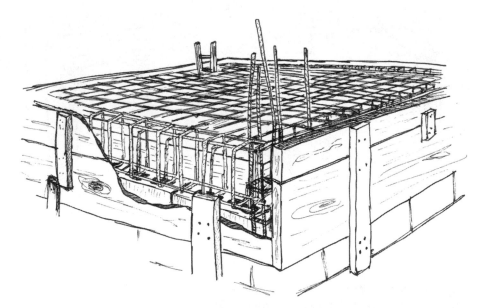

Figure 7-20. Integral Bond Beam and Elevated Roof Slab.

Note: Cut-away view of bond beam and elevated slab forms showing bond beam rebar cage and roof slab rebar grid. Vertical rebar from block cores is bent and tied to the elevated roof slab rebar grid.

Figure 7-21. An Elevated Slab Pour Partially Completed. The Access Hatch Is in the Back Left, and a Concrete Screed Is in the Back Right.

Figure 7-22. Forms and Partial Pour for a Movable Precast Latrine Slab.

Note: No nails were used in forming, and dampened cement bags were used to line the forms. Wood in various sizes has been used to form the boxed slab openings.

Appropriate reinforcing is critical to these slabs. Therefore, design and construct these slabs, as well as embedded items such as handles or locking mechanisms, for movement. Openings in precast concrete slabs (e.g., openings in latrine slabs or access hatches) can be formed in the same manner as those in poured in place slabs (Fig. 7-22).

Access hatch doors or lids can be purchased or constructed. To construct a cover or door, cut a plywood circle 50 mm wider than the outside diameter of the access opening. Cover the plywood with mortar (cement to sand ratio of 1 : 2) and shape it into a dome. Here the plywood serves as a form base. The thickness of the dome should be approximately 75 mm high at the center. Embed two rebar handles (constructed by bending two pieces of rebar), and allow the concrete to cure.

A unique method to cast concrete roof access hatch covers is to pour the roof concrete complete with wood formwork for an opening for the hatch. Install formwork at an angle (like a funnel) so that the access hatch acts as a stopper or cork in the roof. After curing the roof concrete, remove the wood framing and place sheet plastic or another barrier in the opening. Cast the concrete cover (including the reinforcement and cover handle) in the plastic-lined access opening. This method ensures that the hatch cover matches the roof hatch opening. Once the concrete is cured, lift the hatch door and remove the plastic and internal tank formwork.

7.3.7 Wall Construction

Walls may be load bearing or non-load bearing in design. They may be constructed of materials including palm fronds, bamboo, poured concrete, ferrocement, and various types of masonry. This section focuses on poured concrete and masonry walls because local traditions and methods vary for other materials.

Poured Concrete Walls

Poured concrete walls can be used for any application, including tanks, building walls, and foundation walls. One advantage is that they are easy to make watertight. Construct poured concrete walls similarly to columns and grade beams regarding formwork, reinforcing, and placing of concrete. Tie poured concrete walls into the foundation via L-shaped rebar that was previously cast into the foundation, as shown in Fig. 7-23. As in columns, rodding is important to avoid honeycombing.

Masonry Walls

Masonry walls can be used in the same applications as poured concrete walls, including tanks, spring boxes, valve boxes, latrines, or building walls. Workmanship on masonry walls used to contain liquids (tanks) needs to be of high quality to be waterproof.

Masonry walls are built on foundations with vertical rebar, like poured concrete walls. Single-width (wythe) solid brick or rock walls, however, are typically unreinforced. Masonry walls may be constructed with or without corner poured concrete columns, depending on structural design considerations. Also, depending on design and local tradition, walls may be constructed before or after the concrete columns are constructed. Either way, the column should be tied into the masonry walls with reinforcing steel.

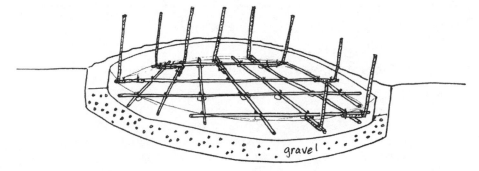

Figure 7-23. Three-Dimensional View of Slab Rebar with L-Shaped Rebar to Tie into Circular Wall Construction.

Brick or Block Masonry Walls

Brick, block, and rock masonry methods vary by country, region, local tradition, and mason, so it is important for local masons to be involved in the design. Each line or layer of brick, block, or rock in a wall is called a *course*.

Reinforced hollow-core concrete block or brick walls are laid using the guidelines listed in Box 7-2. However, for hollow-core block walls, vertical rebar that has been cast into the foundation extends up through the cores of the blocks. The open cores of concrete block are threaded over the vertical rebar that protrudes from the foundation (see Fig. 7-24). The vertical rebar is extended by overlapping and tying on new rebar. (See Chapter 6 for guidelines.) Once rebar is extended and concrete blocks are installed, open block cores are filled with a grout mixture. *Grout* is a wet concrete mixed with small aggregate (no large aggregate is used because it would not pass freely) or a mortar mix of sand, cement, and water.

Rock Masonry Walls

Rock walls can be used for fences, walls around wash areas, valve boxes, and even for spring boxes and small tanks. When constructed with appropriate structural integrity, rock walls are cost-effective solutions for locations rich in rock.

Rock masonry walls are constructed similarly to brick or block walls. However, joints and spacing are less exact and reinforcement is typically not used. Local masons adept in rock wall construction should be included in the project. Rocks in rock walls may be left exposed, or the mason may cover the exterior of the rock with mortar as the wall is constructed. Figure 7-25 shows a masonry rock wall being constructed for a wash area.

7.3.8 Ferrocement Construction

Conventional construction of poured concrete and masonry storage tanks typically involves the use of expensive materials and requires construction and masonry expertise. As an alternative, *ferrocement* or *wire-reinforced cement–mortar* tanks have been successfully used in rural areas to facilitate material storage, most notably drinking water storage.

))) Box 7-2. Guidelines to Follow When Constructing Brick, Block, and Rock Masonry Walls

The term *brick* is used here. However, concrete block or rock may be substituted.

- Arrange the work area (ground level or scaffolding) so that wet bricks, mortar, and water are easily accessible, allowing ample space in which to work safely.
- Install the string (level line) level and taut. Remove any mortar on the string. Attach this line to stationary vertical wood guides or concrete columns when constructing between columns. Then measure and mark this wood to show the height for each course (Fig. 7-A). Another method is to first construct the masonry corners in a pyramid fashion and use the established masonry as a coursing guide (Fig. 7-B). Then stretch a string line between two masonry corners as a level guide for the remaining brick installation.
- Measure and mix dry mortar without adding water.
- Mix mortar with water in a mortar box. For small projects, use a pail, but it is less comfortable and less accommodating. One procedure is to fill the box

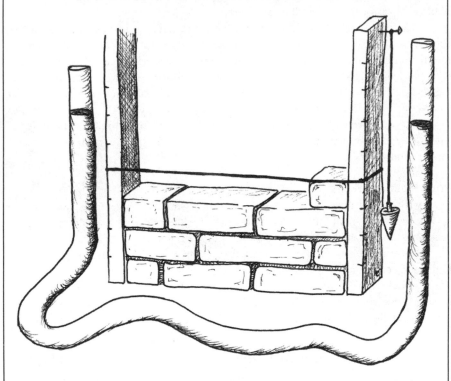

Figure 7-A. Water-Filled Tubing Establishes "Level" Between Two Points. Coursing Is Marked on the Vertical Wood Guides, and the Level Line Is Moved Up After Each Brick Course Is Completed.

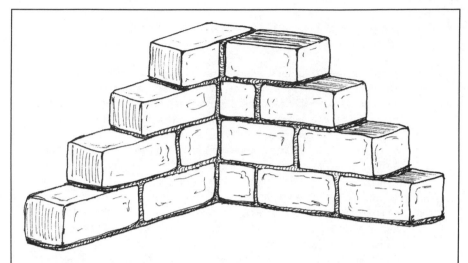

Figure 7-B. A Built-Up Masonry Corner Used to Establish Coursing and Level Line without Wood Guides.

with dry mix, then add water to the mix in one corner. Mix to the proper consistency. When that mortar has been used, mix more water into the remaining dry ingredients in the mortar box. Do not add extra water once the mortar has begun to set, and do not add water to mortar that has been dropped and cleaned from a lower surface.

- Use damp bricks. Either soak the bricks in a large container of water (perhaps a barrel) or pour water over a stack of bricks.
- Lay bricks in a *running bond*. This means that the vertical joints are staggered and do not line up between adjacent courses, which would be structurally unsound.
- When placing bricks on a previously laid course that has dried, wet the top course before laying the next course.
- Cut or break bricks in half for a running bond. Bricks can be scored and then broken with a masonry hammer. Using quarter or eighth bricks is not recommended.
- Using a trowel, lay down a mortar bed on the previous course of brick (or foundation for the first course). Use a sufficient amount of mortar where the brick will be positioned so that the brick can be pushed down into the mortar. Then tap the brick using the trowel handle to the level of the string line, making sure the brick is not touching the line. Scrape excess mortar that is squeezed out by the tapping off both sides and place it in the space between adjacent bricks or use it for the next brick. Do not level the mortar with the trowel or hand before placing the brick on top. Do not lift the brick during the laying process. If for some reason the brick needs to be lifted, use new mortar.
- Fill spaces between adjacent bricks with mortar, and fill voids that may occur below the brick.

- Masonry joints are typically 1 cm ($^3/_8$ in.) thick but may vary when there is irregularity in the bricks. Use thicker joints, not thinner, to compensate for irregularities.
- When concrete columns have been previously poured, tie masonry walls into the columns (or doors, window frames, or hatches). If rebar was not cast into the column, drill holes into the concrete column and insert rebar into the hole. The rebar will extend into the mortar joint about 15–16 cm (6 in.). Tie walls into the columns every 8–10 courses and at critical points.
- After approximately 1 m of wall height has been laid, let the masonry cure for at least 2 h before adding additional courses of masonry. Use this time to clean the bricks, strike the joints, and have lunch.
- Masonry joints can be struck in several manners as shown in Fig. 7-C. Joints can be struck with special tools, a finger, a stick, or a piece of wood with two nails protruding to the depth of the strike. It is important to compress the mortar into the joint for additional strength.
- Clean brick and block with a steel brush or plastic broom. Also clean mortar droppings at the base of the wall before it hardens.
- Once mortar has set, wet the masonry walls several times per day for a few days to improve curing.

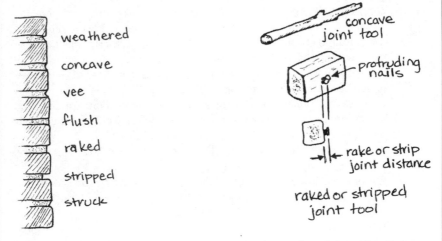

Figure 7-C. Types of Masonry Joints and Tools Used on Joints. Use Only Concave and Vee Joints in Extreme Climates.

Note: Left, Redrawn with permission from Allen and Iano 2003.

However, ferrocement construction techniques can be used for other applications, such as latrine superstructure walls, food or rainwater storage, and depending on the situation, valve or spring boxes.

Ferrocement tanks are usually limited to approximately 5,000 L. These tanks rely on commonly available materials; simple, easily transferable skills; and simple

Figure 7-24. Concrete Block Installed over Vertical Rebar. Workers Are Filling Hollow Cores with Grout to Create a Reinforced Concrete Block Wall.

Figure 7-25. A Masonry Rock Wall Being Constructed as a Wash Area Wall. To Save Purchased Materials, Rocks Are Placed in a Bed of Mortar and Then Covered with Mortar.

construction equipment. Community members can easily contribute labor and local materials, like sand, instead of money. The lifetime of a properly constructed ferrocement tank has been shown to be longer than a masonry tank. Although these structures may be built in many shapes, they are most commonly built as cylinders or jar shapes because the curved walls add strength.

Ferrocement is a thin cement mortar wall reinforced with wire mesh, wire, and rebar. Ferrocement technology takes advantage of the compressive strength of cement and the tensile strength of wire to generate a strong wall. It thereby overcomes the weakness of traditional reinforced concrete construction under tensile stresses. In ferrocement tanks, the mortar contributes to the tensile strength of the composite layer because wire and chicken wire are distributed densely through mortar. This structure allows the load to be taken by the composite layer, rather than concentrating critical stresses in planes of weakness. Wire-reinforced mortar therefore has the ability to resist shrinkage cracking under tensile load. The walls can therefore be thinner and use less cement than traditional reinforced concrete construction (Watt 1978).

Figure 7-26 shows stages of ferrocement tank construction from left to right. Construct ferrocement walls on a poured concrete slab foundation in which L-shaped rebar is embedded along the perimeter, similar to a reinforced masonry wall foundation (Fig. 7-23). Tie the vertical L-shaped rebar wall reinforcement for the tank to the foundation rebar. The vertical portion of the L-shaped bars will eventually reach the top of the tank (after splicing). Attach heavy wire mesh or rebar rings to the vertical rebar. Cover the heavy wire mesh and rebar with two layers of fine mesh or chicken wire, one layer on the inside and another on the outside. Then plaster the walls after installing formwork (if used) and pipe penetrations.

Figure 7-26. A Composite Sketch of the Stages of Ferrocement Tank Construction.

Note: Working from left to right, each step would be completed before the next step began.

Forms to aid ferrocement wall construction may or may not be used. To work without formwork, a coworker must be positioned on the opposite side of the wall holding a concrete float, board or trowel against where the mason is applying the mortar so that it is compacted into the wall and does not fall though the wire structure.

If using formwork, use wood, metal, or plastic pipe (it can be reused for the next tank) for wall forms, depending on project size and conditions (Watt 1978). Use the vertical portion of the L-shaped rebar embedded in the foundation to support the formwork.

Figure 7-27 shows unconventional formwork. Spirally wrap plastic pipe on the inside of the vertical rebar from the bottom of the wall to the top. Tie the pipe to the vertical rebar with pieces of wire. Once the plastic pipe form is secured, wrap two layers of wire mesh on the outside of the tank around the L-shaped rebar and the pipe, extending from the foundation to the top of the wall (Fig. 7-27). The mesh should have holes the size of chicken wire holes or smaller. The holes should not be too small because mortar needs to be able to fill in the holes. At the base of the tank, wrap a 16-gauge tie wire around the outside of the tank (Fig. 7-27) and tie it to itself, forming a single-strand band at the tank bottom. Then tightly wrap the roll end of the wire around the mesh spirally going upward. Because the force of water near the tank bottom is greater than that near the top, the wires will be placed closer together near the bottom and spaced wider proceeding up the tank walls.

(For the following discussion, "square" is used to define the size and shape of the chicken wire hole.) In the bottom quarter of the tank, wrap wire spirally once around the tank for every half square in the chicken wire (about 1 cm). At the one-quarter-height mark on the wall, make four wraps with no spaces. Above the one-quarter mark, maintain one wrap per square of chicken wire (about 2.5 cm) to the halfway mark. At the halfway mark, make four wraps with no spaces. Continue wrapping to the three-quarter mark at a rate of one wrap per 1.5 squares of chicken wire (about 3.5 cm). At the three-quarter mark, make

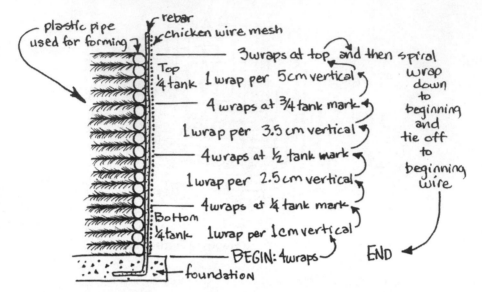

Figure 7-27. Ferrocement Tank Interior Formwork That Consists of Plastic Pipe Is Spirally Wrapped on the Inside of Vertical Wall Rebar.

Note: Reinforcement of chicken wire and 16-gauge tie wire is installed outside the vertical rebar. Alternatively, formwork can be made from wood or metal.

four wraps with no spaces. Then complete each wrap with a vertical rise of two squares of chicken wire (about 5 cm). At the top of the tank, make three wraps with no spaces, and then spiral the wire back down to the base and tie it to the original wire at the bottom. The wire is continuous. If the wire runs out, splice a new roll and continue wrapping.

Once formwork and reinforcing mesh and wire are in place, plaster the exterior of the tank with two layers of cement mortar of $1:2$ ratio cement to sand. The amount of water in the mix is critical to wall integrity. Add too much water, and the wall will be permeable and weak. A dry mix will produce poor bonding, and it is hard to compact as it is being applied. A water-to-cement ratio (by weight) of 0.4 or 0.5 to 1 is recommended. Do not add water if the mortar begins to set. Apply and compact the mortar on the wall in one layer, and once complete, apply the second layer, moistening the surface of the first coat to increase adhesion. Use a concrete finishing float to provide a grainy finish. Once the mortar is applied and compacted to the wire surface, wet-cure it for approximately 2 weeks. After curing, remove formwork (if used) and plaster and parge the inside of the tank to obtain a watertight finish, if required. Apply two layers of mortar to the inside of the tank as well. For the first layer, use the same cement to sand ratio of $1:2$, and then apply a waterproofing parge layer made of cement powder and water (Box 7-3). Walls are usually about 5 cm thick.

After the walls are complete, the roof can be installed. The roof can be made of reinforced concrete, ferrocement, or another material. One simple method is to construct a sloping ferrocement roof by extending and gently arching the vertical rebar that

>>> **Box 7-3. Parging Walls**

Because concrete block and brick are porous, in applications where watertightness is important, it is necessary to *parge* (coat) the inner walls with a cement and sand mortar (cement to sand ratio of 1 : 3 to 1 : 4). This step seals the walls, contains liquid within the tank, and prevents groundwater from seeping into the tank. To parge, first moisten the walls well before applying mortar with a flat trowel. Parging for watertightness is done in three layers. Thinner layers are much better than thick. The first two layers are a cement and sand mix, and the final layer is cement and water that is troweled hard with a steel trowel into the surface. An alternative is to use a chemical waterproofing admixture (such as Sikalite) in the first parge layer, eliminating the following parge layers. Cure each parge layer before applying the next layer. Moisten the wall before applying each coat, and moist cure it after the parge has set. Parge may also be used for aesthetic purposes on exposed concrete or masonry walls or floors (wash areas). In this case, typically only one or two coats are used, or parge is applied as stucco. If the substrate is smooth, chip it to improve adhesion before moistening and installing the parge or stucco.

extend from the foundation through the wall. Place formwork (such as plastic pipe as shown in Figure 7-28) inside the vertical bars of the roof section. Install chicken wire and tie wire, and mortar the roof on the tank exterior and interior in the same manner as the tank walls.

Install an access hatch at the center top of the ferrocement tank. Use a prefabricated hatch, or construct a circular hatch using two rebar rings. The smaller rebar ring is the diameter of the access opening. The larger one is approximately 100 mm larger in diameter. Two 50-mm-long galvanized iron strips connect the two rebar rings. Cast the concentric ring fabrication into the opening at the top of the tank. Construct a hatch cover as outlined earlier, in the section titled "Movable Precast Elevated Slabs."

After construction, if the tank is filled with water too rapidly, there is a risk of cracking and wall failure. Allow the tank to sit for a week with a shallow depth of water before filling it (Watt 1978). If the tank was wet-cured for 2–3 weeks, then it can be filled without the delay.

7.3.9 Pipe

Pipe is necessary in constructing water and wastewater systems and is usually polyvinyl chloride (PVC) or galvanized iron (GI).

Join PVC pipe by gluing. Determine the direction the water will flow before placing and gluing the first PVC pipe joints. Glue PVC pipes so that the bell (female end) is positioned against the flow of water. If placed in the opposite direction, the water will collide with the male end and eventually leak. Be aware of this rule, especially when pipe fittings such as tees and reductions are installed and when pipeline sections are left open for future gully crossings.

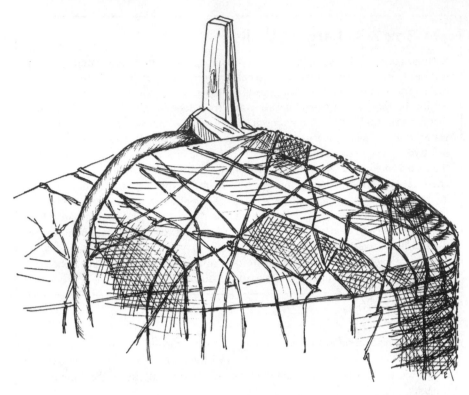

Figure 7-28. Reinforcement of Chicken Wire, Wire Mesh, and Tie Wire Wrapped Around Spirally Wrapped Plastic Pipe (Formwork) for Tank Roof. Wooden Formwork Is Optional.

Connect PVC pipe by first cleaning both the male and female ends of the pipe. To obtain the best bond, use a thinner, which removes grease and is a softener, allowing the glue to weld, rather than merely glue. After removing excess thinner with a rag, apply PVC cement in even, generous coats to both ends of the pipes and insert the pipes straight into each other. Apply a slight twist after the two ends have entered completely. This twist eliminates air passages that may have been created by the presence of tiny debris. Allow the glue to set for 10 s before setting the pipe into final position. In situations where PVC pipe contacts concrete, roughen the surface to allow a better grab between the pipe and the concrete.

Galvanized iron pipe is more expensive than PVC pipe and is generally used when a pipeline travels under roads or crosses streams and ravines. It is also commonly used as the pipe that enters and exits a concrete wall or water tap stand. Connect it by unions or couplings. The required tools for GI pipe connections are a pipe union, two pipe wrenches, and polytetrafluoroethylene (Teflon) tape. Wrap the polytetrafluoroethylene tape around the threads of both pipes, beginning at the end of the pipe, wrapping toward the center of the pipe length. Polytetrafluoroethylene tape creates a watertight seal once the pipes are screwed together. Make connections between PVC and GI pipes with GI adaptors, reducers, and unions when possible, because they are much stronger than PVC and last longer.

⟫⟫⟫ References

Allen, E., and Iano, J. (2003). *Fundamentals of building construction: Materials and methods*, 4th ed. John Wiley & Sons, New York.

Watt, S. B. (1978). *Ferrocement water tanks and their construction*, Intermediate Technology Publications, London.

8

Soil Properties and Identification

))) 8.1 The Need for Soil Identification

Knowledge of soil properties and identification of soil are required for a wide variety of projects. Soil type and properties affect the ease and speed of soil excavation, the extent of erosion under a structure during a heavy rain, the rate of settlement of a structure, the flow of water toward a well, and the flow of wastewater away from a latrine or soak pit.

))) 8.2 Soil Tests

Table 8-1 lists some commonly performed tests and their official American Society for Testing and Materials (ASTM) references. ASTM methods usually require a laboratory with specialty equipment, which is often not available in a field setting. Fortunately, there are several modified field tests using readily available materials that provide similar information. In this section, a brief description of each ASTM method is provided, as well as the corresponding modified field method when available.

8.2.1 Moisture Content

The amount of moisture in a soil affects its unit weight, the ability of water to flow through it, and its ability to support weight. A soil sample with a higher moisture content typically has an increased weight, allows less vertical infiltration of water, and can carry more water horizontally. The removal of water through pumping can cause settling, especially if a building or other weight is placed on top of it. This positioning is not desirable and can lead to the failure of the structure.

The laboratory test for moisture content consists of weighing a soil sample before and after drying. When determining the moisture content of soil, the oven-dry method (ASTM 2008a) is preferred. In the absence of an oven, air-drying may be acceptable. However, in this case, ensure that the sample is completely dry. It is recommended the sample be dried until there is no change in weight over a 6-h period.

John D. Simpson and Marc F. Plotkin contributed to this chapter.

Table 8-1. Soil Laboratory Tests and Their Official ASTM Procedures

Soil Laboratory Test	ASTM Reference
Moisture content	D2216-05 (ASTM 2008a)
Specific gravity	D854-06 (ASTM 2008b)
Sieve analysis (coarse particle size)	D6913-04e1 (ASTM 2008c)
Hydrometer analysis (fine particle size)	D422-63(2002) e1 (ASTM 2008d)
Liquid limit, plastic limit, plasticity index	D4318-05 (ASTM 2008e)
Shrinkage limit	D4943-02 (ASTM 2008f)
Constant head permeability test	D2434-68 (2006) (ASTM 2008g)
Standard proctor compaction test	D698-00ae1 (ASTM 2008h)
Modified proctor compaction test	D1557-02e1 (ASTM 2008i)
Field unit weight of compaction	D1556-00 (ASTM 2008j)
Direct shear test on sand	D3080-03 (ASTM 2008k)
Unconfined compression test	D2166-00e1 (ASTM 2008l)
Consolidation test	D2435-04 (ASTM 2008m)
Triaxial test (compression)	D2850-03a (ASTM 2008n)

8.2.2 Specific Gravity

The *specific gravity* (unitless) of a soil refers to the density of the soil particles themselves compared to the density of water. If the specific gravity of a soil is 2.65 (a typical value for sand), then the soil particles are 2.65 times heavier than water. The specific gravity has implications for soil settling rates in water and other properties described later. Determine the specific gravity of the soil by using ASTM 2008b. There is not an acceptable modified field test to determine specific gravity.

8.2.3 Composition

Generally, soil can be classified into four main groups, based on particle size. These groups are, in order of decreasing size: gravel, sand, silt, and clay. Soils are generally composed of a combination of these four groups in varying percentages. Table 8-2 provides a field method to determine soil composition based on visual observation and lists particle sizes for each type.

Determining the composition, or grain size distribution, of the soil is important because it has implications for the appropriate uses of the soil. For example, larger grained soils such as gravel have a high permeability and allow water to pass quickly. Clays and silts are finer grained and slow the passage of water. A clay would be beneficial when lining a lagoon or landfill but problematic under a soak pit. General soil classifications are provided in Table 8-3, along with some practical characteristics of each soil related to its use.

8.2.4 Texture

The *texture* of the soil is the mixture of the four main soil types presented in Table 8-2. For example, a soil texture that would be good as a building material might be 10% gravel, 75% sand, and 15% silt and clay.

Table 8-2. Visual Field Test for Soil Composition and Associated Particle Size

Visual	Description	Measurement
	Big pieces, very hard	The soil is gravelly. Gravel particle size is 2–20 mm.
	Coarse to small pieces, very rough	The soil is sandy. Sand particle size is 0.02–2 mm.
	Thin powder, soft	The soil is silty. Silt particle size is 0.002–0.02 mm.
	Very thin powder, very soft	The soil is clayey. Clay particle size is 0–0.002 mm.

Source: Adapted from Maini 1997.

The laboratory tests for texture are the sieve analysis and the hydrometer test. The sieve analysis is performed using a set of sieves of different mesh sizes. The sieve with the largest mesh opening size is placed on top, with sieves of progressively smaller mesh sizes below it. The soil sample is placed on the top sieve, and the whole stack of sieves is shaken. When finished, the soil caught by each sieve is weighed. This test provides the fraction of each size in the soil sample. The No. 200 sieve is typically the smallest mesh size, and it corresponds to a particle size of 0.002 mm. This size means that the sieve analysis is acceptable to determine the amount of gravel, sand, and fines (i.e., silt and clay) but will not differentiate between silt and clay particles.

To determine the percent of fine particles (0.001–0.075 mm), a hydrometer test is used. This test requires a hydrometer, which measures the specific gravity of a soil–water mixture. From this test, particle diameters can be calculated.

Although it is a crude method, a "minihydrometer" test can be used to visually estimate the amount of fines versus amount of sand in a sample. To perform the minihydrometer test, place a small amount of material in a clear glass jar with a lid. Fill the jar with water and seal the lid tightly. Shake the combined water and soil sample vigorously for several minutes, and then set it down on a table. Sand and gravel particles will settle to the bottom of the jar within a few minutes; silt and sand particles take many hours or days.

In addition, organic content can be important because the organic matter (e.g., leaves and sticks) will decay with time and leave void spaces in the soil particles. This mixture can cause anything built on top of this soil to settle, and settling may be uneven. A field test for organic content (humus) is to smell it. If the soil smells like decaying material or smells musty, then it has a high organic content (Table 8-4).

8.2.5 Relative Density and Compaction

The density of a soil affects its ability to support structures, its ability to allow water to flow, and the ease or difficulty of excavation. However, to determine how a soil will behave, the in situ density (or in-place density) are typically compared to a standard. Though this test is not typically performed in rural field situations, the terms are explained in more detail in the following paragraphs to provide the reader with background information.

There are two basic standards, the relative density and the percent compaction. The *relative density* is a comparison of a soil's in situ density to its maximum and minimum possible values and is typically used in reference to an undisturbed soil deposit. The relative density can be correlated to the standard penetration test (SPT) (ASTM 2009). The SPT is performed by driving a sampling spoon into the ground by dropping a 140-lb hammer a distance of 30 in. onto the sampler. The penetration resistance (N-value) is the number of drops required to advance the sampler 12 in. There are direct correlations between the N-value and relative density. However, specialized equipment is required for the SPT test.

In the field, it is important to recognize the importance of compacting all fill materials, using an amount of water that allows the soil to be compacted as tightly as possible. It is also important to compact around buried pipes with a fine-grained material to support the pipe. Shovel handles can be used to manually compact material around pipes. In this situation, larger diameter gravels should be avoided when backfilling because they may puncture the pipe.

A different density comparison is typically used when referring to anthropogenic (made by people) fills. How tightly the soil is compacted is referred to as the *percent compaction*. If no mechanical means of compacting is available, manual compaction can be used (see Chapter 7). Percent compaction compares the in situ density of a soil to a laboratory maximum density (proctor) value. This maximum density value may be determined by the standard proctor method (ASTM 2008h) or the modified proctor method (ASTM 2008i). In areas where fills will be used to support structures, the fill should be

Table 8-3. General Classifications and Characteristics of Soils

Major Divisions	Sub-divisions	USCS Symbol	Name	Value as Subgrade
Coarse-grained soils	Gravel and gravelly soils	GW	Well-graded gravels or gravel-sand mixtures with no fines	Excellent
		GP	Poorly graded gravels or gravelly sands, little or no fines	Good to excellent
		GM	Silty gravels, gravel–sand–silt mixtures	Good to excellent
		GC	Clayey gravels, gravel–sand–clay mixtures	Good
	Sand and sandy soils	SW	Well-graded sands or gravelly sands, little or no fines	Good
		SP	Poorly graded sands or gravelly sands, little or no fines	Fair to good
		SM	Silty sands, sand–silt mixtures	Fair to good
		SC	Clayey sands, sand–clay mixtures	Poor to fair
Fine-grained soils	Silts and clays with liquid limit <50	ML	Inorganic silts, rock flour, silts of low plasticity	Poor to fair
		CL	Inorganic clays of low plasticity, gravelly clays, sandy clays	Poor to fair
		OL	Organic silts and organic clays of low plasticity	Poor

Potential Frost Action	Probability of Caving in for Well or Latrine Construction	Risk of Groundwater Contamination for Pit Latrine		Use as Building Foundations
		Water Table (≥10 m Below Grade)	Water Table (5–10 m Below Grade)	
None to very slight	High	High	Very high	Excellent
None to very slight	High	High	Very high	Good to excellent
Slight to medium	Medium-high	Medium	Medium-high	Good to excellent
Slight to medium	Medium	Low	Low-medium	Good
None to very slight	High	High	Very high	Good
None to very slight	High	High	Very high	Fair to good
Slight to high	Medium-high	Low-medium	Medium	Fair to good
Slight to high	Low	Low	Low-medium	Poor to fair
Medium to very high	Medium	Low-medium	Medium	Poor to fair
Medium to high	Low	Low	Low-medium	Poor to fair
Medium to high	Low-medium	Low	Low-medium	Poor

Table 8-3. General Classifications and Characteristics of Soils (*Continued*)

Major Divisions	Sub-divisions	USCS Symbol	Name	Value as Subgrade
	Silts and clays with liquid limit >50	MH	Inorganic silts, micaceous silts, silts of high plasticity	Poor
		CH	Inorganic clays of high plasticity, fat clays, silty clays	Poor to fair
		OH	Organic silts and organic clays of high plasticity	Poor to very poor
Peat	Highly organic	PT	Peat and other highly organic soils	Not suitable

compacted to a minimum of 95% of the maximum dry density value, as determined by the modified proctor method.

8.2.6 Permeability

Soil *permeability* is a measure of how quickly water can flow through a soil sample. It is used in the well-known Darcy's Law, which relates flow to permeability, the slope of the water surface under the ground, and the cross-sectional area of flow. The standard laboratory test to measure permeability consists of running water through the soil sample at a known pressure and measuring how quickly it exits the sample.

Permeability can vary greatly across all soil types. For example, the permeability of gravel can be 10 billion times greater than that of clays. Within individual soils, it can vary significantly as well. The permeability of a medium gravel can be 10,000 times greater than that of a sandy gravel. Verify approximate values shown in critical situations. High

Table 8-4. Smelling Soil to Detect the Presence of Organic Content

If It Smells	There Is	It Is
Rotten	A lot of humus	Not suitable for earth construction
Musty	Humus	Not suitable for earth construction
Agreeable	No humus	Suitable for earth construction

Source: Adapted from Maini 1997.

Note: Heating the soil up will maximize the humus odor.

| Potential Frost Action | Probability of Caving in for Well or Latrine Construction | Risk of Groundwater Contamination for Pit Latrine | | Use as Building Foundations |
		Water Table (≥10 m Below Grade)	Water Table (5–10 m Below Grade)	
Medium to very high	Medium	Low-medium	Medium	Poor
Medium	Low	Low	Low-medium	Poor to fair
Medium	Low-medium	Low	Low-medium	Poor to very poor
Slight	Not suitable	Low	Low-medium	Not suitable

permeable soils may be desirable in soak pits but can quickly allow shallow groundwater to be contaminated. Table 8-5 lists the properties of different soils and their relationship to compressibility and ease of drainage.

8.2.7 Plasticity

The ability of a soil to withstand a deformation without elastic failure such as cracking or disintegration is called the *soil plasticity*. The *liquid limit* is the transition between the fluid and plastic consistencies, and the *plastic limit* is the transition between the plastic and solid consistencies.

Also important is the plasticity index, which is an indication of the range of plastic behavior of the soil. It is the difference between the liquid limit and plastic limit. The plasticity index provides an *elastic behavior,* which means that when the soil is deformed in some way, it will return to its original shape when the load is removed. In contrast, *plastic* means the soil will not return back to its original shape.

8.2.8 Shear Strength

The ability of a soil to resist *shear,* or cutting, stress is important when a soil has to support a heavy structure placed on top of it, or in the situation where it receives lateral loads. The direct shear test involves placing a force on part of the soil sample and measuring how quickly the soil deforms. The triaxial test can determine the stresses in all directions within a soil sample and can only be performed with a special machine.

Table 8-5. Typical Properties of Various Soils and Relationship to Compressibility and Drainage

Major Divisions	Subdivisions	USCS Symbol	Typical Dry Densities		Typical Coefficients of Permeability	Compressibility	Drainage Properties
			lb/ft³	million g/m³	cm/s		
Coarse-grained soils	Gravel and gravelly soils	GW	125–140	2.00–2.24	1–10	Almost none	Excellent
		GP	110–140	1.76–2.24	1–10	Almost none	Excellent
		GM	115–145	1.84–2.32	10^{-6}–10^{-1}	Very slight to slight	Fair to very poor
		GC	130–145	2.08–2.32	10^{-6}–10^{-4}	Slight	Poor to very poor
	Sand and sandy soils	SW	110–130	1.76–2.08	10^{-2}–1	Almost none	Excellent
		SP	105–135	1.68–2.16	10^{-2}–1	Almost none	Excellent
		SM	100–135	1.60–2.16	10^{-5}–10^{-2}	Very slight to medium	Fair to poor
		SC	100–135	1.60–2.16	10^{-6}–10^{-4}	Slight to medium	Poor to very poor
Fine-grained soils	Silts and clays with liquid limit <50	ML	90–130	1.44–2.08	10^{-5}–10^{-2}	Slight to medium	Fair to poor
		CL	90–130	1.44–2.08	10^{-9}–10^{-6}	Medium	Practically impervious
		OL	90–105	1.44–1.68	10^{-6}–10^{-4}	Medium to high	Poor
	Silts and clays with liquid limit >50	MH	80–105	1.28–1.68	10^{-6}–10^{-4}	High	Fair to poor
		CH	90–115	1.44–1.84	10^{-9}–10^{-6}	High	Practically impervious
		OH	80–110	1.28–1.76	10^{-9}–10^{-6}	High	Practically impervious
Peat	Highly organic	PT	NA	NA	10^{-5}–10^{-2}	High	Fair to poor

8.2.9 Cohesion

Soil *cohesion* is the ability of soil to stick together when a tensile stress is applied (i.e., the soil is pulled in some way). Finer soils are generally more cohesive than coarser soils; clay is the most cohesive, silt less cohesive, and sands and gravels noncohesive. This result occurs because of physical and chemical forces that hold fine soil particles together.

Cohesion is determined in the laboratory by the triaxial test. In the field, a *hand penetrometer*, which consists of a calibrated spring-loaded cylinder, can be used to estimate cohesion. Table 8-6 shows a simple field method for determining cohesion.

Table 8-6. Tests for Soil Compaction and Cohesion

Hand Test	Knife Test	Interpretation of Test
		Gravelly soil requires a lot of strength and can be pressed by hand quickly. A knife will penetrate easily into the ball and will come out clean.
		Sandy soil requires some strength and can be pressed by hand for a short duration. A knife will penetrate easily into the ball and will come out mostly clean.
		Silty soil requires little strength and can be pressed by hand for a medium duration. It is more difficult to penetrate the ball with a knife, which will come out slightly dirty.
		Clayey soil requires very little strength and can be pressed by hand for a long duration. The knife is difficult to penetrate and comes out very dirty.

Source: Adapted from Maini 1997.

Note: These two field tests require you to add a little water to make the soil moist and then compress it by hand in an attempt to make a ball. Inserting a knife into the ball provides additional information about the soil's cohesive properties.

Cohesion has important implications for the excavation of wells, latrines, pipe trenches, and other trenches. During excavation, soil collapse may occur unexpectedly and can be fatal. Typically, opening an excavation using a side slope of 3 : 1 (horizontal to vertical) will prevent soil collapse in all types of soils, including gravel and sands. Soils that have a medium cohesion tend to stay more vertical. Such soils include angular gravel, silt loam, and sandy loam. These soils can be excavated safely at the steeper angle of 1 : 1 (horizontal to vertical). For soils with a high cohesive strength, such as clays, silty clays, sandy clays, and clay loams, a safe excavation of 0.5 : 1 (horizontal to vertical) can be used.

))) 8.3 Soil Foundations

Soil foundations are used to support almost every structure, including storage tanks, pipe crossings, houses, roads, and bridges. Foundations are critical because settlement causes many types of failures in structures, from doors and windows closing poorly to complete failure of a water storage tank. Settlement of sands and gravels takes place almost immediately. These are good foundation materials because the settlement will occur during construction. Settlement of clays may take years. These soils should be avoided in foundation construction because future settlement may cause failure of structures.

For large structures, it is important to have a qualified engineer analyze the soil and design a foundation. Table 8-7 provides the bearing capacity values for various soil types. In addition to the presumptive bearing pressures, minimum footing widths should be applied. Specifically, a continuous footing (used for walls) should have a minimum width of at least 60 cm, and isolated column pads should have a minimum width of 75–80 cm. Foundations should normally be taken down to at least 1 m below ground level, or to firm rock at depths less than 1 m. In swelling clays, or in places where frost may occur, foundations should be deeper.

Table 8-7. Bearing Capacity Values for Various Soil Types

Category	Types of Rock and Soils	Presumed Bearing Capacity
Noncohesive soils	Dense gravel or dense sand and gravel	>600 kN/m^2
	Medium dense gravel, or medium dense sand and gravel	<200–600 kN/m^2
	Loose gravel, or loose sand and gravel	<200 kN/m^2
	Compact sand	>300 kN/m^2
	Medium dense sand	100–300 kN/m^2
	Loose sand	<100 kN/m^2
Cohesive soils	Clays and hard clays	300–600 kN/m^2
	Stiff clays	150–300 kN/m^2
	Firm clays	75–150 kN/m^2
	Soft clays and silts	<75 kN/m^2
	Very soft clay	Not applicable
Peat		Not applicable

))) References

American Society for Testing and Materials (ASTM). (2008a). "Standard test methods for laboratory determination of water (moisture) content of soil and rock by mass," D2216-05, <www.astm .org> (Jan. 11, 2008).

American Society for Testing and Materials (ASTM). (2008b). "Standard test methods for specific gravity of soil solids by water pycnometer," D854-06, <www.astm.org> (Jan. 11, 2008).

American Society for Testing and Materials (ASTM). (2008c). "Standard test methods for particle-size distribution (gradation) of soils using sieve analysis," D6913-04e1, <www.astm.org> (Jan. 11, 2008).

American Society for Testing and Materials (ASTM). (2008d). "Standard test method for particle-size analysis of soils," D422-63(2002)e1, <www.astm.org> (Jan. 11, 2008).

American Society for Testing and Materials (ASTM). (2008e). "Standard test methods for liquid limit, plastic limit, and plasticity index of soils," D4318-05, <www.astm.org> (Jan. 11, 2008).

American Society for Testing and Materials (ASTM). (2008f). "Standard test method for shrinkage factors of soils by the wax method," D4943-02, <www.astm.org> (Jan. 11, 2008).

American Society for Testing and Materials (ASTM). (2008g). "Standard test method for permeability of granular soils (constant head)." D2434-68(2006), <www.astm.org> (Jan. 11, 2008).

American Society for Testing and Materials (ASTM). (2008h). "Standard test methods for laboratory compaction characteristics of soil using standard effort," D698-00ae1, <www.astm.org> (Jan. 11, 2008),

American Society for Testing and Materials (ASTM). (2008i). "Standard test methods for laboratory compaction characteristics of soil using modified effort," D1557-02e1 <www.astm.org> (Jan. 11, 2008).

American Society for Testing and Materials (ASTM). (2008j). "Standard test method for density and unit weight of soil in place by the sand-cone method," D1556-00, <www.astm.org> (Jan. 11, 2008).

American Society for Testing and Materials (ASTM). (2008k). "Standard test method for direct shear test of soils under consolidated drained conditions," D3080-03, <www.astm.org> (Jan. 11, 2008).

American Society for Testing and Materials (ASTM). (2008l). "Standard test method for unconfined compressive strength of cohesive soil," D2166-00e1, <www.astm.org> (Jan. 11, 2008).

American Society for Testing and Materials (ASTM). (2008m). "Standard test methods for one-dimensional consolidation properties of soils using incremental loading," D2435-04, <www .astm.org> (Jan. 11, 2008).

American Society for Testing and Materials (ASTM). (2008n). "Standard test method for unconsolidated-undrained triaxial compression test on cohesive soils," D2850-03a, <www.astm.org> (Jan. 11, 2008).

American Society for Testing and Materials (ASTM). (2009). "Standard test method for standard penetration test (SPT) and split-barrel sampling of soils," D1586-08a, <www.astm.org> (Jan. 17, 2009).

Maini, S. (1997). *Soil identification for earth construction: An introduction.* Auroville Earth Institute, Auroville, Tamil Nadu, India.

))) Further Reading

Cedergren, H. (1989). *Seepage, drainage, and flownets.* Wiley Interscience. New York.

Das, Braja M. (2004). *Principles of foundation engineering,* fifth ed. Brooks/Cole, Pacific Grove, Calif.

Day, R. (2000). *Geotechnical engineer's portable handbook.* McGraw Hill, New York.

Holtz, Robert D., and Kovacs, William D. (1981). *An introduction to geotechnical engineering.* Prentice-Hall, Englewood Cliffs, N.J.

Lambe, T. William, and Whitman, Robert V. (1969). *Soil mechanics.* John Wiley & Sons, New York.

Stern, P., and Longland, F. (1983). *Field engineering: An introduction to development work and construction in rural areas.* Intermediate Technology Publications, London.

U.S. Department of Labor. (2007). "Classification of soils for excavation," ID-194. U.S. Department of Labor—Occupational Safety and Health Administration, Washington, D.C.

Water Supply and Treatment

9

Water Use, Access, and Health

))) 9.1 Water Access and Use

9.1.1 Water Scarcity and Locations of Water

Only 2.5% of the world's total water budget is believed to be freshwater, and of this, almost 70% is currently present as glaciers and ice sheets. As shown in Table 9-1, much of the remaining freshwater is available as a groundwater resource, which has an extremely long renewal period. This means that little of the total freshwater budget is readily available as surface water (i.e., lakes, rivers) or as groundwater that is recharged over a short duration.

Water is expected to be one of the most pressing global security problems in the future. *Water scarcity* is defined as insufficient water to satisfy normal human requirements. A country is defined as experiencing *water stress* when annual water supplies drop below 1,700 m^3 per person. When annual water supplies drop below 1,000 m^3 per person, the country is defined as water scarce (UNEP 2002).

Almost 2 billion people currently suffer from severe water scarcity. This number is expected to increase substantially as population increases and as standards of living (and therefore resource consumption) are raised around the world. Furthermore, climate change is expected to have an effect. Although some areas will probably experience increased rainfall, others are likely to suffer from decreased available water (Vorosmarty et al. 2000).

Table 9-2 shows the primary uses of water throughout the world. Globally speaking, most water use is in the energy and agricultural sectors. In higher income areas such as the United States and the European Union, there is a greater use of water for industrial purposes relative to developing economies.

9.1.2 Defining Access to Water

Many development projects focus on improving water supplies, with projects ranging from protecting a water source to building a distribution system. Table 9-3 provides a description of how the World Health Organization (WHO) defines unimproved and improved water supplies.

Paul M. Kennedy contributed to this chapter.

Table 9-1. Percent of World's Total Freshwater in Different Locations

Location	Percent of World's Freshwater
Glaciers and permanent snow cover	68.7
Groundwater	30.1
Lakes	0.26
Soil moisture	0.05
Atmosphere	0.04
Marshes and swamps	0.03
Rivers	0.006
Biological water	0.003

Source: Data from UNESCO-WWAP 2003.

Classifying the type of water source as improved or unimproved by the criteria in Table 9-3 is helpful as an initial observation. However, these criteria reveal no information about how much water is actually used by individuals, or to which uses the water is put. From the perspective of the user, water access should be described based on quantity of water available, distance to the water supply, and time spent collecting water, not just water quality.

WHO defines the term *reasonable access* to a water source as the "availability of at least 20 L per person per day (commonly written as L/capita-day) from a source within 1 km of the user's dwelling" (WHO 2000). Other studies may provide different criteria for what is considered reasonable access. In other words, the volume of 20 L or the distance of 1 km may vary depending on the population and the study in which the definition was created (Fig. 9-1).

By providing a more descriptive definition of access, the living conditions of a village or household can be better understood. Table 9-4 describes four levels of access to water, or service levels, based on the distance the consumer travels to a water supply

Table 9-2. Percent of Water Use Associated with the Agricultural, Industrial, and Domestic Sectors

Location	Agriculture	Industry	Domestic
World	70	22	8
European Union	21	63	16
North America	40	47	13
East Asia and the Pacific	80	14	6
Europe and Central Asia	63	26	11
Latin America and the Caribbean	73	9	18
Middle East and North Africa	90	4	6
South Asia	94	2	4
Sub-Saharan Africa	87	4	9

Source: Data from UNESCO-WWAP 2003.

Table 9-3. Definitions of Improved and Unimproved Water Supplies

Unimproved Water Supplies	Improved Water Supplies
Unprotected wells	Household connections
Unprotected springs	Public standpipes
Vendor-provided water	Boreholes
Bottled water	Protected dug wells
Tanker-truck provided water	Protected springs

Note: Bottled water is considered unimproved because of possible problems of sufficient quantity, not quality.

Source: WHO 2000.

Figure 9-1. Women and Children Collect and Carry Water in Most Parts of the World.

Table 9-4. Service Levels Defined by Distance and Time to Water Source, Quantities of Water Collected, and Level of Health Concern

Service Level	Distance to Source and Total Collection Time	Approximate Quantities Collected	Level of Health Concern
No access	>1,000 m >30 min	Very low Less than 5 L/capita-day	Very high Hygiene not ensured, consumption needs may be at risk. Quality difficult to ensure.
Basic access	100–1,000 m 5–30 min	Low Unlikely to exceed 20 L/capita-day*a*	Medium Not all water needs may be met. Quality difficult to ensure.
Intermediate access	On-plot, e.g., single standpipe on compound or in house	Medium Around 50 L/capita-day	Low Most basic hygiene and consumption needs met. Quality more readily ensured.
Optimal access	Multiple taps in house	Varies Likely to be 100 L/capita-day, possibly up to 300 L/capita-day	Very low All uses met. Quality readily ensured.

aLaundry or bathing may occur at a source with additional quantities of water used.

Source: From Howard and Bartram 2003.

and the time spent collecting water. Different quantities of water used are associated with each of the service levels. *Collection time* to a water source is the amount of time it takes for a person to travel from the home to the water source, collect water, and return home. Table 9-4 also lists the level of health concern associated with each service level. The health concern is based on the amount of water collected and available for use. Using Table 9-4 to describe households or a village can provide information that is useful when planning a project. However, it does not differentiate between improved and unimproved sources.

One concern in comparing data from multiple sources is that definitions of the same terms may vary by organization. Country censuses, health organizations, and nongovernmental organizations may have their own definitions of *unimproved water source*, *safe water*, or *basic sanitation* (see Ahrens and Mihelcic 2006). Moreover, data can be interpreted differently. For example, a study may count everyone in a village as having access to an improved water source, although not all community members actually use the hand pump in question.

))) 9.2 Categories of Water Use

In Table 9-4, the degree of health concern for each service level was derived from the fulfillment of the four different uses of water, which are described in Table 9-5 (Howard

Table 9-5. Categories of Water Use and Examples of Each Use

Water Use Category	Examples
Consumption	Drinking and cooking
Hygiene	Personal and domestic cleanliness (i.e., bathing, laundry, washing floors, dust suppression)
Productive	Gardening, brewing, animal watering, construction (e.g., manufacturing concrete)
Amenity	Washing a vehicle or motor scooter, lawn watering

and Bartram 2003). Water used for consumption and hygiene affects health by fulfilling physiological needs and controlling water-related diseases. Productive uses affect the livelihood of a household and thus indirectly affect human health, whereas amenity uses may not affect human health at all (Howard and Bartram 2003). As an example of how total water use is broken down at the community level, a study performed in Kenya found the total village water use to be 17 L/capita-day: consumption accounted for 21%, hygiene for 51%, and productive uses for 28% (Kennedy 2006).

9.2.1 Water for Consumption

A person not receiving enough water for consumption can be affected in many ways, depending on the magnitude of dehydration. Susceptible groups may experience increased risks of urinary stones, oral health problems, coronary disease, and even certain types of cancer.

Many factors influence the amount of water the human body needs, such as body weight, climate, and physical activity. Water requirements also vary between subsections of a population, such as pregnant women or children. Published values suggest that a 60-kg adult requires between 2 and 3 L/day (White et al. 1972; WHO 2000). For vulnerable populations in tropical climates, the amount of consumptive water required to maintain health can be as high as 4.5 L/capita-day (Howard and Bartram 2003).

When measuring the total volume of water consumed over a day, the water used in the preparation of food should be added to the amount that is drunk. The amount of water used in cooking varies with the type of food being prepared and the role of water in the preparation. In most cases, 2 L/capita-day should be available. Depending on the location and types of food consumed, the quantity of water required for food preparation can be up to 4 L/capita-day. The quantities of water used to develop Table 9-4 include a total water requirement for drinking and food preparation of approximately 7.5 L/capita-day (Howard and Bartram 2003).

9.2.2 Water for Hygiene

Because many diseases are *water-washed*, domestic water use must include sufficient water for hygiene, in addition to consumption (see Chapter 2). The amount of water used

Figure 9-2. After Water Is Collected in Clean Containers, Cover Water Containers to Prevent Contamination. Sometimes Cultural or Religious Symbols Are Placed on Storage Containers to Protect the Water or to Identify Clean Containers.

for bathing must be sufficient to remove dirt and soap (White et al. 1972). If not enough water is available, bathing may be less frequent, or water may be used from an unsafe source. The frequency of washing clothes and eating utensils may also suffer without sufficient water.

The quantity of water used is only one important component of maintaining personal hygiene. In addition to quantity, timing and availability of water are other factors that influence hygiene behaviors. For example, the effectiveness of hand washing after using a latrine can be diminished if there is not enough water to wash the hands thoroughly or if timing is delayed, allowing for possible transmission of pathogens before the hands are clean (Fig. 9-3). Quantities of water needed to maintain hygiene vary by region due to differing customs.

The type of sanitation technology used can result in an additional water requirement that ranges from 0 to 75 L/capita-day (Table 9-6). Although composting and pit latrines have minimal water requirements, pour flush latrines and sewers create an increased demand for water. This situation is especially important for areas of the world where climate change will result in increased water scarcity or where sewage is unlikely to be treated and will thus contaminate nearby surface water (Fry et al. 2008).

Figure 9-3. Use of Water and Soap Is Key to Reducing the Incidence of Diarrheal Disease.

Note: The *tippy tap* pictured here is an appropriate technology for washing hands that can be used in the absence of a piped water supply. You can learn how to make one at www.cdc.gov/safewater/publications_pages/tippy-tap.pdf.

9.2.3 Water for Productive and Amenity Uses

Howard and Bartram (2003) define *productive uses* as activities at the domestic level only (home gardening, brewing, and home construction), instead of community-level productive uses such as industry or energy production. Productive uses of water influence the well-being of household members in many ways. These small-scale activities may provide better food security or additional income, helping to improve a household's living conditions.

As accessibility to water increases, the collection time for a household may decrease. The time now made available can be used for other purposes, such as productive uses, or for hygiene activities. This change is especially important to women and children, who are usually responsible for collecting water.

With increased service levels, the health risks of the members of the household are expected to be reduced. As the service level and associated quantity of water available

Table 9-6. Comparison of Water Use for Water Supply and Sanitation Service

	Service Level	Access Measure	Water Use (L/capita-day)
Water supply	No access	More than 1,000 m or 30 min total collection time	5
	Basic access	Between 100 m and 1,000 m or 5 to 30 min total collection time	20
	Intermediate access	One tap on-plot, or within 100 m or 5 min total collection time	50
	Optimal access	Multiple taps with continuous supply	100
Sanitation	Unimproved	Excreta is not contained to prevent contamination of the environment	0
	Intermediate	Pit latrine	1–2
	Optimal	Pour flush toilets	6–10
	Optimal in water scarce world	Vault toilets and cartage	3–6
		Sewerage	50–75
		Compost	0

Source: Definitions of service level are based on Howard and Bertram (2003) for water supply and Gleick (1996) for sanitation.

continue to increase for a specific household, water may become available for amenities. However, if water access does not increase uniformly among all households in a community, water used for amenities by some households may prevent more vulnerable people from fulfilling their basic needs of consumption and hygiene.

))) 9.3 Relationship of Access to Water Usage and Health

9.3.1 Improvements to Water Quality and Quantity

Water quality improvements can be accomplished in a number of ways and can increase the amount of water available that is suitable for consumption and hygiene purposes. Examples of these projects include constructing a springbox or building a raised lip around a well that prevents the entrance of contaminated runoff and debris. Constructing a distribution system may prevent some contamination that occurs during the transport of water between the source and the point of use.

Variations in the quality of water can also affect usage patterns. For example, community members may use a river to collect water for drinking only in the morning. They are aware that water quality decreases in the river as it is used upstream for activities such as washing laundry, bathing, and watering livestock.

Communities can gain access to increased quantities of water not only through developing new sources (e.g., drilling a new well), but also by increasing water storage capacity. This strategy will allow water to be collected during off-peak hours, providing greater quantities of water during the day.

9.3.2 Improvements to Accessibility

The link between access to water and the levels of health concern described in Table 9-4 occurs because of the relationship between the time of collection or distance to a water source and the quantity of water used in the household. Some health behaviors, such as hand washing, occur more often when water is more accessible (shown by Curtis et al. 2000). The occurrence of water-washed diseases (e.g., trachoma) may also be greatly affected by significant changes in distance to the water source. Lack of water for sufficient bathing can also allow for more disease transmission.

The time and energy required to collect water also influence health. Large reductions in collection time will increase the use of water and provide more time that can be spent on other activities, such as childcare, food preparation, and income-generating activities. This time savings is especially important to women and children, who are usually responsible for collecting water. A reduction in collection time from 5 h to 15 min was found to increase the quantity of water used for child hygiene by a factor of 30 (Prost and Négrel 1989).

Figure 9-4 shows that the amount of water collected decreases with an increase in collection time. The amount collected decreases significantly between 0 and 3 min travel time. This scenario is equivalent to a collection time when the water source is located in the household compound or near the point of use. For example, water used from a standpipe located in the compound would be greater than water collected from a short distance away, such as from a neighbor's compound. Once a user is collecting water from a location no longer in the household compound or nearby, the quantity of water that is collected remains about the same until about 30 min travel time. The third section of Fig. 9-4 shows that as collection time increases past 30 min, the amount of water collected begins to decrease once again.

The exact values in Fig. 9-4 vary depending on the circumstances. The plateau in the middle section, where quantity collected does not vary with an increasing collection time, is generally close to a service level of 15–20 L/capita-day, not including water used for bathing or laundry. In a study in Mozambique, water consumption was approximately 12 L/capita-day for a village with a collection time of 15 min and approximately 3 L/capita-day for a village with a collection time of 5 h (Cairncross 1987). Kennedy (2006) found that the people of Muthara (Kenya) collected 12.5 L/capita-day (not including water used for laundry). Interestingly, as collection time continued to increase, the quantity of water collected did not decrease. This result suggests that there may be some minimum value to Fig. 9-4, where individuals collect an amount of water they see as necessary for their lifestyle.

Access to water can be improved by developing water sources closer to the community, making piped distribution systems do the work of transporting water, or allowing more water to be carried at one time with inventions such as the Hippo Water Roller (see www.hipporoller.org for more information).

Figure 9-5 shows an example of an expansion path for reducing the risk of waterborne diarrhea. At first glance, Fig. 9-5 suggests that more disability adjusted life years (DALYs) can be averted by investing in household water treatment versus improvements at the source. There is recent evidence that suggests water quality interventions that use point-of-use treatment are more effective at reducing diarrhea morbidity than improve-

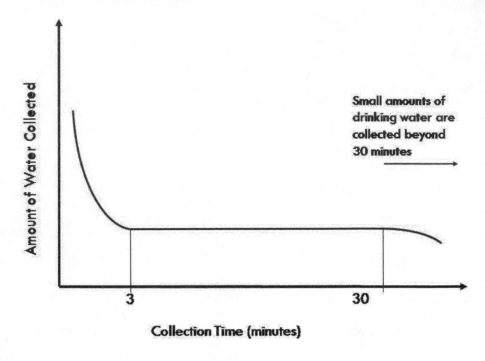

Figure 9-4. General Relationship Between Volume of Water Collected and Collection Time.

Source: Adapted from WHO and UNICEF 2005.

ment of water quality at the source (Clasen et al. 2006). Source intervention also reduces the risk of diarrhea. It is, however, different in some ways to household interventions that use appropriate point-of-use treatment technologies. Issues of convenience, adoption, and access to a sufficient quantity of water are all important to community members and can be addressed by intervention at the source. An additional consideration for determining whether to use point-of-use or source interventions would be the objective of the intervention. For example, if reduction of a water-washed disease was the main objective, then improvements at the source and improved access to water (e.g., via distribution) may be a preferred intervention.

⟩⟩⟩ 9.4 Source of Water

9.4.1 Available Water Sources

There are four main sources of water: groundwater, surface water, rainwater, and greywater. Greywater should not be ignored as a source of water, especially in what is becoming a water-scarce world. *Blackwater*, which contains excrement, is not discussed in this chapter, although it is used in some locations to irrigate and fertilize fields. As shown in

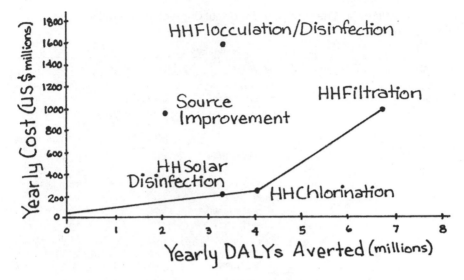

Figure 9-5. Example of an Expansion Path for Reducing Risk of Diarrhea.

Note: This graph shows the annual cost for averting a specific number of DALYs. The graph is specific to particular countries in Africa for reasons that include geographical considerations of health, costs, population, and the specific technologies that are appropriate for either improving a water source or used for household (HH) point-of-use treatment. Source: Adapted from information provided by Thomas Clasen, London School of Hygiene and Tropical Medicine.

Table 9-7, these four sources have different advantages and disadvantages relating to their location, ease of access, water quality, and seasonality.

Surface Water

Where it exists, surface water is not only the most accessible source and available in the largest quantities but also the most readily polluted. It is open to contamination from humans, animals, and agricultural runoff. Microbial contamination is often extremely high, as is turbidity. Increased runoff associated with the rainy season(s) increases the turbidity. Surface water is almost never suitable to drink without treatment. However, it can be used as-is for many purposes, and with adequate treatment can also be used for drinking and cooking.

Groundwater

Because soil functions as a filter, groundwater is generally less prone to microbial contamination than surface water. However, many exceptions exist in the vicinity of latrines, septic tanks, cemeteries, and other sources of pollution.

The chemical quality of groundwater should be considered when it is used for drinking or cooking. Naturally occurring arsenic is found throughout the world, including Argentina, Bangladesh, Chile, Croatia, Hungary, India, Mexico, Serbia, and the United States. Fluoride concentrations in groundwater have also been found to exceed

Table 9-7. Microbial, Chemical, and Seasonal Characteristics of Different Water Sources

Parameter	Surface Water	Groundwater	Rainwater	Greywater
		Water Source		
Microbial quality	Poor	Good unless near a latrine, cemetery, dump, etc.	First flush is poor. After first flush, depends on roofing material.	Poor. Risk comes from the extent of fecal cross-contamination
Chemical quality	Varies. Affected by agricultural runoff and upstream human uses	Arsenic and fluoride can be local problems. Depending on individual preference, may have aesthetic issues associated with hardness and other dissolved metals	Good	High in suspended solids. Can be high in phosphorus if phosphorus-containing soaps are used for washing
Seasonality	During dry season, certain water bodies dry up completely. Rainy season can increase runoff of pathogens and other chemical constituents	Shallow water tables may decline during dry season. Deeper wells may remain productive throughout the year	Often available only during rainy season(s), so storage is required	Based on availability of other water sources

recommended levels in many parts of the world, including parts of South Asia, North and East Africa, the Middle East, and China. Other dissolved constituents (e.g., hardness, iron) may be viewed in some cultures as an aesthetic problem.

Switching to a water source that contains safe levels of fluoride or arsenic is generally the preferable option when faced with a contaminated supply. Wells can be painted red (high arsenic content) or green (safe levels of arsenic) to allow users to take their drinking and cooking water only from noncontaminated wells. An increased level of calcium consumption, along with several other vitamins and minerals, can help to mitigate the effects of fluorosis. If the concentration of fluoride-contaminated groundwater is known, it can be diluted with rainwater to safe levels.

Rainwater

In many developing countries, the majority of annual rainfall occurs during a few months of the year. This leads to periods where more water is available than is needed, followed by months of inadequate water supply. Rainwater harvesting, combined with storage, can cover all or most water needs in the rainy season and extend rainwater availability into the beginning of the dry season. Rainwater can also be viewed as a supplementary source to increase the quantity of water. Because rainwater is typically stored for longer periods of time relative to other sources of water, care should be taken to design and maintain hygienic storage facilities.

When not affected by localized industrial air pollution, rainwater itself is safe to drink. However, before it is stored, it passes over a roof and through a gutter. These surfaces become dirty between rainy seasons and between rain events. Gutters can become clogged with leaves and other organic material, and fecal matter from birds and other animals may have been deposited on a roof. Rainwater harvesting systems can be designed to remove this contaminated water, referred to as the "first flush" (see Chapter 17).

Greywater

Greywater is water that has been used for cleaning, bathing, or cooking. It does not include water used to convey excrement (*blackwater*) or water used for anal ablution. Greywater usage has the double benefits of increasing the amount of water available for use (or, more accurately, increasing the benefits derived from each available unit of water that is collected) while decreasing the total volume of wastewater that must be disposed of. The main risks of using greywater are associated with fecal cross-contamination. Other constituents of concern are detergents and grease. Greywater can become anaerobic and develop an odor after 24–48 h, so minimizing the storage time is important when designing a greywater storage tank.

9.4.2 Matching Water Source and Use

Humans require water for a variety of purposes, and it is important to match water usage with water quality. In contrast, in many Western countries, all domestic water, whether it is for drinking, bathing, or washing a car, is treated to drinking water quality standards.

In settings where water is scarce and providing safe drinking water is a challenge, a different approach to water supply should be taken. The goal should be for people to have access to adequate amounts of water of appropriate quality for each individual purpose. Table 9-8 shows how the use of water can be matched with a specific source. When viewed in this context, water supply becomes a collection of solutions to a variety of problems. As an example, briny water that is not suitable for drinking may be used for cleaning tile flooring.

Many communities already make this differentiation. For example, people may do their laundry and bathing in a river, while collecting water from a well for drinking and cooking. As part of this process, the benefits and costs of collecting water from various sources are weighed. For example, women might choose to collect drinking water from a pathogen-contaminated lake that is a 10-min walk away instead of from a pathogen-free well that is 30 min away. In their calculations, the health benefits from the well water may be less than the cost associated with the increase in collection time. It is possible that with increased health education, the situation would be reversed. However, even extensive knowledge about the importance of safe drinking water might not change the fact that the time required to collect the well water is needed for other purposes.

When working with a community to develop a water supply plan, it is crucial to gain an understanding of what uses people actually consider to be priorities. Development workers may make the mistake of focusing solely on water for drinking, cooking,

Table 9-8. Water Sources That Are Appropriate for Specific Uses of Water

	Water Source					
Water Use	Surface Water (Untreated)	Surface Water (Treated)	Groundwater[a]	Rainwater (First Flush)	Rainwater (After First Flush)	Greywater
Drinking	X	A	A	X	A	X
Cooking	X	A	A	X	A	X
Bathing, hand washing, anal ablution	A[b]	U	A	A	A	X
Irrigation	A	U	A	A	A	A
Latrine flushing	A	U	A	A	A	A

Note: A = Appropriate, U = Unnecessary expense, and X = Inappropriate.

[a]Care must be taken in areas where arsenic, fluoride, or microbial contamination of groundwater is a concern. Sanitary surveys should be used to determine the risk of microbial contamination.

[b]In areas where schistosomiasis is endemic, people's skin should not come into contact with untreated surface water. Infected snails release a larval form of the parasite that can burrow through skin.

and hygiene, whereas community members might be more interested in water for productive activities, such as gardening. This misunderstanding can lead to different decisions regarding which water source to use, as well as an underestimation of potential demand. Both women and men, who may have contrasting ideas about which water-using activities are most important, must take part in the prioritization of a community's water needs.

))) References

Ahrens, B. T., and Mihelcic, J. R. (2006). "Making wastewater construction projects sustainable in urban, rural, and peri-urban areas." *J. Engineering for Sustainable Development: Energy, Environment, Health*, 1(1), 13–32.

Cairncross, S. (1987). "The benefits of water supply." *Developing world water*. J. Pickford, ed., Grosvenor Press, London.

Clasen, T., Roberts, I., Rabie, T., Schmidt, W., and Cairncross, S. (2006). "Interventions to improve water quality for preventing diarrhoea (A Cochrane Review)." In *The Cochrane Library*, Issue 3.

Curtis, V., Cairncross, S., and Yonli, R. (2000). "Domestic hygiene and diarrhoea—Pinpointing the problem." *Tropical Medicine and International Health*, 5(1), 22–32.

Fry, L. M., Mihelcic, J. R., and Watkins, D. W. (2008). "Water and non-water-related challenges of achieving global sanitation coverage." *Envir. Sci. and Technol.*, 42(4), 4298-4304.

Gleick, P. H. (1996). "Basic water requirements for human activities: Meeting basic needs." *Water International*, 21, 83–92.

Howard, G., and Bartram, J. (2003). *Domestic water quantity, service level and health*. World Health Organization, New York.

Kennedy, P. M. (2006). *An analysis of the relationship between water accessibility, use and health in Muthara, Kenya.* <http://cee.eng.usf.edu/peacecorps/Resources.htm> (Jan. 18, 2009).

Prost, A., and Négrel, A. D. (1989). "Trachoma and conjunctivitis." *Bulletin of WHO*, 67, 9–18.

United Nations Educational, Scientific, and Cultural Organization (UNESCO) and World Water Assessment Programme (UNESCO–WWAP). (2003). *Water for people, water for life: The United Nations world water development report*, United Nations Educational, Scientific, and Cultural Organization (UNESCO) and Berghahn Books, Oxford.

United Nations Environment Programme (UNEP). (2002). *Global environmental outlook 3*, Earthscan Publications, London.

Vorosmarty, C. J., Green, P., Salisbury, J., and Lammers, R. B. (2000). "Global water resources: Vulnerability from climate change and population growth." *Science*, 289, 284–288.

White, G., Bradley, D., and White, A. (1972). *Drawers of water: Domestic water use in East Africa.* University of Chicago Press, Chicago.

World Health Organization (WHO). (2000). *Global water supply and sanitation assessment 2000 report.* World Health Organization and United Nations Children's Fund, New York.

World Health Organization (WHO) and the United Nations Children's Fund (UNICEF). (2005). *Water for life: Making it happen.* World Health Organization and the United Nations Children's Fund, Geneva, Switzerland.

))) Further Reading

Esrey, S. A., Potash, J. B., Roberts, L., and Shiff, C. (1991). "Effects of improved water supply and sanitation on ascariasis, diarrhoea, dracunculiasis, hookworm infection, schistosomiasis, and trachoma." *Bulletin of the World Health Organization*, 69(5), 609–621.

Morel, A., and Diener, S. (2006). *Greywater management in low- and middle-income countries: Review of different treatment systems for households or neighbourhoods.* Swiss Federal Institute of Aquatic Science and Technology (EAWAG), Dübendorf, Switzerland.

World Health Organization (WHO). (2004). *Water, sanitation and hygiene links to health.* <http://www.who.int/water_sanitation_health/facts2004/en/index.html> (March 20, 2006).

10

Watersheds: Hydrology and Drainage

))) 10.1 Ecological Capital

The many benefits of natural and well-managed watersheds include water supply, drainage, flood protection, erosion control, wastewater treatment, wildlife habitat, and other biodiversity functions. In developing countries, and especially in rural areas, watershed conditions tend to affect human health and livelihood more directly than in urban areas or industrialized countries. This situation occurs because of the absence of infrastructure for freshwater, wastewater, and storm water conveyance and treatment—in these locales, the watershed itself must provide conveyance and treatment functions, essentially serving as ecological capital. Ecological capital is recognized to provide a "fundamental stepping stone in the economic empowerment of the rural poor" (WRI 2005). When watershed degradation occurs, beneficial functions are impaired or lost, and the ecological capital on which low-income communities depend is depleted.

Watershed degradation can occur for many reasons, some of which are illustrated in Fig. 10-1. Most notably, human activities can change land cover and lead to increased surface runoff and soil erosion, decreased spring output, and degraded water quality. It is also common for natural wetlands to be drained to increase arable land. Grazing activities affect vegetation cover and soil compaction, leading to reduction in evapotranspiration and infiltration rates and increased runoff. Animal wastes may also affect water quality. Deforestation due to agriculture, grazing, or firewood collection will also affect evapotranspiration and infiltration, but the magnitude of these effects depends on soil properties and subsequent land use. Any consumptive use of water, whether for domestic, industrial, or irrigation purposes, will affect the water budget as well (e.g., manifested in reduced stream flows or lower water tables) and could affect water quality through the resulting wastewater discharge.

Hydraulic structures (e.g., dams, culverts) or other stream channel modifications will affect the volume and timing of stream flows. Any construction activity can potentially lead to increased runoff and soil erosion. Construction over or along stream channels, such as roads and bridges, may also affect stream flow hydraulics.

Finally, regional climate change, whether because of natural cycles or anthropogenic effects, may affect watershed functions as vegetation and stream channels adjust to

---

This chapter was written by David W. Watkins, Jr.

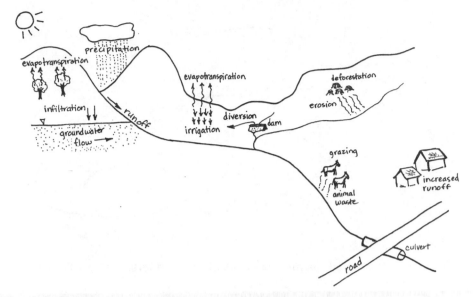

Figure 10-1. Examples of Human Activities Affecting the Hydrologic Cycle and Water Quality.

changing seasonal patterns of precipitation, temperature, and humidity. These effects can adversely affect human health through diminished access to water supplies or through the introduction of pathogens to water supplies. Environmental systems can be adversely affected through any changes to the natural hydrologic cycle.

Efforts to maintain beneficial watershed functions and protect human and environmental health can be made at a range of spatial and temporal scales. Ideally, sustainable development would entail long-term, holistic, consensus-based planning at the watershed scale (as well as other pertinent socioeconomic, political, and ecosystem scales). In practice, engineers find themselves arriving on the scene of a watershed in transition, often observing unintended or incompletely understood consequences of past and present activities, and with a limited budget with which to correct problems. Thus, practical, small-scale, and sometimes temporary solutions must be found, but these solutions should take into account larger space and time scales to the extent possible.

⟫⟫ 10.2 Watershed Management Principles

A *watershed* is defined as the land area that drains to a particular water body or, in the case of a river or stream, to a particular location along the channel. To delineate a watershed manually, a topographic map (showing elevation contours) is used to locate high points around the water body or outlet point. These points are then connected with lines sketched approximately perpendicular to the contours (Fig. 10-2).

Many organizations around the world have access to geographic information systems (GIS) software, such as ArcGIS. Watershed delineation is easily performed using GIS

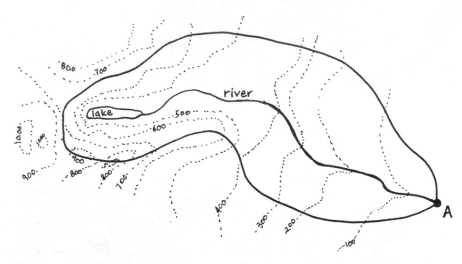

Figure 10-2. Watershed Delineation Overlaid on a Topographic Map. The Area Outlined Drains to Point A.

elevation data, such as a digital elevation model or elevation contour data. Free GIS data from various organizations are available online for many parts of the world. Online help is available for ArcGIS software, with instructions for watershed delineation at www.esri. com. Chang (2006) provides an introduction to GIS topics, such as coordinate systems, data models, data analysis, spatial interpolation, watershed analysis, and other useful topics for working with spatial data.

For many practical purposes, the watershed provides a physically meaningful boundary within which to carry out water resources assessment and planning activities. However, groundwater levels do not necessarily follow land surface elevations.

Within any watershed, it is important to recognize the social, ecological, and economic value of the landscape and its water resources. Although certain economic uses of the watershed (agriculture, grazing, timber, roads) may be easily observed, and human health benefits from clean and plentiful water supplies are apparent, many environmental functions may not be widely recognized. Denny (1997) lists a number of environmental functions of watersheds (Box 10-1).

The stewardship of watershed lands is often neglected. Water supplies and other environmental functions may deteriorate rapidly if contributing areas are not protected from point and nonpoint source pollution or from damaging hydrologic effects of land use changes or structural modifications. Furthermore, because land use change is usually gradual, the effects of pollution may be delayed, and these effects are often distant from the point of water delivery, there may be a "psychological disconnect" in which residents fail to recognize the effects of their land use, water use, and waste disposal decisions (Lee 2000).

Among the most serious of these impacts are the public health effects of poor drainage and standing water, which can promote transmission of a host of water-borne and mosquito-borne diseases, including bacterial, viral, and parasitic infections. Of primary

))) Box 10-1. Environmental Functions of Watersheds

Climatic effects
  Carbon fixation and CO_2 balance
  Rainfall and humidity improvement
  Microclimate influences
Water quality functions
  Particulate filtration and settling
  Nutrient stripping
  Biodegradation of toxic chemicals
  Pathogen reduction
  Heavy metal sequestration
  Wastewater treatment
Biodiversity and habitat functions
  Landscape and ecosystem diversity
  Ecosystems transition zones
  Water supply
  Wildlife habitat
  Genetic biodiversity
Hydrologic and hydraulic functions
  Drainage
  Erosion protection
  Flood mitigation
  Water supply
  Groundwater recharge

From: Denny 1997.

importance are water-borne diseases (including cholera, typhoid, and many common diarrheal diseases), which are spread through fecal contamination of water supplies. This contamination may infect peoples' hands, utensils, food, or drinking water. There are also water-related diseases, such as schistosomiasis, which are acquired though skin contact with contaminated water. Finally, mosquito breeding in standing water can promote the spread of parasitic infections, such as malaria and filariasis, as well as viruses, such as those that cause yellow fever and dengue fever.

To the extent possible, the water supply, health, and environmental functions of watersheds must be valued explicitly, and a community-based watershed plan should be developed to evaluate trade-offs and promote watershed stewardship. Eight basic steps to developing a community-based watershed plan are illustrated in Fig. 10-3. These steps range from involving the community and gaining an understanding of the watershed and its functions, to identifying problems, establishing goals, identifying alternatives, and evaluating performance. As indicated in Fig. 10-3, sustainable watershed planning should be an ongoing process—as populations, technology, economics, and climate conditions change—rather than a linear process with an end point.

Figure 10-3. Steps in Developing a Community-Based Watershed Plan.

⟫⟫ 10.3 Hydrologic Analysis and Design Methods

Historically, the ubiquitous lack of watershed data has led to the development of relatively simple, empirical methods for performing hydrologic analysis and design calculations. This section outlines a few standard methods that are widely used to assess hydrologic effects of land use change and to design solutions for drainage and flooding problems.

10.3.1 Water Budget Method

One simple analysis, in concept at least, is to develop a water budget for the watershed or subarea of the watershed to be analyzed. Considering the hydrologic processes shown in Fig. 10-1, a water budget may be represented as

$$\Delta S = P - I - ET - R \tag{10-1}$$

where ΔS is the change in surface storage (amount of ponded water), P is precipitation, I is infiltration, ET is evapotranspiration (which may also include "interception" of rainfall by plants), and R is runoff. While simple in concept, an accurate water budget over any given time period may be elusive in practice because of uncertainties in measuring each of the components.

Precipitation may be measured at one or possibly a few gauges throughout the watershed but can vary significantly even over short distances. Runoff may be estimated as the increase in stream flow volume (over some "base flow," which is the flow during

))) Box 10-2. Thornthwaite-Type Monthly Water Balance

A Thornthwaite-type monthly water balance model is a simple and relatively reliable method for simulating a steady-state climatic average or continuous water balance in a watershed. The method assumes that if potential evapotranspiration (PET) is less than or equal to water input, then evapotranspiration (ET) is equal to the potential evapotranspiration. Otherwise, ET is the sum of water input and an increment removed from soil storage. PET was defined by Penman (1956) as "the amount of water transpired . . . by a short green crop, completely shading the ground, of uniform height and never short of water." It is best estimated by the Hamon method, which is based only on temperature and day length.

A typical Thornthwaite-type model requires the following input:

- Field capacity, θ_{fc} (moisture content of soil after excess water has drained away)
- Vertical extent of the root zone, Z_{rz} (cm)
- Total monthly precipitation, P_m (cm)
- Average monthly temperature, T_m (°C)

1. Calculate the maximum soil-water storage capacity:

$$\text{Soil}_{max} = \theta_{fc} \times Z_{rz}$$

2. Calculate the monthly water input:
   a. Divide precipitation into rain and snow:

$$\text{Rain}_m = F_m \times P_m$$
$$\text{Snow}_m = (1 - F_m) \times P_m$$

   The melt factor, F_m, depends on temperature:

$$F_m = 0 \text{ for } T_m \leq 0\ °C$$
$$F_m = 0.167 \times T_m \text{ for } 0\ °C < T_m < 6\ °C$$
$$F_m = 1 \text{ for } T_m \geq 6\ °C$$

   b. Calculate monthly snowmelt.
   The monthly snowmelt depends on snowfall and snowpack. The monthly snowpack, Pack_m is calculated using the snowpack from the previous month:

$$\text{Pack}_m = (1 - F_m)^2 \times P_m + (1 - F_m) \times \text{Pack}_{m-1}$$

   Monthly snowmelt is then:

$$\text{Melt}_m = F_m \times (\text{Pack}_{m-1} + \text{Snow}_m)$$

   c. Calculate monthly water input, W_m:

$$W_m = \text{Rain}_m + \text{Melt}_m$$

3. Calculate the evapotranspiration and soil moisture:
   a. Calculate PET_m. PET has been estimated by Hammon (1963) as

$$PET = 29.8D \frac{e_a^*(T_a)}{T_a + 273.2}$$

Where PET is in mm/day, D is day length in hours, and $e_a^*(T_a)$ is the saturation vapor pressure at the mean daily temperature, T_a:

$$e_a^*(T_a) = 0.611 \exp\left(\frac{17.3T_a}{T_a + 237.3}\right)$$

b. If $W_m \geq PET_m$, then

$$ET_m = PET_m$$
$$Soil_m = \min\{[(W_m - PET_m) + Soil_{m-1}], Soil_{max}\}$$

c. If $W_m < PET_m$, then

$$ET_m = W_m + Soil_{m-1} - Soil_m$$

where the decrease in soil moisture is found using:

$$SOIL_{m-1} - SOIL_m = SOIL_{m-1}\left[1 - \exp\left(-\frac{PET_m - W_m}{SOIL_{max}}\right)\right]$$

This type of model can be used for either continuous monthly data or climatic monthly averages. If the model is used with climatic monthly averages, then at month $m = 1$, the previous month, m, is $m - 1 = 12$.

Source: Dingman 2002.

dry periods), with stream flow calculated as the cross-sectional area of the flow times the average velocity (see Fig. 10-4). Infiltration and evapotranspiration (ET) are difficult to measure directly, however. This problem prevents closure of the water budget. For even rough estimates of these components of the hydrologic cycle, simplified methods may be needed. One approach is to rely on tabulated "crop coefficients" to estimate ET. Another approach is to use simple equations, as in a Thornthwaite-type water balance model (Dingman 2002).

10.3.2 Analyzing Large Rainfall Events

In reality, the water budget of a watershed varies continuously in time, with significant changes throughout each day, even if no precipitation occurs, because of diurnal changes in solar radiation, temperature, wind, and humidity. Because of the difficulties of continuously tracking water as it moves throughout the watershed, a number of simplified methods (or models) have been developed specifically for analyzing large rainfall events that may lead to flooding or drainage problems. Two of these are the rational method and the Natural Resources Conservation Service (NRCS) runoff curve number method.

Rational Method

One common method for estimating peak runoff rates in urban settings or small watersheds (less than 200 acres) is the rational method. This method computes a peak runoff

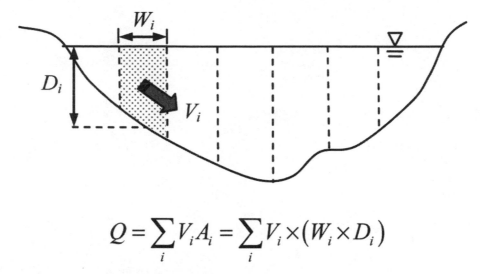

$$Q = \sum_{i} V_i A_i = \sum_{i} V_i \times (W_i \times D_i)$$

Figure 10-4. Illustration of Flow Measurement Technique.

Note: The average velocity at each section may be roughly estimated by measuring the distance a small float, such as a leaf, travels in 10 s and multiplying this surface velocity by 0.85.

rate, Q (ft³/s), from a storm based on a given rainfall intensity, i (in./h), the drainage area, A (acres), and a runoff coefficient, C, that depends on land use, terrain, and soil conditions. The equation is

$$Q = C \times i \times A \qquad (10\text{-}2)$$

No unit conversions are required in Eq. 10-2 because 1 acre-in./h is approximately equal to 1.008 ft³/s.

Infiltration rates are higher for sandy soil than for clay soil. Infiltration may be minimal for rocky or compacted ground and is essentially zero for impervious surfaces, such as rooftops and paved surfaces. Also, infiltration decreases on steep slopes because the water has less time to infiltrate than when it moves slowly or ponds over flat surfaces. Land use also affects infiltration rates because vegetation loosens the soil. Runoff coefficients, as shown in Table 10-1, are therefore higher in areas of clay soil or rock, on steep slopes, and in urbanized areas.

Two important concepts must be understood to select the proper rainfall intensity, i: the duration of the rainfall and the return period of the storm. The duration is important because rainfall intensity can vary significantly during a storm, with intense bursts for short periods but a much lower average intensity over the course of a long event.

The rational method assumes that the critical duration of a storm is equal to the time it takes for the entire watershed to be contributing runoff to the watershed outlet (design point), which is called the *time of concentration* of the watershed. Alternatively, the

Table 10-1. Runoff Coefficients for Use with the Rational Method

Land Use or Land Cover	Recommended C Value (McCuen 2005)
Business	
Downtown	0.85
Neighborhood	0.60
Residential	
Single-family	0.40
Multifamily, detached units	0.50
Multifamily, attached units	0.70
Industrial and commercial	
Light	0.65
Heavy	0.75
Parks, cemeteries	0.20
Unimproved (fields, woods)	0.20
Pavement[a]	
Asphalt and concrete	0.85
Brick	0.80
Roofs[a]	0.85[b]
Lawns, sandy soil[a]	
Flat (slope <2%)	0.08
Average (2–7%)	0.13
Steep (>7%)	0.18
Lawns, heavy soil[a]	
Flat (<2%)	0.15
Average (2–)	0.20
	0.30

Note: Values shown are recommended values for return periods of 25 years or less (ASCE 1969). For storms with longer return periods, larger runoff coefficients should be used (see McCuen 2005).

[a]These values may be used to develop a composite runoff coefficient based on the percentage of different types of surfaces in the watershed.

[b]A representative value for metal, concrete, asphalt, and built-up roofs. Thatch roofs have much lower runoff coefficients (perhaps as low as 0.2).

time of concentration may be considered as the travel time from the most hydraulically distant point of the watershed to the watershed outlet. Example 10-1 outlines a method for estimating the time of concentration based on the slope and land cover of the watershed. For drainage areas less than 1 acre, a time of concentration of 15 min may be assumed.

⟩⟩⟩ **Example 10-1.** Estimating Time of Concentration

Estimate the time of concentration for a 5-acre watershed, where the longest drainage path is identified as 1,200 ft in total length, including 300 ft of pasture (overland flow) with a 1% slope, 300 ft of pasture (small gullies) with a 1.5% slope, 450 ft of grass-lined waterway with a 2% slope, and 150 ft of concrete gutter with a 2.5% slope.

Solution

Estimate the velocity of flow and time of travel along each segment, with velocity (ft/s) depending on slope and land cover according the function:

$$V = K \times S^{0.5}$$

where K is an empirical constant and S is the slope (ft/ft). For this example:

Pasture (overland flow): $V_1 = K_1 S_1^{0.5} = 22(0.01)^{0.5} = 0.22$ ft/s

Pasture (gullies): $V_2 = K_2 S_2^{0.5} = 25(0.015)^{0.5} = 3.1$ ft/s

Grass channel: $V_3 = K_3 S_3^{0.5} = 20(0.02)^{0.5} = 2.8$ ft/s

Concrete gutter: $V_4 = K_4 S_4^{0.5} = 42(0.025)^{0.5} = 6.6$ ft/s

$$\text{Time of concentration} = \sum_i \frac{L_i}{V_i} = \frac{300 \text{ ft}}{0.22 \text{ ft/s}} + \frac{300 \text{ ft}}{3.1 \text{ ft/s}} + \frac{450 \text{ ft}}{2.8 \text{ ft/s}} + \frac{150 \text{ ft}}{6.6 \text{ ft/s}}$$

$$= 1{,}650 \text{ sec.} = 28 \text{ min.}$$

The time of concentration of a storm is the time it takes for the entire watershed to be contributing runoff to the watershed outlet.

Other useful K values are

Forest (overland flow): $K = 0.7 - 1.5$, for dense to no underbrush

Grass (overland flow): $K = 1.0 - 1.8$, for very dense or tall to thin or short

Paved area (sheet flow): $K = 9$

Alluvial fans (gully flow): $K = 10$ ⟫⟫

The *return period* is an estimate of how many years, on average, there would be between storms of a given severity. For instance, a 10-year, 24-h storm is the maximum rainfall occurring in a 24-h period that can be expected once every 10 years, on average, or with a 10% chance of occurring in any given year. Designing for a longer return period (more severe and rare storm) provides higher reliability, but also adds to the cost of the design. In residential areas of low-income communities, it may be necessary to design for a very short return period (perhaps even less than one year), to use scarce funds in an equitable manner. For road crossings, where flooding could lead to costly damage and have a severe effect on social and economic activities, one may want to design for a longer return period (perhaps 5–10 years or more, depending on the importance of the infrastructure).

Once the critical storm duration (time of concentration) is determined, and an appropriate return period is selected, historical rainfall data must be analyzed to determine the corresponding rainfall intensity. In many urban areas, analysis of historical rainfall data has been done, and the results are summarized in the form of intensity–

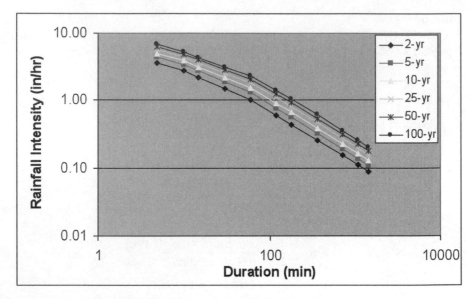

Figure 10-5. Example of Site-Specific Rainfall IDF Curves.

duration–frequency (IDF) curves, as shown in Fig. 10-5. If such data do not exist nearby, one may use—with caution—IDF curves from a distant location with similar climate and rainfall patterns. For instance, IDF curves for southern Florida, an area of low relief, may be used in low-relief regions of the Caribbean if no other data records are available.

))) **Example 10-2.** Determining Peak Runoff

Here, we will calculate the 10-year peak runoff for a 5-acre watershed (assuming we can use the data in Fig. 10-5). The land use for 4.5 acres of the watershed is 50% pasture and 50% single-family residential, and there is an impervious parking lot on the remaining 0.5 acre. The time of concentration for the watershed was provided in Example 10-1.

Step 1
With a duration equal to the time of concentration (28 min, which is approximately 30 min), use Fig. 10-5 to determine the design rainfall intensity. For a return period of 10 years, the value is $i = 2.2$ in./h.

Step 2
Because there are different land uses in the watershed, compute a weighted (composite) runoff coefficient.

Pasture: $C_1 = 0.2$

Single-family residential: $C_2 = 0.4$

Parking lot (asphalt): $C_3 = 0.85$

$$C_w = \frac{\sum_i C_i A_i}{\sum_i A_i} = \frac{0.2(2.25) + 0.4(2.25) + 0.85(2.25)}{5} = 0.36$$

Step 3

Calculate design discharge using the rational method (Eq. 10-2) with the composite runoff coefficient:

$$Q = C \times i \times A = 0.36 \times (2.2 \text{ in./h}) \times (5 \text{ acres}) = 4.0 \text{ acre-in./h} \approx 4.0 \text{ ft}^3/\text{s}$$

》》》

Natural Resources Conservation Service Runoff Curve Number Method

Another method for runoff estimation is the Natural Resources Conservation Service (NRCS) runoff curve number method (USDA 1972). This method was developed primarily for agricultural watersheds on the order of 10 to 1,000 acres, but it has been applied to larger watersheds with different land uses. Its use should be limited to watersheds smaller than 20 mi².

The key parameter in the NRCS method is the runoff curve number (CN), which is a function of land use and soil type, as shown in Table 10-2. For a precipitation event of a total given depth P (in.), the total runoff depth R (in.), is calculated as follows:

$$R = \frac{(P-0.2S)^2}{P+0.8S}, \text{ for } P \geq 0.2S \ (R=0 \text{ otherwise}) \tag{10-3}$$

where the potential abstraction (infiltration plus other losses) is $S = (1,000/CN) - 10$, and CN is the runoff curve number determined as a function of land use, soil type, and antecedent soil moisture conditions. Equation 10-3 can also be represented as a set of curves, indicating total runoff depth as a function of total precipitation depth and CN value, as shown in Fig. 10-6. To obtain an estimate of runoff volume from a storm event, one would multiply the runoff depth R by the watershed area.

All curve numbers listed in Table 10-2 are based on average antecedent soil moisture conditions. If the antecedent conditions were dry or wet, however, the curve numbers must be adjusted according to the chart shown in Fig. 10-7. Dry conditions are defined as less than 0.5 in. of rain in the last 5 days in the dormant season, or less than 1.4 in. in the last 5 days in the growing season. Wet conditions are assumed if heavy rainfall (1.1 in. in the dormant season or 2.1 in. in the growing season), or light rainfall and low evapotranspiration, have occurred within the past 5 days. Discussions with community members may help to determine typical conditions before large runoff events.

The peak discharge from a storm event may be calculated using the curve number method by estimating the time of concentration for the watershed and assuming

Table 10-2. Runoff Curve Numbers for Various Land Uses and Soil Types, Assuming Average Antecedent Soil Moisture Conditions

Land Use	Soil Group[a]			
	A	B	C	D
Lawns, open spaces, parks, cemeteries				
Good condition (grass cover 75% or more)	39	61	74	80
Fair condition (grass cover 50–75%)	49	69	79	84
Poor condition (grass cover less than 50%)	68	79	86	89
Paved parking lots, roofs, driveways	98	98	98	98
Streets and roads				
Paved with curbs and storm sewers	98	98	98	98
Paved with open ditches	76	85	89	91
Gravel	72	82	87	89
Dirt	83	89	92	93
Commercial and business districts (85% impervious)	89	92	94	95
Industrial districts (70% impervious)	81	88	91	93
Residential, average lot size:				
1/8 acre or less	77	85	90	92
1/4 acre	61	75	83	87
1/2 acre	54	70	80	85
1 acre or more	51	68	79	84
Cultivated agricultural land[b]				
Fallow, straight row or bare soil	77	86	91	94
Fallow, conservation tillage	74	83	88	90
Row crops, straight row	67	78	85	89
Row crops, conservation tillage or contoured	64	75	82	85
Noncultivated agricultural land				
Pasture or range (groundcover > 50%)	39	61	74	80
Pasture or range (groundcover 25–50%)	49	69	79	84
Pasture or range (groundcover < 25%)	68	79	86	89
Meadow	30	58	71	78
Forest, condition:				
Good (groundcover density > 70%)	32	58	72	79
Fair (groundcover density 30–70%)	44	65	76	82
Poor (groundcover density < 30%)	55	73	82	86

[a]Soil groups: A, Deep sand, deep loess, aggregated silts; B, Shallow loess, sandy loam; C, Clay loams, shallow sandy loam, soils low in organic content; and D, Heavy clays, soils that swell when wet.

[b]Assumes good condition (more than 20% residue cover). Increase slightly if less cover.

Source: McCuen 2005.

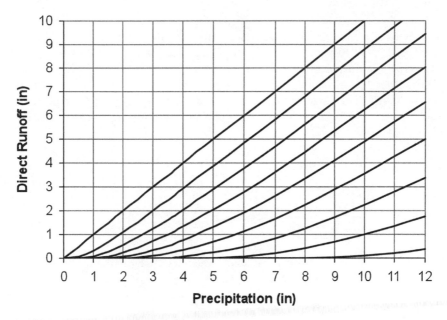

Figure 10-6. Runoff Depth as Calculated by the NRCS Curve Number Method for the Case Where Runoff Occurs Only When $P > 0.2\ S$.

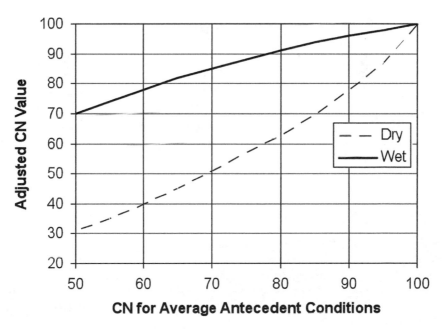

Figure 10-7. Adjustment of Runoff Curve Numbers for Dry or Wet Antecedent Conditions.

particular time distributions of precipitation and runoff within the storm event (Bedient and Huber 2002; McCuen 2005). Equations 10-4 and 10-5 provide approximate peak discharge Q_p (ft³/s) per unit area A (mi²), for two particular storm distributions (the NRCS Type II and III distributions, respectively) and a particular relationship between 24-h runoff and peak runoff. The Type II distribution is applicable in most of the central and eastern United States, where peak annual runoff typically results from summer thunderstorms. The Type III distribution is recommended for coastal locations affected by tropical storms.

$$\text{Type II: } Q_p = 150 \left(\frac{0.2S}{P} \right)^{-0.314} T_c^{-0.581} A \tag{10-4}$$

$$\text{Type III: } Q_p = 115.6 \left(\frac{0.2S}{P} \right)^{-0.319} T_c^{-0.481} A \tag{10-5}$$

These equations should only be used if the time of concentration is between 0.1 and 10 h and the curve number is greater than 55. Also, the parameter $((0.2\,S)/P)$ should be between 0.1 and 0.5. If the value of $((0.2\,S)/P)$ is outside of this range, the minimum or maximum value should be used. For other storm distributions (e.g., Types I and IA, developed for the Pacific maritime climate with wet winters and dry summers), a graphical peak discharge method is recommended (USDA 1986). A description of this method is available by contacting the authors.

Choosing Between the Rational Method and the NRCS Runoff Curve Number Method

Like the rational method, the NRCS curve number method also considers land use, soil type, and time of concentration to estimate runoff volumes and peak discharge rates. Although in some cases this method may be considered an alternative to the rational method, it is applicable to larger watersheds, and it also may be used to calculate the total runoff volume (or depth over the watershed area) from a storm, whereas the rational method only determines a peak discharge. Both methods were developed for runoff estimation for large rainfall events (1- to 2-year return periods or longer) and thus may not be accurate for smaller, more frequent storms.

10.3.4 *Manning's Equation*

Once a given volume of runoff or peak discharge is estimated, conveyance or storage facilities may be designed to prevent or reduce flood effects. By far the most common equation used for designing conveyance structures (e.g., unlined ditches, lined drains, culverts) is Manning's equation, which relates the discharge capacity of the structure to cross-sectional geometry, longitudinal slope, and lining material of the drain channel as follows:

$$Q_p = \frac{1.49}{n} A R_h^{2/3} S^{1/2} \tag{10-6}$$

where Q is the discharge capacity (ft^3/s); A is the cross-sectional area of the flow (ft^2), R_h is the hydraulic radius (ft), defined as the area divided by the wetted perimeter, A/P_w; and S is the longitudinal slope (ft/ft). Manning's roughness coefficients (n) are listed in Table 10-3. An example calculation for drainage channels is provided in Example 10-3.

Table 10-3. Manning's n Coefficients for Various Materials

Material	Manning's n (Typical)
Unlined earth channels	
No vegetation	0.020
Grass, some weeds	0.024
Dense weeds	0.030
Natural, unmaintained channels	
Clean bottom, brush on sides	0.065
Dense weeds, high as flow	0.10
Dense brush, high as flow	0.12
Lined channels	
Concrete, smooth	0.013
Brick, plastered	0.015
Gravel bottom, concrete sides	0.020
Loose rubble, rip-rap	0.032
Culverts, pipes	
Concrete	0.013
Corrugated metal	0.024
Cast iron	0.013
Smooth plastic, PVC	0.010

Note: Values assume straight, uniform sections and good construction. For nonuniform sections, channels with bends, or poor construction, use larger values of n.

⟫⟫⟫ Example 10-3. Sizing Drainage Channels

The maximum depth of flow in the trapezoidal channel shown below is 1.2 ft. The longitudinal slope of the channel is 0.005 ft/ft. Determine the discharge capacity if the channel is lined with smooth concrete ($n = 0.013$). What would be the required bottom width to carry the same discharge at a depth of 1.2 ft if the channel were unlined ($n = 0.024$)?

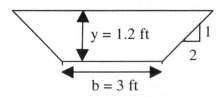

$y = 1.2$ ft

$b = 3$ ft

For the lined channel, compute the flow area, wetted perimeter, and hydraulic radius:

$$A = yb + 2y^2 = (1.2)(3) + 2(1.2)^2 = 6.48 \text{ ft}^2$$

$$P_w = b + 2\sqrt{y^2 + 4y^2} = 3 + 2\sqrt{(1.2)^2 + 4(1.2)^2} = 8.37 \text{ ft}$$

$$R_h = A/P_w = 6.48/8.37 = 0.774 \text{ ft}$$

The discharge capacity is determined from Manning's equation (Eq. 10-6):

$$Q = \frac{1.49}{n} AR_h^{2/3} S^{1/2} = \frac{1.49}{.013}(6.48)(.774)^{2/3}(.005)^{1/2} = 44.3 \text{ ft}^3/s$$

To determine the bottom width required for an unlined channel to have the same capacity, apply Manning's equation with b as the unknown variable:

$$\frac{Qn}{1.49\,S^{1/2}} = A^{5/3} P^{-2/3} \Rightarrow \frac{(44.3)(.024)}{1.49(.005)^{1/2}} = 10.09 = (1.2b + 2.88)^{5/3}(b + 5.37)^{-2/3}$$

Solve iteratively for b: ($b = 6.6$ ft). The rougher unlined channel requires a wider width.

》》》

For any drain pipe or culvert, as long as it is not placed at too steep a slope (so that entrance conditions constrict the flow), it may be assumed that the circular cross section flows full, so that

$$R_h = \frac{A}{P_w} = \frac{\left(\dfrac{\pi D^2}{4}\right)}{\pi D} = \frac{D}{4} \qquad (10\text{-}7)$$

To choose the proper culvert diameter (ft) for a given discharge (ft³/s), Manning's equation may be rearranged as follows:

$$D = \left[\frac{2.16\,nQ}{S^{1/2}}\right]^{3/8} \qquad (10\text{-}8)$$

This calculation assumes that entrance and exit conditions do not constrict the flow. If these conditions do block the flow, please consult a reference for more details on culvert hydraulics (e.g., Wurbs and James 2002).

Manning's equation of the form

$$V = \frac{1.49}{n} R_h^{2/3} S^{1/2} \qquad (10\text{-}9)$$

may also be used to estimate flow velocity (ft/s) in a channel to determine if scour and channel erosion will occur. Table 10-4 provides maximum velocities to prevent scour and erosion of channel linings.

))) 10.4 Flood Storage and Drainage for Small Areas

Natural or constructed storage may also be used to reduce flooding and protect property and infrastructure such as roads. The concept is to allow runoff to pond temporarily in a low-lying area, which is then drained at a more gradual rate, as illustrated in Fig. 10-8. Given a volume of runoff (say, calculated from the NRCS curve number method), a peak inflow rate to the storage area, and a peak discharge rate (calculated from Manning's equation), the volume of storage required can be estimated from Fig. 10-9. Alternatively, for given runoff and storage volumes, the ratio of the peak inflow to the peak outflow discharge can be estimated from this graph.

The goals of a drainage system are to reduce runoff and ponding, slow runoff and reduce peak discharges, and allow peak discharges to pass safely. All system designs begin with a specified design discharge, which is essentially a policy decision. For instance, should the system be relatively inexpensive and designed only to pass a discharge that is expected to occur a few times each year, or should additional funds be spent to pass a discharge that is expected to occur only every 5–10 years? In the United States, it would not be unusual for a small drainage system to be designed for the 10-year flood event, but this decision may not be practical in low-income communities. For low-income communities, it may be more feasible to allow some temporary ponding of runoff from frequent rainfall events rather than to construct a system that operates at capacity only once every few years.

10.4.1 Infiltration Systems

Reducing impervious surfaces or increasing vegetative cover can serve to increase infiltration. Rock infiltration galleries or rain gardens can also be constructed to capture runoff

Table 10-4. Maximum Permissible Velocities in Channels Susceptible to Erosion

Material	Clear Water (ft/s)	Water Transporting Silts (ft/s)
Fine sand	1.50	2.50
Sandy loam	1.75	2.50
Silt loam	2.00	3.00
Stiff clay	3.75	5.00
Fine gravel	2.50	5.00
Coarse gravel	4.00	6.00
Cobbles, shingles	5.00	6.00
Grass cover, erodible soils	4.00	4.00
Grass cover, stable soils	6.00	6.00

Note: Values are for straight channels with small slopes, after aging.

Source: Fortier and Scobey 1926; Chow 1959.

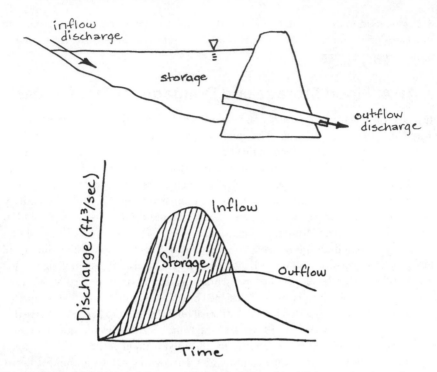

Figure 10-8. Field Use of Storage (Top) and Graphical Representation of Storage (Bottom) to Reduce Peak Discharge.

and allow it to soak into the ground. As shown in Fig. 10-10, these filters may be either low-lying or excavated areas that are back-filled with more porous materials, such as rock, gravel, or sand.

There are two basic design considerations for these types of infiltration systems: storage volume and infiltration capacity. For belowground storage, the available storage volume (in ft³) is calculated as

$$V_s = n \times H \times W \times L \tag{10-10}$$

where n is the porosity of the fill material, and H, W, and L are the depth, width, and length, respectively (in ft), of the trench or rain garden.

If the rate of rainfall plus runoff routed to the system exceeds the infiltration capacity of the fill material, then the depth required is

$$H = \frac{f_c \times T_d}{n} \tag{10-11}$$

where f_c is the (saturated) infiltration capacity of the soil (ft/min) and T_d is the duration of the runoff event (min) (Davis and McCuen 2005). In this case, water will pond on the surface if there is a natural or constructed berm around the system. Although such a berm

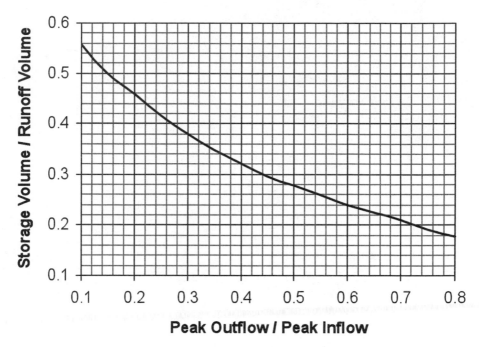

Figure 10-9. Approximate Storage–Discharge Relationship for Rainfall Distribution Types II and III.

Source: Based on data from McCuen 2005.

may provide additional storage, any ponding should be temporary so as not to enhance breeding of mosquitoes. Table 10-5 lists typical porosity values and saturated infiltration rates for various soils.

10.4.2 Open Drainage Channels

In most cases, the three main options for conveyance of storm water runoff or wastewater will be open drainage channels, covered drains, or culverts. Examples of open drainage channels, either lined or unlined, are shown in Fig. 10-11. Obviously, culverts or covered drains are required wherever flow is to pass underground or beneath existing infrastructure, such as roads. For all forms of conveyance, Manning's equation may be used to estimate the discharge capacity for a given slope, cross section, and lining material or to determine proper values of the design variables to convey a design discharge.

Unlined channels are the least expensive option for conveyance and may be constructed by hand. The sides of a channel should be sloped no more than 1:2 to ensure stability. The longitudinal slope may vary according to topography, but if the slope is greater than 1%, there may be potential for scour and erosion of the channel bottom. For slopes between 1% and 5%, partial lining, a grass lining, or even a compacted gravel lining may be sufficient.

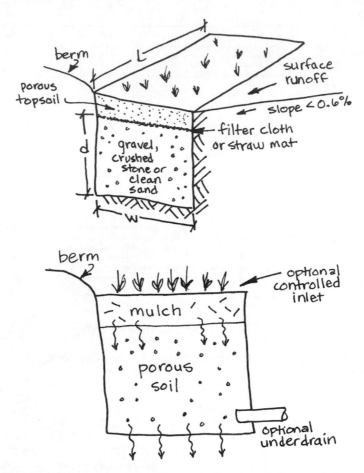

Figure 10-10. Infiltration Systems: Top, an Infiltration Gallery with Aggregate Media, and Bottom, a Rain Garden That Attempts to Mimic More Natural Hydrology.

Source: Davis and McCuen 2005.

Grass-lined channels have the advantage of providing some water quality improvement through filtration, sedimentation, and uptake of nutrients. However, more maintenance may be required to keep the grass healthy and at a proper length. For steep channels (slopes of 5–10%, or even greater), solid linings consisting of concrete, masonry, or rock *gabions* (packs of rocks held together by a wire mesh) may be needed to prevent scour. These linings may be constructed or poured in place, or they may consist of precast concrete elements. Precast elements are advantageous over masonry or poured concrete because they can be laid quickly and do not require several days curing time once they have been put in place. Precast channels should be laid on a bed of compacted sand, preferably 1.5 ft thick.

Concrete or masonry-lined channels may fail if water pressure is allowed to build up or flow occurs on the outside of the lining. These problems can be mitigated by provid-

Table 10-5. Values of Saturated Infiltration Capacity and Porosity for Typical Soils

Material	Infiltration Capacity (cm/s)	Porosity
Well-sorted gravel	10^{-2}–1	0.25–0.50
Sand and gravel mixed	10^{-3}–10^{-1}	0.20–0.35
Well-sorted sands	10^{-3}–10^{-1}	0.25–0.50
Silty sands, fine sands	10^{-5}–10^{-3}	0.25–0.45
Sandy silts, clayey sands	10^{-6}–10^{-4}	0.30–0.50
Silt	10^{-6}–10^{-4}	0.35–0.50
Clay	10^{-9}–10^{-6}	0.33–0.60

Source: Data from Freeze and Cherry 1979.

ing weep holes, perhaps 0.5-in. in diameter, to allow water from the ground to seep into the channel. Short lengths of pipe may be embedded in the concrete or masonry at 1- to 3-ft intervals to allow sufficient seepage through the lining.

In urban areas with narrow streets, the road surface itself may function as a drainage channel, provided that the flow does not interfere with traffic or flood adjacent buildings.

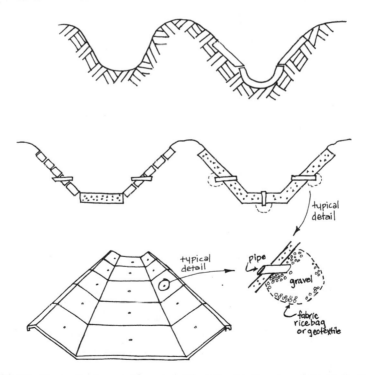

Figure 10-11. Open Drainage Channels That Are Unlined and Partially Lined (Top), and Lined with Brick or Concrete (Middle and Bottom). The Middle Right Cross Section Is Shown in Isometric View at the Bottom Left.

To prevent erosion of the road surface, the slope should be less than 5%, and the road surface should consist of compacted gravel, stone, or pavement. Example 10-4 illustrates the application of Manning's equation to estimate flow depth and velocity on a road surface.

))) **Example 10-4.** Estimating Flow Depth and Velocity on a Road Surface

A design discharge for a compacted gravel road is computed to be 60 ft³/s. The road is 16 ft wide, with 1-ft-high berms on each side and a longitudinal slope of 1.5%. Compute the flow depth and velocity and determine if the road surface is vulnerable to scour.

Solution
Manning's equation can be used to determine the flow depth y, assuming that the road has a rectangular cross-sectional area. Because the flow area will be much wider than it is deep, we can approximate $R_h = y$. Then, with $n \times 0.025$ for gravel,

$$\frac{Qn}{1.49 S^{1/2}} = AR_h^{2/3} \Rightarrow \frac{(60)(.025)}{1.49(.015)^{1/2}} = 8.22 = (16\,y)(y)^{2/3} \Rightarrow y = 0.67 \text{ ft}$$

Next, determine the velocity:

$$V = \frac{Q}{A} = \frac{60}{16(0.67)} = 5.6 \text{ ft/s}$$

From Table 10-4, we see that this number exceeds the permissible velocity for clean water conditions for both fine and coarse gravel, indicating that scour may occur. The fact that the gravel is compacted may alleviate this problem, but alternative measures should be taken to prevent scour of the road surface. If possible, the runoff should be reduced, or some of the flow on the road surface should be diverted to storage or infiltration areas.

)))

Drainage on steep slopes (typically greater than 5%) can be particularly troublesome and may require some form of terracing. Various terracing methods exist for controlling erosion on agricultural land, and these methods may be adopted for drainage systems. Figure 10-12 illustrates four designs: check dams, turnout drains, stepped drains, and baffled drains.

Turnout drains are gently sloped channels that divert a portion of the flow from the main drainage channel and direct it to an infiltration trench or pit. Check dams are relatively inexpensive and may be used in unlined drainage channels. Ponded water deposits silt behind each check dam, forming a stepped channel. Check dams may be constructed of wood, stones, gabions, or even concrete, but regardless of the materials used, they should be set into the ground so that water does not cut a path beneath or around them. In particular, the foundation of each dam should not be at a higher elevation than the crest of the dam downstream. Baffled drains use concrete blocks, bricks, stones, or boards to deflect the flow and thereby reduce the velocity and scour potential. Stepped drains

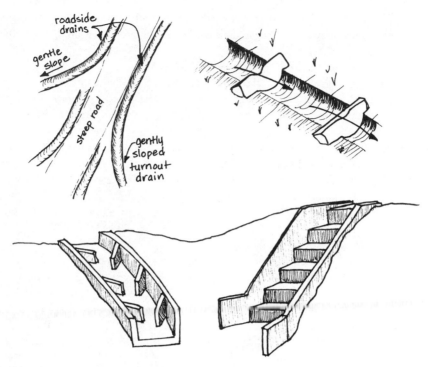

Figure 10-12. Options for Drainage Channels on Steep Slopes, Including (Top Left) Turnout Drains, (Top Right) Check Dams, (Bottom Left) Baffled Drains, and (Bottom Right) Stepped Drains.

Source: WHO 1991.

may be practical if the slope is very steep (perhaps 20–30% or greater), but otherwise are likely to be much more expensive than alternatives.

Open drainage channels are generally preferred to closed or covered drains for a number of reasons, including the following:

they are less expensive, easier to construct, and require less excavation;

defects and accumulation of debris are easier to monitor;

mosquito breeding is easier to control; and

corrosive gases do not build up, as may occur in poorly ventilated closed drains.

The primary advantage of closed drains is that they do not take up surface area, and they may be required whenever drainage must pass through an embankment, such as at a road crossing.

10.4.3 Closed Drains and Culverts

Closed drains with diameters of 6–12 in. are commonly constructed from prefabricated sections of concrete pipe, though other materials, such as galvanized iron or polyvinyl-chloride (PVC), may be used. Culverts used under embankments tend to have larger diameters (up to several feet) and are primarily constructed of prefabricated reinforced concrete, corrugated metal, or corrugated plastic. Table 10-3 provides Manning's *n* values for these materials.

Construction of closed drains requires excavation of a trench (Fig. 10-13). The trench may be approximately 2 ft wider and 5 ft deeper than the diameter of the pipe. This design allows 1.5 ft of compacted and carefully leveled sand to be placed beneath the pipe (set at the proper slope), and at least 3 ft of soil to cover the top of the pipe to protect it from heavy vehicles passing over it. Inlets must also be provided to allow water to drain from the surface, and these should be spaced approximately 100–150 ft apart, depending on the slope and design rainfall intensity. The inlets should be covered with iron grates to prevent debris from entering the drains. Finally, manholes should be provided to allow for cleaning and maintenance of the drains. The recommended manhole spacing is approximately 350–500 ft for drains larger than 2 ft in diameter, and 200–300 ft for smaller drains.

Additional care should be taken in construction of closed drains and culverts to prevent blockage from trash, accumulation of sediments, and standing water, which can promote mosquito breeding. Iron grills or racks can be placed at the openings of closed drains to keep out debris, but these racks must be cleaned regularly to prevent blockage and reduction of the drain's capacity. Dips, or low spots, should not exist in a drainage system, nor should culverts be placed at a lower level than the downstream streambed, to prevent standing water and sediment accumulation. Small basins, or silt traps

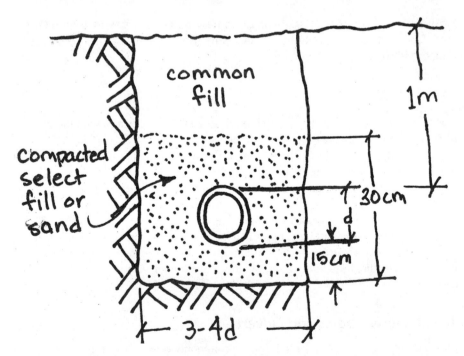

Figure 10-13. An Example Design for a Closed Drainage System with a Pipe Laid in a Trench.

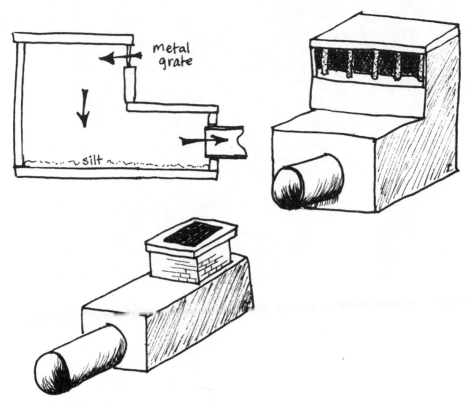

Figure 10-14. Example Design for a Silt Trap for Closed Drainage.

(Fig. 10-14), may be placed at the entrance to closed sections. However, these traps can quickly fill with sand or trash and become breeding grounds for mosquitoes if they are not cleaned out regularly.

10.4.4. Maintenance

Proper maintenance of drainage systems is essential to ensure that they function as designed. In fact, a well-designed but poorly maintained drainage system can lead to more severe flooding, greater public health concerns, and worse watershed degradation than no system at all. Drainage systems concentrate storm water and wastewater flows and accelerate flow through the watershed, so that areas in the lower part of the watershed may be more vulnerable to flooding if downstream drains become blocked or storage areas become filled with sediment. In most cases, regular cleaning (removal of debris and sediments) prevents these problems, but occasionally design flaws or changing watershed conditions need to be identified to make corresponding adjustments.

  Erosion of channels and surrounding areas is another serious concern. This problem can result whenever conveyance and storage capacities are exceeded or if flow velocities are simply too high. If erosion is occurring in an existing drainage channel, four possible solutions exist: reduce peak discharge in the channel, line the channel, widen the channel, or reduce the channel slope. Rocks may also be placed over localized areas prone to erosion. For larger channels needing erosion protection, rock gabions may be used to line the channel.

))) References

American Society of Civil Engineers (ASCE). (1969). *Design and construction of sanitary and storm sewers,* ASCE Press, New York.

Bedient, P., and Huber, W. (2002). *Hydrology and floodplain analysis,* Prentice Hall, Upper Saddle River, N.J.

Chang, K. (2006). *Introduction to geographic information systems,* 3rd ed., McGraw Hill Higher Education, New York.

Chow, V. T. (1959). *Open-channel hydraulics,* McGraw-Hill, New York.

Davis, A. P., and McCuen, R. H. (2005). *Stormwater management for smart growth,* Springer, New York.

Denny, P. (1997). "Implementation of constructed wetlands in developing countries." *Water Sci. and Technol.,* 35(5), 27–34.

Dingman, S. L. (2002). *Physical hydrology,* 2nd ed., Prentice Hall, Upper Saddle River, N.J.

Fortier, S., and Scobey, F. C. (1926). "Permissible channel velocities." *Transactions of the American Society of Civil Engineers,* 89, 940–956.

Freeze, R. A., and Cherry, J. A. (1979). *Groundwater,* Prentice Hall, Englewood Cliffs, N.J.

Hamon, W. R. (1963). "Computation of direct runoff amounts from storm rainfall." *Symposium on Surface Waters, Publication 63,* International Association of Scientific Hydrology, Gentbrugge, Belgium, 52–62.

Lee, M. D. (2000). "Watershed protection challenges in rapidly urbanizing regions: The case of Tegucigalpa, Honduras." *Water International,* 25(2), 214–221.

McCuen, R. H. (2005) *Hydrologic analysis and design,* 3rd ed., Prentice Hall, Upper Saddle River, N.J.

Penman, H. L. (1956). "Evaporation: An introductory survey." *Netherlands J. Agricultural Science,* 193, 120–145.

U.S. Department of Agriculture (USDA). (1972). "Hydrology, Sec. 4" *National engineering handbook,* Soil Conservation Service, Washington, D.C.

U.S. Department of Agriculture (USDA). (1986). "Urban hydrology for small watersheds, Soil Conservation Service, *Technical Release 55,* Washington, D.C.

World Health Organization (WHO) in collaboration with United Nations Environment Programme. (1991). *Surface water drainage for low-income communities,* World Health Organization, Geneva, Switzerland.

World Resources Institute (WRI) in collaboration with United Nations Development Program, United Nations Environment Programme, and World Bank. (2005). *World resources 2005: The wealth of the poor—Managing ecosystems to fight poverty.* World Resources Institute, Washington, D.C.

Wurbs, R. A., and James, W. P. (2002) *Water resources engineering,* Prentice Hall, Upper Saddle River, N.J.

))) Further Reading

Food and Agriculture Organization (FAO). (1986). "Strategies, approaches and systems in integrated watershed management." *Conservation Guide No. 14.* FAO, Rome.

Food and Agriculture Organization (FAO). (1998). "Developing participatory and integrated watershed management." *Community Forestry Case Study Series No. 13.* FAO, Rome.

Stone, L. J. (2003). *Earthen dams for small catchments: A compilation of design, analysis, and construction techniques suitable for the developing world.* <http://cee.eng.usf.edu/peacecorps/Resources.htm> (Jan. 17, 2009).

Tawney, E. J. (2006). *Watershed management and planning in the Tagabe River watershed catchment area in and around Port Vila, Vanuatu.* <http://cee.eng.usf.edu/peacecorps/Resources.htm> (Jan. 18, 2009).

11

Gravity-Fed Water Supply Systems

))) 11.1 Fluid Mechanics and Pipe Networks

11.1.1 Quantity of Flow in a Pipe

As water flows in a pipe, a certain volume of water passes through per unit of time. This rate is referred to as the *flow rate* or *discharge*, and it is in units of volume per time (e.g., L/s, m³/s, or ft³/s). For a given flow rate, if the pipe has a small diameter, then the water has a high velocity, and conversely, if the pipe is large, then the velocity is low. This notion can be quantified by Eq. 11-1, which is referred to as the continuity equation:

$$Q = V \times A \tag{11-1}$$

where Q is the volumetric flow rate (length cubed/time), V is the velocity (length/time), and A is the cross-sectional inside area of the pipe (length squared).

11.1.2 Pressure (or Head) at Various Locations along the Pipe

A pipe will carry its maximum discharge as long as the pressure anywhere in the pipe does not fall below atmospheric pressure. If the pressure does go below zero, then a *siphon* will be established. This is a negative pressure in the pipe that can result in contamination being drawn into the pipe if a crack or hole develops. Negative (or even low) pressures are therefore to be avoided. For this reason, it is important to be able to determine the pressure at all locations in the pipe network.

Assuming no change in elevation of a pipe, pressure diminishes in the pipe as water flows downstream. This pressure loss (also called *head loss*) occurs because of friction with the pipe walls and also because of the swirling flow patterns found at the entrance to pipes, valves, pipe bends, and where the flow exits the pipe.

The pressure (or head) loss due to friction can be determined as follows, assuming fully turbulent flow:

$$h_L = 16 f \frac{LQ^2}{2g\pi^2 D^5} \tag{11-2}$$

Matthew A. Niskanen, Nathan Reents, John D. Simpson, and Stephen P. Good contributed to this chapter.

In Eq. 11-2, h_L is the frictional head loss (length), f is a friction factor, L is pipe length, Q is flow rate in pipe (length cubed/time), g is the gravitational constant (length per time squared), π equals 3.1416, and D is the pipe diameter. Friction factors can be determined for different relative roughness and pipe diameters, as provided in Tables 11-1 and 11-2.

In addition to losing energy because of friction with the pipe walls, losses occur in any kind of pipe fitting because of swirling flow patterns. These are called *minor losses* and can be quantified as follows:

$$h_{L(minor)} = 16K \frac{Q^2}{2g\pi^2 D^4} \tag{11-3}$$

where K is a function of pipe geometry and type of fitting. Values of K can be found in Table 11-3.

11.1.3 The Energy Equation

Using Eqs. 11-1 to 11-3 and Tables 11-1 to 11-3, it is possible to calculate the amount of energy (or pressure) that is present at any location along any of the pipes in a water distribution system. Pressure is sometimes conveniently expressed in terms of *head*, which is the pressure divided by the specific weight of water, γ. The pressure at any point in a pipe can be thought of as how far the water would rise if a vertical tube was connected to the pipe at that point. The energy equation is used for this purpose and is written as follows:

$$\frac{p_1}{\gamma} + \frac{8Q_1^2}{g\pi^2 D_1^4} + z_1 + H_p = \frac{p_2}{\gamma} + \frac{8Q_2^2}{g\pi^2 D_2^4} + z_2 + \Sigma h_L$$

$$\uparrow \qquad \uparrow \qquad \uparrow \qquad\qquad\qquad\qquad \uparrow$$
$$\text{Term 1}\ \ \text{Term 2}\ \ \text{Term 3} \qquad\qquad\qquad \text{Term 4}$$

$$\tag{11-4}$$

In Eq. 11-4, p is pressure (force/length squared), γ is the specific weight of water (force/length cubed), Q is the flow rate (length cubed/time), g is the gravitational constant (length per time squared), D is the pipe diameter (length), z is the height above some reference elevation (length), H_p is the pressure head supplied by a pump (length), and Σh_L is the total head loss (length). In SI units, g is equal to 9.81 m/s^2, and in English units,

Table 11-1. Roughness (k_s) of Various Pipe Materials

Pipe Material	k_s (ft)
Polyvinyl chloride	3.3×10^{-7}
Copper, brass	4.9×10^{-6}
Steel	1.5×10^{-4}
Galvanized iron	4.9×10^{-4}
Cast iron	8.5×10^{-4}
Concrete	9.8×10^{-4} to 9.8×10^{-3}
Riveted steel	3.0×10^{-3} to 3.0×10^{-2}

Source: Crowe et al. 2001.

Table 11-2. Friction Factor for Use in Eq. 11-2 for Various Relative Roughness and Pipe Diameter Ratios (k_s/D), Assuming Fully Turbulent Flow

k_s/D	Friction factor (f)
0.00001	0.008
0.00005	0.011
0.0001	0.012
0.0002	0.014
0.0004	0.016
0.0006	0.018
0.0008	0.019
0.001	0.02
0.002	0.024
0.004	0.029
0.006	0.032
0.008	0.036
0.01	0.038
0.015	0.044
0.02	0.05
0.03	0.056
0.04	0.065
0.05	0.07

Source: Crowe et al. 2001.

32.2 ft/s^2. The specific weight of water is a function of temperature. At 10 °C (50 °F), the specific weight of water is 9.804 kN/m^3 (62.41 lb/ft^3).

The subscript 1 in Eq. 11-4 denotes an upstream point in the pipe, and the subscript 2 refers to a downstream point. Term 1 in Eq. 11-4 is called the *pressure head*; term 2, the *velocity head*; and term 3 the *elevation head*. Term 4 is the *head loss* term. The energy equation can be solved for D and is used to size pipes, as illustrated later in Example 11-1.

11.1.4 Use of the Hydraulic Grade Line

The hydraulic grade line (HGL) is a plot of $z + p/\gamma$ (i.e, the elevation head plus the pressure head). This is terms 3 and 1 in the energy equation (Eq. 11-4). The ground elevation is first plotted, and then a line of p/γ is plotted above that. Figure 11-1 shows an HGL in relation to the topography for the situation that will be examined in Example 11-1.

An easier and more intuitive method of plotting the HGL is to start at the upstream water surface (i.e., at the source) and then subtract head loss as you go downstream. There are minor losses in the inlet to the pipe at the source (term *a* in Fig. 11-1), then friction losses along the pipe. When pipes are joined, the head at the downstream end of the upper pipe is the head available at the upstream end of the lower pipe. This process can be continued along the entire pipeline. The slope of the HGL will be steeper as D is decreased or

Table 11-3. Minor Loss Coefficient (K) Values for Various Fittings

Fitting Shape	Description of Fitting	Representation of Pipe Geometry	Minor Loss Coefficient	Minor Loss Coefficient
Pipe Entrance		r/d		
		0.0	$K = 0.50$	
		0.1	$K = 0.12$	
		> 0.2	$K = 0.03$	
Sudden Contraction		D_2/D_1	K for $\theta = 60°$	K for $\theta = 180°$
		0.0	0.08	0.50
		0.20	0.08	0.49
		0.40	0.07	0.42
		0.60	0.06	0.27
		0.80	0.06	0.20
		0.90	0.06	0.10
Sudden Expansion		D_1/D_2	K for $\theta = 20°$	K for $\theta = 180°$
		0.0		1.00
		0.20	0.30	0.87
		0.40	0.25	0.70
		0.60	0.15	0.41
		0.80	0.10	0.15
90° Bend			$K = 1.1$	
Pipe Fittings	Globe valve—wide open		$K = 10.0$	
	Angle valve—wide open		$K = 5.0$	
	Gate valve—wide open		$K = 0.2$	
	Gate valve—half open		$K = 5.6$	
	Return bend		$K = 2.2$	
	Tee			
	straight-through flow		$K = 0.4$	
	side-outlet flow		$K = 1.8$	
	90° elbow		$K = 0.9$	
	45° elbow		$K = 0.4$	

Source: Crowe 1981. Reproduced with permission of American Society for Heating, Refrigerating, and Air-Conditioning Engineers via Copyright Clearance Center.

Q is increased because both situations will result in higher velocity and therefore higher head loss, as can be seen in Eq. 11-2.

Using the energy equation (Eq. 11-4), it is possible to plot the head at all points along the pipe, thereby allowing any points of excessively high or low head to be detected for later correction. (When determining whether the head is too high or too low, the pressure head is of concern. Visually, this is the elevation difference between the hydraulic grade line and the ground surface.) As stated above, low-pressure regions may not be able to provide the needed discharge, and high pressures can burst the pipe. This situation is illustrated later in Example 11-1.

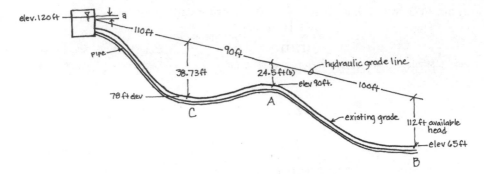

Figure 11-1. Topography and Hydraulic Grade Line (HGL) from a Water Storage Tank (at 120 ft Elevation) to a Water User (at 65 ft).

Note: The drop in the HGL represented by the letter *a* after the water tank is the minor loss caused by the pipe inlet.

Each type of pipe can withstand a different amount of pressure (or head). This maximum allowable pressure for a certain type of pipe can be found from the pipe manufacturer and is provided in Table 11-4. If you cannot find this information, then an approximate value to use is 150 ft of head for polyvinyl chloride (PVC) pipe and 600 ft for galvanized iron pipe. These are low estimates. Actual values depend on the specific material, wall thickness, and operating temperature.

))) 11.2 Components of a Gravity-Fed Water Supply System

A gravity-fed water system consists of many components as described in detail in this section. The hydraulic design process is demonstrated later in Example 11-1.

Table 11-4. Standard Values for Typical Pipe Materials

Material	Hazen-Williams C	Pressure Limit (Bars)	Pressure Limit (m of Head)
PVC (Class B)	150	6	61
PVC (Class C)	150	9	91
PVC (Class D)	150	12	122
PVC (Class E)	150	15	125
Polyethylene (medium density)	150	12.5	127
Polyethylene (high density)	150	16	163
Iron tubing	120	40	407
Prestressed concrete	120	20	204

Notes: 1 bar equals 10.197 m of water. C is the Hazen-Williams roughness coefficient.

Source: Purcell 2003; Menon 2005.

))) **Box 11-1. Eight Items to Consider When Determining Whether a Gravity Flow Water Project Is Feasible**

1. Ensure that water at the source does not appear turbid, does not have an odor, does not carry a lot of sediment, and is not contaminated with pathogens or agricultural chemicals.
2. The owner of the land where the source is located must be trustworthy and must provide written permission to use the source. Educate the owner on how his or her activities could influence the water source.
3. For gravity systems, locate the community houses at an elevation below the source.
4. The minimum flow of the source (during the dry season) must cover the needs of the community or it must be supplemented with other sources of water.
5. If the source is located below the community, a pump system will be required, and thus the community needs access to power (human, renewable, or fossil fuel) to lift the water. Also consider the availability of parts to repair a pumping device.
6. There needs to be a suitable place for a storage tank located above the community.
7. There must be funds available from nearby organizations, government, and community members to construct the system initially.
8. The members of the community must show interest and be willing to provide labor and the funds to construct and maintain the system.

11.2.1 The Source

Communities that build gravity-fed water supply systems typically begin by identifying an adequate and permanent source of safe water in an area of higher elevation than the community. This water could be surface water collected behind a dam or groundwater collected by means of a spring box. Water at the source must not appear turbid, have an odor, carry a lot of sediment, or be contaminated. If the flow of the source does not cover the needs of the community during the dry season, it will need to be supplemented with other sources of water. The owner of the land where the source is located must be trustworthy, understand how his or her activities could influence the water source, and provide written permission for use of the source.

Dams are used when the available water source is a stream or river. The advantage of constructing a dam is that during the rainy season, the water will easily pass over the structure and not damage it. The main disadvantage is that this water source is more susceptible to contamination and generally carries more sediment than the groundwater of a natural spring. Dams can be built out of many materials, depending on availability in the region.

Sedimentation tanks (Fig. 11-2) are often needed when using dammed surface water, particularly because of the high turbidity of water during the rainy season. The tank is built to slow the flow of the water, causing particles to settle out by gravity.

For the tank to work properly, a plumber must periodically open the clean-out valve to release sediment in the bottom of the tank. Sediment in the conduction line can cause obstructions and unnecessary wear along the inner walls of the pipe. A sedimentation tank also serves as a break-pressure tank (discussed later), because the pressure is returned to atmospheric pressure. However, they are typically placed near the source to remove sediments as early as possible so as to minimize damage to pipes caused by suspended particles.

11.2.2 The Conduction Line

A conduction line transmits water from the elevated source to a storage tank. The tank then stores the water accumulated during periods of lower demand for use during periods of higher demand. Pipe material is typically PVC or galvanized iron. Table 11-5 compares the advantages and disadvantage of each and provides an explanation of where they are typically used. As this table shows, the choice of pipe material affects the roughness of the pipe and therefore the head loss because of friction.

Water velocity through pipe is typically maintained between 0.5 and 3 m/s. If the velocity is lower than 0.5 m/s, suspended solids in the water may settle and collect in the pipe. This slowness can increase head loss and lead to clogging in the pipe. Sediment clean-out valves (or washouts) that are strategically placed at low points allow removal of accumulated sediment; however, a lapse in maintenance will lead to problems. If a low velocity cannot be avoided, frequent line flushing needs to take place, or a sedimentation tank may have to be installed at the beginning of the pipeline, or the spring box can be sized to provide for sedimentation. If the velocity is greater than 3 m/s, the interior of the

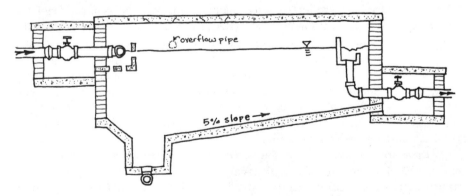

Figure 11-2. Cross-Sectional View of a Horizontal Settling Tank and Its Components.

Source: Redrawn from SANAA (1999).

Table 11-5. Typical Pipe Material Sizes, Advantages, and Disadvantages

Pipe Material	Lengths and Types[a]	Advantages	Disadvantages	Typically Applied
PVC pipe (polyvinyl chloride)	Typically 10- or 20-ft lengths 1/4"–4" diameter	Low cost Flexible, easy to place in a trench Also available in long spiral sections, which require fewer joints Modification by cutting or burning Smoother, therefore causes less head loss and allows a smaller diameter to be used	Limited allowable pressure Possible damage from blunt force Easier to vandalize Becomes brittle when exposed to sunlight for an extended time	Lower pressure areas Areas not vulnerable to impact
GI pipe (galvanized iron)	Typically 20-ft lengths 1/4"–4" diameter	Very durable Allowable pressure is much higher	Higher cost Added labor to bend or modify Additional tools required to install Requires pipe threading Can experience corrosion and blockage due to calcium deposits	High-pressure areas Stream or gorge crossings and other locations where pipe is exposed Another option for this scenario is to have a larger diameter PVC pipe surrounding the smaller diameter pipe that actually carries the water, to protect the pipe.

[a]Subject to specific availability in each country.

Table 11-6. Pipeline Flow Rate Design Limits (*L/s*) for Different Pipe Sizes

	Pipe Diameter, mm					
	25	32	40	50	80	100
Minimum	0.35	0.60	0.90	1.4	3.5	6.0
Maximum	1.4	2.0	3.5	5.0	NA	NA

pipe can be eroded, leading to increased frictional losses and a reduction in the design life. A larger pipe size reduces the velocity and also leads to lower frictional losses. Table 11-6 provides velocity limits for different pipe sizes.

Pipe sizing is based not only on survey data but also on peak flow from the water source. A conduction line is typically designed to carry the maximum daily flow from the source to the storage tank. The tank then acts to store the water carried during the periods of lower demand for use during periods of higher demand. Pipe diameters can change going from one pipe segment to another. Pipe sizes usually decrease as a pipeline progresses downstream from the first pipe segment when decreasing discharge is required (Box 11-2). Include the head loss in the contraction fitting from a large to smaller diameter pipe in calculations. Table 11-3 provides the coefficient to be used for contractions.

Pipe size can be calculated by hand if necessary (Reents 2002; Simpson 2003). Software programs such as *GoodWater* (Good 2008) are also available to assist the design of gravity-fed water systems. *GoodWater* is unique because it also incorporates principles of sustainability. A copy can be obtained by contacting this book's lead author, James R. Mihelcic. Commonly used pipe diameters are 0.5, 1, 1.5, 2, 3, and 4 in. Larger, more expensive pipes are rare in rural water projects. GI pipe is usually sold in 6-m (20-ft) sections, and PVC pipe is usually solid in 19-ft sections. The actual pipe inner diameter is slightly different than the nominal size.

⟫⟫⟫ Example 11-1. Using the Energy Equation to Solve for Pipe Size

Suppose the discharge from a pipe by a group of users is 0.7 ft³/s. We want there to also be at least a small amount of head available everywhere in the pipe, 10 ft, for example. This measure allows for sufficient flow and takes care of possible errors (e.g., in the land survey). PVC is readily available in this instance, so it is specified as the pipe material. Water enters the pipe from an inlet tank in which the water surface is 5 ft above the ground level. (We are designing the pipe that comes after the storage tank, the distribution line. The same process would also be used for the conduction line.)

The task here is to choose a suitable pipe diameter and then plot the head along the pipe to make sure that all points have sufficient pressure to allow the design flow rate (0.7 ft³/s) to pass, where no points have pressure high enough to cause the pipe to burst. The ground levels are shown in Fig. 11-1. There is a high point in the ground 200 ft downstream of the intake (Point A), and the end of the pipe (Point B) is another 100 ft from the high point. There will be a small amount of head loss at the entrance (indicated as *a* near the storage tank in Fig. 11-1), but that will be considered later.

The allowable head loss because of friction in the pipe until Point A is calculated by taking the water surface elevation at the source (120 ft above mean sea level) and subtracting the ground elevation at Point A (90 ft) but then adding in the 10 ft of head assumed to be available as a safety factor. This step results in a head loss (h_L) because of friction of 20 ft. Equation 11-2 is then solved for D, using $h_L = 20$ ft.

$$D = \sqrt[5]{16f\frac{LQ^2}{2g\pi^2 h_L}} = \sqrt[5]{16(0.008)\frac{(200 \text{ ft})(0.7 \text{ ft}^3/s)^2}{2(32.2 \text{ ft}/s^2)\pi^2(20 \text{ ft})}} = 0.25 \text{ ft} = 3.0 \text{ in}$$

>))) **Box 11-2. Community Understanding of Pipe Size**
>
> You may observe that some individuals do not understand why the diameter of the pipe changes as the distribution system gets closer to the community. These individuals may suggest, for example, using 2-in. pipe on the entire system instead of a smaller size pipe. They may initially look in disbelief when you explain that the same pipe size will not be used for the whole transmission line.
>
> In this situation, the individual believes that the smaller 1.5-in. pipe is too small to provide water for the entire community and the system will run dry. They may visually show the difference in size of the diameters of the two pipes with their thumb and finger and explain how the larger, 2-in. pipe will carry more water. Their belief may be based on an observation that a larger river carries more water than a small stream. Careful planning and integration into a community will allow you to earn the trust and respect of community members so that you can then explain to them the fundamentals of fluid mechanics and the design in their terms.

The idea here is to find a value of pipe diameter D that satisfies both Eq. 11-3 and Table 11-2. The value of $f = 0.008$ was guessed because we do not know D, which is required for the k_s/D ratio listed in Table 11-2. We will now check the value of f by using the value of D (0.25 ft) that we got when using Eq. 11-3.

$$\frac{k_s}{D} = \frac{0.0001 \text{ mm}}{(308.5 \text{ mm/ft})(0.25 \text{ ft})} = 1.3 \times 10^{-6}$$

This value of k_s/D is less than the lowest value (0.00001) in Table 11-2. In this case, the value of f could be lowered, but we don't know how much because it is below the range of data used to make Table 11-2. Maintain the value of $f = 0.008$ to be conservative. A high value of f results in more head loss, so if the water will flow with a higher head loss, then it will surely flow if the real head loss is less than we calculated.

If this value of k_s/D hadn't matched, then we would have needed to use the value of D obtained to get another value of the friction factor, f, from Table 11-2, calculate D again and check if k_s/D matches that of Table 11-2. This process is continued until Table 11-2 and Eq. 11-3 are both satisfied. It may take three or four iterations of this procedure to converge on an answer.

If the resulting diameter is available as commercial pipe, we can stop here. If it is not commercially available, we shall round up to the next largest commercially available pipe size.

Let's assume that 3-in. pipe is not available and that the next largest size is 4 in. This would lead to a head loss of

$$h_L = 16f\frac{LQ^2}{2g\pi^2 D^5} = 16(0.008)\frac{(200 \text{ ft})(0.7 \text{ ft}^3/s)^2}{2(32.2 \text{ ft}/s^2)3.14^2(0.33 \text{ ft})^5} = 5.0 \text{ ft}$$

Rounding up to a 4-in. diameter reduced the head loss because of friction from about 12 ft down to 5 ft.

Now we can consider the head loss at the inlet calculated from Eq. 11-3, with a value of 0.5 for K from Table 11-3, assuming a square-edged entrance with $r = 0$.

$$h_{L(\text{minor})} = 16K\frac{Q^2}{2g\pi^2D^4} = 16(0.5)\frac{(0.7\,\text{ft}^3/s)^2}{2(32.2\,\text{ft}/s^2)3.14^2(0.33\,\text{ft})^4} = 0.52\,\text{ft}$$

The minor head loss is thus 0.52 ft, which corresponds to *a* in Fig. 11-1. The total head loss is the summation of the frictional head loss and the minor head loss:

$$H_{L\text{-total}} = 5.0\,\text{ft} + 0.52\,\text{ft} = 5.52\,\text{ft}$$

Using the energy equation (Eq. 11-4) with the water surface in the tank as the upstream point and Point A as the downstream point, we can simply subtract the total head loss, 5.52 ft, from the water surface elevation. This equation yields a value of 120 ft − 5.52 ft = 114.48 ft. This value is above the ground level at that point by a value of 114.48 ft − 90 ft = 24.48 ft, which is above our minimum specified value of 10 ft, so the correct discharge can go through this pipe.

If this value were not above the minimum value, then we would have had to increase the pipe diameter by one pipe size and check again. We would continue to increase the pipe size until we found the smallest commercially available pipe that would leave at least 10 ft of head at all points along the pipeline.

To design the pipe size for the downstream section of pipe, we repeat the process, but using a head of 114.48 ft as the upstream value. If 4-in. diameter pipe is maintained all the way until the downstream end of the pipe, the hydraulic grade line at the end of the pipe can be calculated by subtracting the frictional head loss from the value of head at Point A. The frictional head loss can be calculated as follows:

$$h_L = 16f\frac{LQ^2}{2g\pi^2D^5} = 16(0.008)\frac{(100\,\text{ft})(0.7\,\text{ft}^3/s)^2}{2(32.2\,\text{ft}/s^2)3.14^2(0.33\,\text{ft})^5} = 2.5\,\text{ft}$$

Therefore, the value of the hydraulic grade line at the end of the pipe is 114.48 ft − 2.5 ft = 112 ft. Because the ground elevation here is 65 ft, that means the head is 112 ft − 65 ft = 47 ft. This 47 ft of head is now the inlet value for any pipe(s) attached here that continue downstream (not shown). Additional pipes can be calculated in the same manner, always using the outlet value from the pipe upstream as the inlet value of the pipe being calculated.

The maximum pressure occurs at the location where the difference between the hydraulic grade line and the ground elevations is the largest. This location occurs at Point B in Fig. 11-1. This 47 ft of head is not more than the maximum allowable pressure for PVC (150 ft), so the pipe will not burst.

However, if the pressure is too high at any point in the pipe, you can decrease the pipe diameter upstream to increase the amount of head loss or you can put galvanized iron pipe in the spots with excessively high pressure. An additional option is a break-pressure tank, discussed in Section 11.2.3. If a smaller diameter pipe or a break-pressure tank is used, make sure you have enough head to get the water back up any hill downstream of this point. If the head is greater than 600 ft, the maximum allowable pressure for galvanized iron pipe, then you must decrease the pipe diameter(s) upstream. If a large amount of pressure is needed to cross a valley and then carry the water back up the other side and no pipe is strong enough to withstand the pressure, the pipe can be suspended across the valley.

》》》

11.2.3 Break-Pressure Tanks

Static head is the difference in elevation from the source to any point on the ground profile. Static head must always be considered because excessive pressure in the pipe from elevation head can rupture the pipe or pipe connections. Piping is rated by its ability to withstand an internal pressure without breaking. This ability is a function of the type of pipe material and the thickness of the pipe wall. Table 11-4 provides information on the pressure limits of particular piping materials.

Break-pressure tanks are strategically placed along the pipeline to eliminate excessive pressure that will rupture the pipe or cause failures at the joints. The function of a break-pressure tank is to allow the flow to discharge into the atmosphere, thereby reducing its hydrostatic pressure to zero, and establishing a new static level. Figure 11-3 depicts a break-pressure tank and its components. In Fig. 11-3, as the entry pipe turns vertically downward, there are holes placed in its sides to help dissipate the pressure and protect the inside of the box from damage. A storage tank also functions like a break-pressure tank.

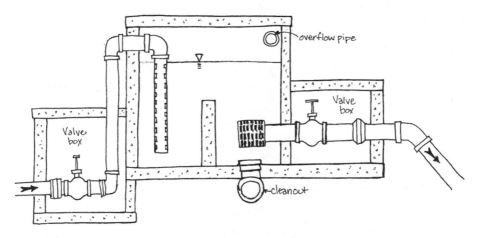

Figure 11-3. Cross-Sectional View of a Break-Pressure Tank and Its Components.

Note: Float valves can be installed to prevent the overflow of water.

Source: Redrawn from SANAA (1999).

A general rule of thumb is that static head at any given point within the conduction line (with a good factor of safety) should not exceed 100 m of head. This limit requires that a break-pressure tank be installed approximately every 100 m of change in elevation from the source yet still leaving enough pressure to flow to the storage tank.

There is no minimum required capacity for a break-pressure tank, as long as water is able to drain from it as quickly as it is discharged. The dimensions of the tank are influenced more by the size of the fittings (such as control valves or float valves) that must fit inside it and the size of the pipe wrenches, which must be able to be rotated inside as well (Jordan 1980). When adding a break-pressure tank, similarly to when reducing pipe size, it is important to check that the head that will be lost is not needed farther downstream to bring the water back up to a localized high point.

Figure 11-4 shows what happens to the HGL of a system when a break-pressure tank is added to the system and the frictional pipe loss is plotted over the ground profile. Proper location of break-pressure tanks is an important part of a successful design. The

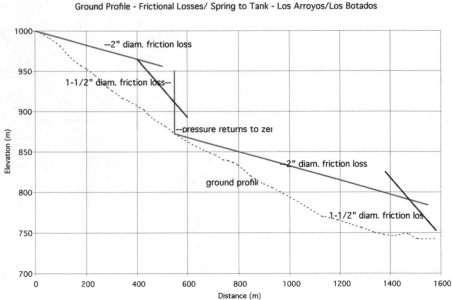

Figure 11-4. Pressure Head Plotted over Ground Profile (Determined by the Topographic Survey) to Create the HGL.

Note: The top solid line shows the frictional head loss, which is determined by the source flow and the size of pipe. Here the difference in head from the source to the storage tank is approximately 260 m, therefore one or two break-pressure tanks might have been recommended. In the real scenario depicted here, only one break-pressure tank was recommended for the midpoint of the elevation change. This decision was made considering issues of land use and project costs. The static head at the midpoint is 130 m, which is greater than the 100-m rule of thumb mentioned in the text. However, if PVC RD17 pipe was used, the allowable pressure would be 176 m (Table 11-4), so the system should not fail from pipe rupture (Niskanen 2003).

ideal placement is a location where the exiting flow will have an immediate drop in elevation to regain energy. In general, the top of a hill is a good location for a break-pressure tank, as long as the next high point along the conduction line is at a lower elevation.

There is sometimes a float valve (similar to that in a western-style toilet) located inside the break-pressure tank that helps regulate flow inside the tank. In this case, the float valve shuts off when the break-pressure tank is full (meaning that water is not being used downstream). However, if the float valve does not shut off, water is lost in the overflow tube. Most float valves have much lower pressure head limits (around 60 m of head) than standard tubing materials; if you are installing break-pressure tanks, double-check the final pressure at the tank entrance. The other option is that the break-pressure tank will overflow when the water demand is not great, which is allowable when the water source provides more water than necessary. Placing rocks or something hard where the water falls will disperse energy and avoid erosion at this spot. Float valves break fairly often. Therefore, it is important to minimize the number of break-pressure tanks and make sure that the water committee has replacements and knows how to install them.

11.2.4 Clean-Out Valves and Air-Release Valves

Sediment and air in the pipe are the two most common obstructions found in water systems that result in partial or complete blockage of the pipeline. Figure 11-5 shows a typical profile of a water supply system and the proper placement of air-release and sediment clean-out valves. Air-release valves are best installed at high spots along the conduction

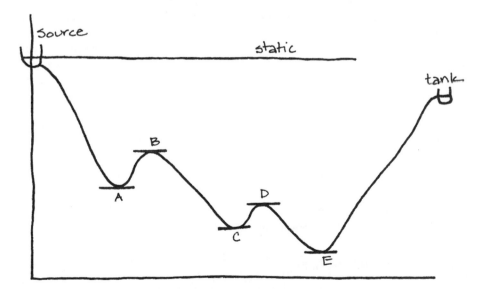

Figure 11-5. Locations of Sediment Clean-Out Valves (Points A, C, and E) and Air-Release Valves (Points B and D).

Source: Adapted with permission from Jordan (1980)

line, marked here as B and D. Not all high points will require air-release valves, however, and once the system is completed, air valves can be installed starting from the source, working downward until air blockages are not a problem. Install sediment clean-out valves at low spots along the pipeline, which are marked as A, C, and E in this figure. During construction, install sediment clean-out valves at the low points in the terrain and the air-release valves at the low points of the terrain. This installation may require that a section of pipe be cut.

The minimum size for sediment clean-out valves is typically a 1-in. gate valve. For larger main lines, a relationship of $D/3$, where D is the inner diameter of the main line, can be used to calculate the size of the required valve.

The purpose of an air-release valve is to release this trapped air. Because air is less dense than water, it sometimes is retained in the higher areas of the pipeline and thus blocks the movement of water at the design flow. The location of air-release valves is especially important where the hydraulic grade line is close to the level of the terrain. Air-release valves that work by a spring attached to a ball that allows air to escape if it builds up to enough pressure to push the ball away from the opening and seal itself again when the pressure reduces are available commercially. Blockages because of air typically occur after pipes have been emptied for cleaning or service. Standard, nonautomated valves can also be installed if the cost of automatic air-release valves is too high or if they are not available in the area. In this case, the people servicing the pipeline open all the air-release valves while walking up the pipeline to the area to be cleaned. After the work is completed, these people can return the same way, closing the air-release valves manually once water has arrived at each location and all air has been forced out. These valves are typically enclosed in a concrete box to provide easy access and prevent damage by humans and animals.

Installing too many clean-out and air-release valves not only increases the cost but also makes the system more susceptible to leaks and vandalism during the operation of the system. Financially, the engineer must review the cost effect of adding clean-out and air-release valves because they can become a very expensive portion of the system cost. The maintenance and upkeep of these types of valves are important to prolonging the operating life of the system.

11.2.5 The Storage Tank

Storage tank design is covered in Chapter 14. Locate the tank to allow for at least 10 m of head at all points in the distribution network (some people suggest that this value can be 5–10 m of head). It is also preferable that it be located close to the community for easy maintenance. Because storage tanks act as break-pressure tanks, locate them where the exiting flow will have an immediate drop in elevation to regain energy.

11.2.6 Distribution Lines

Distribution networks consist of the pipe and accessories required to connect the storage tank to the users. Design distribution lines similar to the conduction line, as in the process described in Example 11-1, or use software such as *GoodWater* (Good 2008), which was mentioned earlier.

Branched Networks

A *branched system* is one in which there are no loops in the network. In other words, water can reach each tap stand by only one path. Many water systems in the developing world are branched systems, except perhaps in large cities.

The flows required for any pipe in a branched network (not a looped one) are determined by the user demand at the end of that pipe. The flow of any pipe coming into a junction is the sum of all the flows going out of that junction. Figure 11-6 illustrates this point.

The hydraulic grade line of each pipe can be determined using the value at the downstream end of the pipe just upstream of the pipe in question as the new upstream head value. HGL calculations proceed from the source to the end of each branch line once the pipe diameters have been selected.

Looped Systems

If there is a tap stand in a system that water can reach by more than one path, then the system has one or more pipe loops. This system provides the advantage of adding

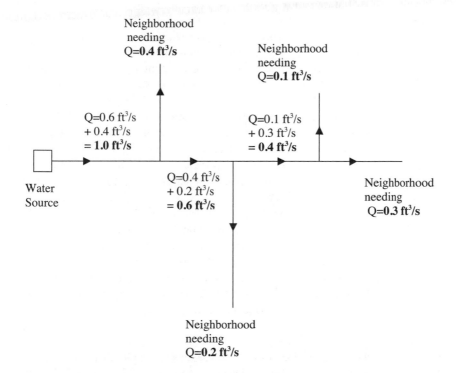

Figure 11-6. Determination of Flow Rate in Each Pipe of a Branched System.

Note: Determine neighborhood needs by adding individual user needs. In this system, the flow rates would be calculated by starting at the right and working backward to the left.

redundancy to the design. If there is a blockage or break in a pipe, water can still go to the downstream taps through an alternate path. It adds cost to the design, however, because of the need for more piping. It also makes it harder to calculate the flow and pressure at points throughout the system. For analyzing looped networks, there are computer programs as well as the classical Hardy–Cross method, which can be done by hand (although it takes a while) or a spreadsheet (Crowe et al. 2001).

11.2.7 Tap Stands

Tap Stands

Water is distributed from the distribution line to a user via a tap stand. Base tap stand design on community needs, especially the type of container commonly used for collection. Also consider what activities (e.g., washing dishes, laundry) will take place near the tap stand in addition to water collection.

Typically, one-half-in. PVC piping is used to connect each tap stand to the distribution line. Piping within the community is not buried as deep as in the main conduction line. It is usually 60 cm deep for normal terrain and 80 cm deep to cross roads. In areas of erosion, GI pipe may be used to cross roads.

A water distribution system can be based on community tap stands, or separate tap stands can be provided for each participating household. The collection and sanitary disposal of greywater is a design consideration as well. Figure 11-7 shows the components of a typical tap stand. Figure 11-8 depicts various types of tap stands, and Fig. 11-9 shows the integration of greater collection, drainage, and shutoff valves. Concrete skirts not only inhibit erosion at the tap stand but can allow a user to direct excess water to nearby gardens.

A larger PVC pipe (approximately 10 cm in diameter) that is cut in half lengthwise and held together with hose clamps or bailing wire can work as a mold for the taps shown in Fig. 11-7 and the top illustration of Fig. 11-8. For this particular design, the piping and fittings that exit the top and bottom of the tap stand are placed inside the mold, which is filled with concrete and then cured. Encasing the tap standpipe in concrete protects it from large animals, children, and carts.

A minimum head of 10 m pressure is recommended at the tap to ensure the delivery of sufficient flow from the tap. When determining the minimum head required at the tap, it is important to consider the frictional losses at the elbows and gate valves shown in Fig. 11-7. Standard taps purchased at a hardware store can generally withstand approximately 60 m of head before they are damaged. If the static head is more than 60 m before the placement of community tap stands, a break-pressure tank with a float valve or a break-pressure valve can be installed (Reents 2003).

Valves

Valves are typically placed outside of each house, enabling the plumber to control access to the system. Enclose the valves in a valve box, which may be locked so that only the plumber can adjust them. If the water fee is not paid, water supply can be terminated. Lockable faucets can also be purchased. Locate valve boxes on public property close to the road for ease of access. Install additional valves to isolate sections of the water system

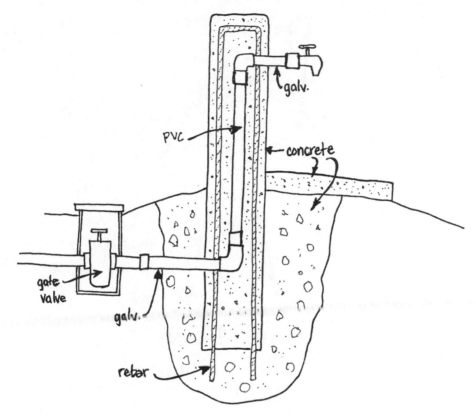

Figure 11-7. Components of a Typical Tap Stand Design.

for repairs. With strategically located valves, access to the system can be given for part of the day to each section of the town. Close valves slowly to prevent damage from water hammer.

11.2.8 Other System Components

The components discussed in previous sections form the basic parts of a water system. Other nontraditional items may also be included. Consider other components to improve the quality of the water in the system. These components can add significant cost and maintenance constraints to a system, therefore, carefully consider their additions. Installing aeration units in the line can improve the taste and smell of water. These are simple tanks where water enters from above and falls through the air to an outlet at the bottom of the tank.

An aeration unit will not remove pathogens. In this case, different filters may be installed along the pipeline. A simple rock or sand filter can be installed at the intake if water is of poor quality. These filters will require maintenance, though, and cannot be guaranteed to remove pathogens. Slow sand filtration can also be used to improve water

Figure 11-8. Tap stands with Concrete Post and Concrete Erosion Protection (Top), Concrete Post and Underground Pipe to a Soak Pit to Minimize Erosion (Middle), and Unattached Wooden Post and Large Stones to Provide Erosion Protection (Bottom).

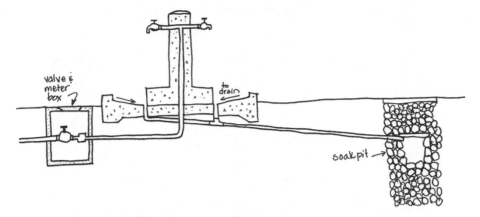

Figure 11-9. Tap Stand Design Integrated with Greywater Evacuation, Drainage, and Optional Shutoff Valves and Meters.

Source: Redrawn with permission from Pickford (1991).

quality. These filters can require large amounts of space and can also be complicated to install. Water may also be treated at the home with point-of-use filtration systems.

To manage water in the network, splitter boxes can be installed. Installing these boxes can be much more cost effective than valves, and they can more precisely control the distribution of water. They consist of a simple box where water enters, flows over a partition, and exits into different tubes. These boxes are particularly important where water must be divided into specific fractions, for example, when one community must receive exactly half or a quarter of the water from a source. The partition divides the box into two halves, and the half where the water exits may be subdivided into as many sections as the designer requires. The placement of these subsections corresponds exactly to the ratio of water that that subsection is to receive. For instance, if the user wants the water distributed into three pipes, one with 50% of the water and two with 25% of the water each, install a splitter box with three subsections (Fig. 11-10). Arrange the partitions of the subsections so that water flows over the main partition in the correct ratios.

In addition to the standard tap stand, various other types of extraction points can be included in a water system. Special needs may need to be addressed at locations such as schools, churches, and community centers. Public bathing or showering areas may also be a priority in the community. A specially designed area for washing clothes that has multiple faucets can also be considered in communities that have a traditional washing area.

))) 11.3 Construction Tips

Typically, the community digs the trench and installs the pipeline. They will bury it about 1 m below ground because it may cross agricultural fields that are plowed on a regular basis or that may be burned to clear brush. The erosion of soil on hilly terrain can also expose piping that has not been buried deeply enough. In addition, burying the conduction line reduces illegal tapping for household or agricultural use. One-meter sticks can be

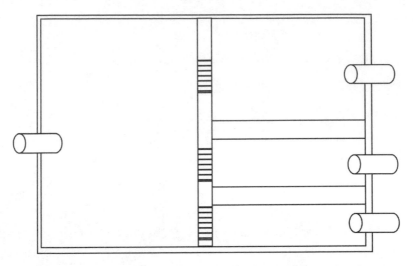

Figure 11-10. Splitter Box That Divides Water into 50%, 25%, and 25% Pipes.

cut and provided for workers to measure consistent depth along the trench. Small marks can also be notched on picks and shovels to identify trench depth.

Connect PVC pipe as described in Chapter 7. Glue PVC pipes so that the bell end (i.e., the female end) is positioned against the flow of water to prevent future leakage. Do not accidentally introduce sediment into the pipes during joining. Take care when connecting pipes of different materials, especially PVC to galvanized iron. These areas are problematic because the connection sleeve is often short. Glue a longer section of PVC the night before into the adaptor so that the connection is stronger.

Experience suggests that 20 to 30 pieces of 6-m PVC pipe can be laid through rocky soil during a normal workday with a work brigade of 15 people. The count can increase to approximately 40 pieces of pipe if the soil is sandy and free of rocks. When installing sections of galvanized iron piping, install universal unions every 8 to 10 sections of pipe. Having an ax available to cut through roots and narrow shovels as well can reduce the time required to dig a wide trench. Box 11-3 gives a typical water project construction timeline.

))) 11.4 Operation, Maintenance, and Security

The operation and maintenance of a water supply system is vital to its long-term sustainability. The most important asset of a well-operated water supply system, aside from its design, is the formation and management of a community water committee (see Chapter 3 and Schweitzer 2009). This group of elected individuals has the task of ensuring that the water system will last as long as possible. They collect fees to maintain the system, determine what to do if users do not pay their fees, and arrange necessary system repairs and maintenance.

The second most important asset of operating and maintaining a gravity-fed water supply system is to have several trained plumbers who reside in the community. Training local plumbers starts during the project's conceptual design and begins in earnest on the

>>> **Box 11-3. Typical Water Project Construction Timeline**

Water supply projects require coordination of laborers, materials, weather, and other factors, and the order of construction activities can affect progress. Below is a suggested project plan and justification for its ordering.

1. First Third of the Main Pipeline
Starting with the conduction line is easier than starting with a complicated intake structure, which often necessitates hauling cement and other heavy materials. Use this time to organize labor brigades and work out any problems. Connect pipes to bring water where people are working.

2. Intake Structures
Interest in the project will be highest near the beginning. Take advantage of this time to work on sections farthest from the community.

3. Finish Main Pipeline
Once the conduction line is finished, water can be brought to the tank location. This step will excite community members and provide further motivation.

4. Storage Tank
Working on the storage tank once the conduction line has been completed allows water to be available to mix concrete without hauling water long distances.

5. Entire Distribution Network
The distribution network will proceed rapidly as community members are working near their homes and can see daily progress.

6. All Tap Stands
Connect all tap stands at the same time. Workers are liable to stop working on the project once water arrives near their own homes. To be fair to all community members, ensure that all tap stands are installed over a short time frame.

first day of construction. During the construction, these individuals learn how to install and replace every element of the system. Additionally, they need to be trained to
- walk the conduction line;
- inspect the line for leaks;
- release air from the system, if present;
- clean sediment valves;
- inspect and clean an emptied storage tank once a year;
- maintain bleach in a system chlorinator, if required; and
- understand how to purchase parts that need replacement or repair.

Compensation for these individuals is a decision made by the community water committee and can be monetary or in kind (e.g., rice, beans). Box 11-4 shows some common water myths.

⟫⟫ Box 11-4. Common Water Myths

Decreasing pipe size increases pressure at the tap.	The energy equation (Eq. 11-4) shows that if you decrease the pipe size, there is more head loss, meaning that more pressure is burned off because of friction by the time the water reaches the end of the pipe.
If there is not enough pressure at the tap, then the control valve just has to be opened more.	Opening the valve may help, but if there are other causes of head loss (e.g., too small a pipe, blockage by sediment), then you still won't get enough flow.
Water does not flow uphill.	Water can flow uphill if there is enough pressure to push it up the hill. This situation can be seen in the energy equation (Eq. 11-4) by solving for all of the pressure terms on one side of the equation. All that is left on the other side of the equation is head loss and elevation difference. If there is enough pressure difference to overcome the head loss and elevation difference, then water can indeed flow uphill. The needed pressure can be supplied by a pump or, for gravity systems, by a sufficient difference in elevation.
Having a sudden drop in the pipeline is not good.	Having a sudden drop is okay if the pressure stays below the pressure that would cause the pipe to rupture.

A final aspect of water supply system design is security. While designing a system, provide secure boxes or hatches to access the valves, tanks, and other places where a system has potential for tampering. The hatches on valve boxes and tanks, whether concrete or steel, can be designed with a padlock for security. This system also includes protecting vulnerable areas, such as the spring box or intake dam from animals.

⟫⟫ References

Crowe, C. T., Roberson, J. A., and Elger, D. F. (2001). *Engineering fluid mechanics*, 7th ed. John Wiley and Sons, New York.

Good, S. (2008). *Development of a decision support system for sustainable implementation of rural gravity flow water systems*, <http://cee.eng.usf.edu/peacecorps/Resources.htm> (Jan. 17, 2009).

Jordan, T. D., Jr. (1980). *A handbook of gravity-flow water systems*, Intermediate Technology Publications, London.

Menon, Shashi. 2005. *Piping calculations manual*. McGraw-Hill, New York.

Niskanen, M. A. (2003). *The design, construction, and maintenance of a gravity-fed water system in the Dominican Republic*, <http://cee.eng.usf.edu/peacecorps/Resources.htm> (Jan. 17, 2009).

Pickford, J. (1991). *The worth of water: Technical briefs on health, water and sanitation*. IT Publications, London.

Purcell, Patrick. 2003. *Design of water resource systems*. Thomas Telford, London.

Reents, N. (2003). *Designing water supply systems in rural Honduras*, <http://cee.eng.usf.edu/peacecorps/Resources.htm> (Jan. 17, 2009).

Schweitzer, R. N. (2009). *Community managed rural water supply systems in the Dominican Republic: Assessment of sustainability of systems built by the National Institute of Potable Water and Peace Corps, Dominican Republic,* <http://cee.eng.usf.edu/peacecorps/Resources.htm> (May 21, 2009).

Simpson, J. D. (2003). *Improvement of existing gravity-fed rural drinking water systems in Honduras,* <http://cee.eng.usf.edu/peacecorps/Resources.htm> (Jan. 17, 2009).

Servicio Autónomo Nacional de Acueductos y Alcantarillados (SANAA). (1999). *Normas de diseño para acueductos rurales V.1.0,* Tegucigulpa, Honduras.

))) Further Reading

American Society of Civil Engineers (ASCE). (1969). *Design and construction of sanitary and storm sewers,* ASCE Press, New York.

Annis, J. (2006). *Assessing progress of community managed gravity flow water supply systems using rapid rural appraisal in the Ikongo District, Madagascar.* <http://cee.eng.usf.edu/peacecorps/Resources.htm> (Jan. 17, 2009).

McConville, J. R. (2006). *Applying life cycle thinking to international water and sanitation development projects: An assessment tool for project managers in sustainable development work,* <http://cee.eng.usf.edu/peacecorps/Resources.htm> (Jan 17, 2009).

McConville, J. R., and Mihelcic, J. R. (2007). "Adapting life-cycle thinking tools to evaluate project sustainability in international water and sanitation development work." *Environ. Eng. Sci.,* 24(7), 937–948.

12

Increasing Capacity of Existing Gravity-Fed Water Systems

))) 12.1 Methodology for Problem Identification

A typical gravity-fed water supply project consists of the source, conduction line, storage tank, and distribution line (see Chapter 11). Problems encountered in these systems, their causes, and suggested solutions are listed in Table 12-1. Difficulties can be associated with environmental, mechanical, or behavioral issues.

Figure 12-1 provides a flowchart that can be used to identify problems in a gravity-fed water supply system. Even if only one problem is initially observed, analyze the entire system to correctly increase capacity. For example, an overflowing source may indicate that the conduction line needs to be improved. However, if the existing storage tank is also overflowing, improving the conduction line will make the situation worse. You may need to repeat the process shown in Fig. 12-1 if after the first attempt at increasing capacity, problems still remain.

A basic understanding of Chapter 11 will assist you in finding the correct method to increase capacity. For example, the most common error made when trying to improve existing systems is to use smaller diameter pipes in an attempt to increase pressure. However, a smaller pipe actually increases velocity, which leads to additional head loss and thus decreased pressure.

12.1.1 General Tank Observations

Some problems can be identified by investigating water levels in the storage tank. Interview community members, inspect the inside of the tank for water marks, and observe the water level for a 24-h minimum.

If the tank never overflows, then the troubleshooting process should continue to the next section. If the tank overflows intermittently, an increase in the tank volume will allow for the currently wasted water collected during nonpeak hours to be used during peak-hour demands. If the tank always overflows, then the problem lies in the distribution line.

John D. Simpson contributed to this chapter.

12.1.2 Tank Inflow Measurement

Use engineering judgment to determine the required water demand of the community. A value of 25 gal/capita-day (95 L/capita-day) is used in Fig. 12-1. A community analysis may prove beneficial in determining exact water demand. Include any productive uses of water, such as the irrigation of small gardens, that may have emerged in the community since the system was built.

Table 12-1. Typical Problems, Causes, and Solutions in Gravity-Fed Water Systems

Problem	Cause	Solution
Insufficient supply	Faulty catchment structure	Repair structure or build new structure
	Capacity of source diminished because of environmental factors	Reforestation, excavate source deeper
Piping system unable to deliver sufficient water	Undersized pipes or excessive leakage	Replace pipes or install additional pipes
	Blockage in pipes	Open clean-out valves, inspect pipes
Increased demand	Community development (population or businesses)	Educate community on water conservation
	Source is seasonally dry, causing increased demand on other water sources	Educate community on water conservation
	Misused water (lack of valves at each house)	Repair leaking pipes, replace valves

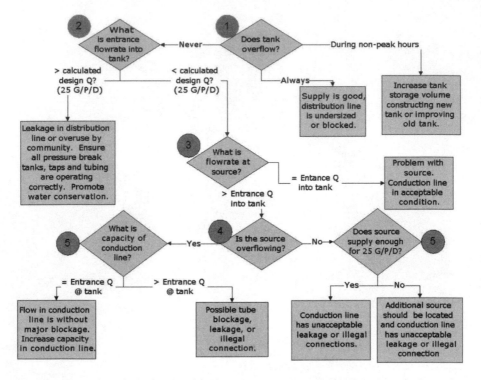

Figure 12-1. Flowchart for Troubleshooting a Gravity-Fed Water Supply System.

Source: Simpson 2003.

After the required demand is calculated, determine the flow rate into the tank. This measurement indicates whether enough water is being produced by the source and transported through the conduction line. The flow rate for small systems can be determined by measuring the time it takes to fill a 5-gal bucket. If the flow rate entering the tank is too large for the bucket method, close the outlet of the tank and measure the time required to fill the tank. Calculate the flow rate by dividing the volume of the container by the time required to fill it.

If the measured flow rate entering the tank is greater than the calculated demand, the problem lies within the distribution system. If the measured flow rate entering the tank is less than the calculated demand, then the troubleshooting process should continue to the source.

12.1.3 Source Flow Rate Measurement

To determine if the source is contributing to the failure or poor performance of a system, measure the flow rate at the source. If physically possible, as when measuring flow rate into the tank, this step is most easily accomplished by measuring the time required to fill a container of a known volume. Do this measurement at the end of the dry season because flow rate is the lowest at this time. If the community is experiencing problems in the wet season as well, then a flow rate measurement in the wet season can also be useful.

This measurement will also provide information on whether the conduction line needs to be improved. If the measured flow rate at the source equals the flow rate entering the tank, then the conduction line is in an acceptable state, given the current condition of the source. The source would have to be improved to see an increase in the amount of water delivered to the tank. If the measured flow rate at the source is greater than the flow rate entering the tank, then the troubleshooting process continues to the next section.

12.1.4 Overflowing Source Observation

During normal operating conditions, observe whether all of the water at the source enters the intake to the conduction line or if some of the water bypasses the intake. No immediate actions to improve capacity can be taken from this observation, which is used in the following section.

12.1.5 Conduction Line Analysis or Demand Analysis

Depending on the results of the previous section, a final analysis is made. This analysis corresponds with the bottom row of the Fig. 12-1 flowchart.

Source Overflowing
If the source is overflowing, calculate the capacity of the conduction line. This may involve performing a topographical survey (see Chapter 5) of the existing conduction line to determine elevation changes. Survey data, along with known pipe sizes and the locations of break-pressure tanks, can be used to calculate the hydraulic grade line (HGL) along the length of the water distribution system. The capacity of the conduction line is the maximum flow rate that can be achieved before the head drops below the required minimum at any point in the system.

If the calculated capacity of the conduction line is approximately equal to the flow rate entering the tank, then the conduction line is operating without major blockages or leakage. To supply more water to the users, the conduction line would need to be upgraded, as discussed later in the chapter. If the calculated capacity of the conduction line is greater than the flow rate entering the tank, then the conduction line is failing. Possible problems are blockage of the pipes with air or sediment, excessive leakage, or illegal connections.

Source Not Overflowing
If the source is not overflowing, determine if the supply flow rate is sufficient for the needs of the community. If the source does not supply the required demand, find an additional source or improve the existing source. If the source does in fact supply the required demand, then the conduction line is experiencing excessive losses or illegal connections that divert water and prevent it from entering the tank.

⟫⟫ 12.2 Source Improvement Techniques

12.2.1 Improvement of an Existing Source

The most widespread environmental circumstance that leads to a decrease in flow rate at the source is deforestation. In the future, climate change will become increasingly important in affecting flow in some parts of the world. Another common situation is the construction of other water catchment structures upstream of the source. Spring boxes can be improved by methods such as creating infiltration trenches upstream, reforesting the watershed, or excavating deeper to expose groundwater that was previously unused.

Over time, catchment structures may lose their ability to capture all of the available flow. Visually inspect the source to determine if some flow is bypassing the catchment structure. Often, improvements such as the placement of rocks and mortar to channel water directly to the catchment structure may be sufficient to increase the flow.

12.2.2 Addition of a New Source

If a community has access to a new source, perform an analysis to determine if the storage tank is large enough to store approximately 8–11 h of water from the combined sources. This is an approximation; for more detailed water storage tank sizing techniques, see Chapter 14.

If the tank cannot accommodate additional water without excessive overflowing, size a new conduction line to deliver only the quantity of water that the tank can handle. This method avoids spending resources that will not lead to a significant improvement in the system. Alternatively, loss of water from the tank during nonpeak hours may be considered acceptable to deliver a higher quantity of water for the times of the day when demand is highest.

A new source can be connected to the conduction line in a number of ways, but only after a careful hydraulic analysis of the proposed system. If the new source is located at a higher elevation than the old source, it can be made to flow directly into the original source. Analyze the conduction line to ensure that it has the capacity to carry the extra water.

Figure 12-2 depicts two methods to add a new source to an existing system when it is located at a lower elevation than the original source. The original system, before the addition of the new source, is shown in Fig. 12-2a.

It is possible to join the two sources with a "T" connection, as shown in Fig. 12-2b. However, in this example, water from the original source will flow through the "T" and back up to the new source. This backflow will occur because the HGL at the intersection is higher than the elevation of the lower source. If the upper source is flowing into the lower source, the storage tank will actually receive less water than it did before. Analyze systems on an individual basis to prevent this unfortunate situation.

In this case, the use of a junction box (Fig. 12-2c) prevents any backflow. The location of the new source may serve as the junction box if local conditions warrant. Because the junction box is open to the atmosphere, the HGL is lowered to the elevation of the junction box, and the capacity of the conduction line is decreased after that point. In other

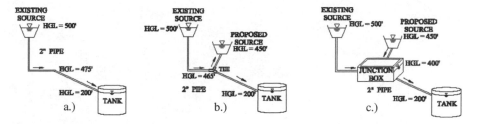

Figure 12-2. Example of Adding a Source to the Existing Source (a) That Backflows into the New Source When Joined by a "T" Connection (b), But Not When Joined with a Junction Box (c).

words, a junction box acts as a break-pressure tank. Therefore, analyze the conduction line to ensure that it would still have the capacity to carry sufficient water to the tank.

If the existing conduction line does not have the capacity to carry the combined flow, particularly after the HGL has been lowered, add a new section of pipe (Fig. 12-3). This example shows how existing components of a system can be integrated with new components to improve performance

Figure 12-4 shows a scenario similar to that in Fig. 12-2, but with the original pipe size decreased from 2 in. to 1 in. The HGL is therefore different at the point of intersection

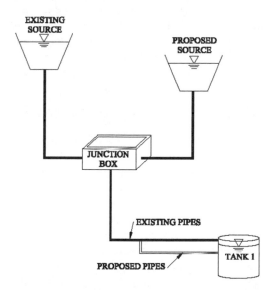

Figure 12-3. A New Source Can Be Added to an Existing System by Constructing a Junction Box to Unite Flows from Two Sources.

Note: In this example, an additional pipe (labeled "proposed pipes") was deemed necessary in the lower reaches of the conduction line before the tank to carry the additional flow from the new source. This solution is used partially because the junction box acts as a break-pressure tank.

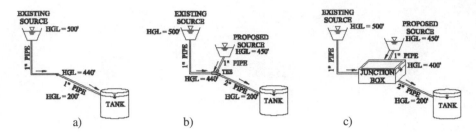

Figure 12-4. Example of Adding a Source Showing (a) Original Conditions, (b) Use of a "T" Connection, and (c) Use of Junction Box to Unite Flows.

of the two sources. In this case, the use of a "T" connection will prove to be beneficial (Fig. 12-4b). There will not be any backflow of water to the new source because the HGL at the "T" is lower than at the new source.

Figure 12-4c shows that joining the two sources with a junction box will also ensure that water flows from both sources. The pipe after the junction box will have an increased carrying capacity compared to Fig. 12-4a because of the increase in pipe size. In this example, the increased pipe diameter increased the capacity of the system, despite the energy lost at the junction box.

A third (and more costly) option is to construct a completely new, independent conduction line that starts at the new source and ends at the tank. This method ensures that there is no interference between the two sources. As can be seen from the extensive options available when adding a new source, an iterative design process must occur to ensure that water will flow in the direction desired.

))) 12.3 Conduction Line Improvement Techniques

A conduction line is typically sized to carry the maximum daily flow from the source to the tank for the estimated future population. The tank then acts to store the water carried during the periods of lower demand for use during periods of higher demand. It is possible that the conduction line is not sufficiently sized to carry the necessary amount of water, assuming the capacity of the line is not impeded by problems such as air or sediment deposition in the pipes or leakage along the line. An undersized conduction line can be diagnosed by a source that is overflowing. An undersized conduction line should not, however, be confused with material failure or construction practices in which losses within the pipes do not allow for a sufficient supply of water to the tank.

When the measured flow rate is compared to the theoretical flow rate, the problem can be pinpointed. If the measured flow rate equals the theoretical flow rate, it can be assumed that the conduction line is undersized. If the measured flow rate is less than the theoretical flow rate, then most likely the problem lies within construction practices, maintenance, or material failure. However, if the source does not supply at least a quantity of water equal to the theoretical capacity of the conduction line, then the observed flow rate entering the tank will be below the theoretical flow rate. The flow rate entering the tank should then be equal to the yield at the source.

If the troubleshooting process in Section 12.1 established that there are physical problems with the conduction line, open sediment clean-out valves in an attempt to flush out any sediment trapped in the system. Check air-release valves to make sure that they are functioning. Also inspect the system to verify that there are no illegal connections drawing water from the conduction line.

If it is determined that the conduction line is undersized, analyze the system to determine what measures are needed to increase the flow. This analysis is most easily accomplished by installing pipe with a larger diameter in some or all of the conduction line. New pipe may be installed next to the old pipe, and both can be used to carry water. This, of course, assumes that the old pipe is still in good condition.

Exercise caution in adding new pipes because the HGL will change in all sections, and you must avoid problems such as air blocks and unacceptable negative pressure. The installation of larger diameter pipes in the upper reaches of the conduction line versus the lower reaches will create a situation in which the HGL becomes flatter in the upper reaches and steeper in the lower reaches. This difference can be an advantage if you expect low or negative pressures but a drawback if the lower reaches are already under excessive pressure.

⟩⟩⟩ 12.4 Tank Improvement Techniques

The two parameters to observe at the storage tank are the water level and the entering flow rate. A tank is useful only if it is constantly changing levels, storing water in nonpeak hours and allowing for an increased flow to the community during peak hours. Measuring the flow rate entering the tank is necessary to ensure that it is the same as the flow rate measured at the source.

Tank size does not affect pressures within the conduction or distribution lines. A larger tank stores otherwise wasted water, allowing more water to be available during peak hours. It is common for the outflow of a tank to exceed the inflow of the tank for a long period of time, completely emptying the tank if it is not large enough.

If you determine that the tank is undersized, you can encourage storage of the water within the community, such as filling containers at each house during the night. As long as water is stored in clean, covered vessels, this can be an inexpensive and efficient method of increasing storage.

If an additional tank needs to be built, the two tanks should either have their overflows at the same level, or the lower tank should be equipped with a float valve that will stop flow when the tank is full. Without a method to stop the flow from the upper tank to the lower tank, the upper tank will serve no purpose. It will always empty to the lower tank, and the water will be lost in the overflow of the lower tank. Building a second tank with its overflow at the same elevation as that of the first is the preferred option because it eliminates the need for a float valve. The bottom elevation of the tanks is not as critical because the HGL is based on the elevation of the water surface. However, ensure that both tanks can drain completely if the bottoms are at different elevations, as described in the following sections.

12.4.1 Equal Overflow Elevation Tanks

Figure 12-5 shows different ways to connect two tanks with the overflows located at the same elevation. A typical arrangement is to have distribution lines joined immediately upon exiting the two tanks, creating only one main distribution line (Figs. 12-5a and 12-5c). There may be a "T" connection installed in the conduction line to fill the two tanks separately (Figs. 12-5a and 12-5b), or the conduction line may only enter one tank (Figs. 12-5c and 12-5d). If the conduction line enters both tanks, it is not necessary to interconnect the two tanks, assuming both tanks are connected to the distribution line (Fig. 12-5a). However, if only one tank is connected to the distribution line, then the two tanks must be interconnected (Figs. 12-5b and 12-5d).

If the bottoms of the tanks are at different elevations, the lower tank must be the tank that feeds the community. This method prevents stagnant water from building up in the lower tank and also maximizes available storage.

In Fig. 12-5c, the conduction line only enters tank 1, but both tanks are connected to the distribution line. If the two tanks are not directly connected to each other, water will flow out of tank 1, into the distribution line, and back into tank 2. However, we recommend adding the optional connection depicted in the figure. The common practice of closing the outlet valve of a tank at night would not allow tank 2 to fill if the valve were located before the distribution lines are joined.

Another typical arrangement is to have only one tank connected to the conduction line and to the distribution line. In this situation, there must be a connection between the bottoms of the two tanks (Fig. 12-5d). The tanks will fill and empty simultaneously.

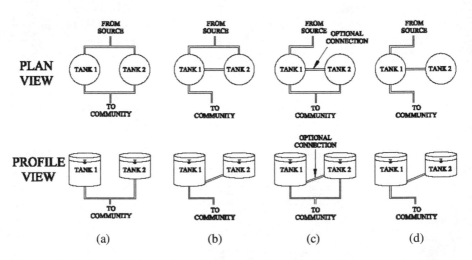

Figure 12-5. Several Methods to Connect Tanks with Overflows at an Equal Elevation. Shown Are Plan and Profile Views.

12.4.2 Unequal Overflow Elevation Tanks

Figure 12-6 shows various arrangements for a new tank added with its overflow at a different elevation than the original tank. In all the cases depicted in Fig. 12-6, the lower tank must be equipped with a float valve. The addition of a float valve, however, increases required maintenance. When the lower tank is full, this valve closes the entrance to the lower tank, allowing the tank at the higher elevation to fill. Any water that cannot be stored in the two tanks then exits through the overflow of the higher tank. The lower tank should have an overflow pipe at an elevation higher than the entrance for emergency purposes, but if the tank system is operating correctly, overflows should only occur in the upper tank. For the scenario shown in Fig. 12-6d, the pipe connecting the two tanks should be sufficiently sized to carry the maximum hourly flow. In all cases shown, this pipe can technically be sized separately from all other parts of the system because the head difference is small. In addition, the pipe connecting the two tanks is short; therefore, the cost differential is not great for placing in a larger diameter size.

Interconnecting the two-tank system is more complicated when the overflows are located at different elevations. If the outlet of the upper tank (located near the floor) is at an elevation below the overflow of the lower tank, both tanks must be connected to the distribution line so that all available storage in the upper tank is obtained (Figs. 12-6a, 12-6b, and 12-6c). A *check valve* that allows flow in only one direction must then be installed on the outlet of the lower tank to prevent water from flowing from the upper tank into the lower tank, reducing the storage capacity of the upper tank.

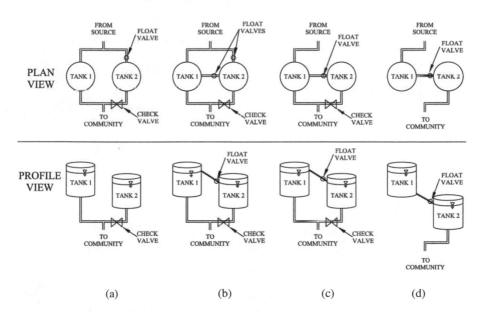

(a)   (b)   (c)   (d)

Figure 12-6. Connecting Tanks with Overflows at an Unequal Elevation. Shown Are Plan and Profile Views.

If the outlet of the upper tank is at an elevation higher than the entrance to the lower tank, then the distribution line should only be connected to the lower tank (Fig. 12-6d). The conduction line will first enter the upper tank, and a float valve installed on the line from the upper tank to the lower tank will allow the upper tank to fill after the lower tank has filled.

))) 12.5 Distribution Line Improvement Techniques

It is common to experience discrepancies in flow at houses located either higher or lower in the distribution system. For homes located in the higher elevations of the distribution system, this discrepancy may occur because the hydraulic grade line is at, or near, the elevation of the homes. The high flow rate to the lower houses causes a drop in the hydraulic grade. As the use at the lower houses increase, the hydraulic grade line drops and the higher houses receive less and less water, until the higher houses receive no water at all. One option is to install *control valves* (i.e., gate valves and globe valves) in the lower sections of the system to control the quantity of water that can flow to the lower houses. Case-by-case examination of each system must be made for the proper installation of valves.

Proper sizing of pipes may help prevent this problem as well. For example, if the lower houses lie farther from the tank than the upper houses, a smaller pipe can be used to provide the proper quantity of water. In effect, the smaller pipe creates the head loss that a globe valve would create, thereby reducing the flow rate.

Use these suggestions with caution. If the HGL falls below the elevation of houses because of increased head loss, those houses will not receive water (Fig. 12-7). Intention-

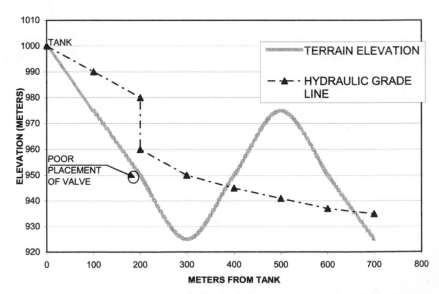

Figure 12-7. Example of Poor Placement of a Control Valve That Shows Negative Pressures in Higher Reaches Because of Head Loss Created by the Valve.

ally decreasing the HGL is only recommended in the lowest section of the system, where it does not continue to other houses located at a higher elevation.

12.5.1 Closing the Tank Outlet at Night

The most common occurrences of water loss in a community water system are houses without valves, houses with faulty valves, and households that leave a valve open. Houses without valves can easily be located and provided with a new valve. Houses with faulty valves can also be located and the valves replaced or repaired. The other commonly encountered problem is severe leakage in the distribution system. Locating major leakage can be done by walking the length of the conduction line.

While these improvements are in the process of being made, it may make sense for the plumber to close the outlet to the tank at night. This process allows the tank to fill and prevents wasted water at night. However, it requires someone to travel to the tank every day to open and close the valve if a valve is not installed close to the community.

Simple modifications can be made to the tank to obtain maximum storage and prevent losses of water if the designated individual does not reopen the tank before overflowing occurs. One method is to connect the overflow pipe to the distribution line. This method allows water to be supplied to the community once the tank is full. In this case, an additional overflow pipe with the inlet located at a higher elevation must be installed as an emergency overflow, in case the regular overflow pipe connected to the distribution line is full or cannot carry the flow entering the tank from the conduction line.

12.5.2 Engineering Practices to Improve Performance

Keep the use of moving parts to a minimum because they can fail and are expensive to replace. This admonition includes all valves, such as float valves in break-pressure tanks, clean-out valves, and air-release valves.

Clean-out valves are useful to remove sediments and other debris from the system. However, if the valve or the connection fails, water is lost through the clean-out valve and decreases or prevents flow. Valves often fail because of breakage of the mechanical parts from normal use. Valves are also vulnerable to intentional or unintentional damage. The most critical points for a clean-out valve are at the first few low points after the source and after the storage tank.

Break-pressure tanks in the conduction line can be equipped with operating float valves so that they never overflow. Once all houses downstream close their valves, the tank will fill, raising the float and closing the valve, and allowing for increased pressures upstream. The storage tank will then start to fill as all the houses close their valves. However, if a faulty valve does not prevent water from entering the break-pressure tank after it is full, water is lost in the overflow pipe of the break-pressure tank, and the storage tank may never fill. For this reason, it is best not to overuse break-pressure tanks. The elimination of a break-pressure tank in the distribution system after the storage tank to the community can only occur by carefully locating the storage tank at the lowest elevation possible while still providing sufficient pressure to all houses.

12.5.3 Sizing of Pipes

The sizing of pipes is the most important factor in equal distribution of water. Using pipes with a larger diameter improves pressures for systems in which sufficient pressure is not obtained because larger diameter pipes create smaller head loss for a given flow rate.

12.5.4 Sectorization

Sectorization, or manually opening and closing valves to parts of the community at various times of the day, is widely practiced in an attempt to create a more equal distribution of water. Typical sectorization can range from a house receiving water once a week to every day for only a few hours.

During the morning hours as people are preparing for work, many faucets may be open, but during the night the valves are closed. Usage will also vary between households. Every family needs to have access at one point during the day, although lower pressures may be noticed during peak hours. Sectorization creates a situation in which every house feels they must have the faucets fully open at all times to receive water. Typical sectorization can range from a house receiving water once a week to every day for only a few hours. Sectorization with periods of days shows the greatest problems, as a house will try to store a week's worth of water in a short time. This type of sectorization increases the peak hourly factor many times, and therefore the losses created with high flow rates drop the HGL line, causing homes at higher elevations to receive limited or no water.

In a correctly operating system, all lower elevation houses should close faucets when not in use, in which case a system control valve is not necessary. This concept is closely related to community education and emphasizing to those who live at lower elevations the importance of careful water management.

The use of control valves can help in cases in which houses at lower elevations receive water and houses at higher elevations do not. The topography and system layout is important, though, when considering installation of valves. For a system in which the higher elevation homes are located farther from the tank than the lower elevation houses, a control valve installed on the main line will not accomplish the goal of raising the HGL in the higher elevations. Figure 12-7 depicts this problem. Only in the situation where the lower elevation houses are located farther from the tank than the higher elevation houses should a control valve be used.

There is one case in which sectorization, or completely closing the discharge on the tank so nobody receives water, may benefit the community. This is when household faucets cannot close or are voluntarily left open. In this case, closing the storage tank every night will prevent losses.

⟫⟫⟫ Reference

Simpson, J. D. (2003). *Improvement of existing gravity-fed rural drinking water systems in Honduras*, <http://cee.eng.usf.edu/peacecorps/Resources.htm> (Jan. 17, 2009).

13

Pipeline Crossings

))) 13.1 Constructing a Crossing

Suspension bridges and cable structures in general can be complicated to design. In this chapter, a few dimensions have been defined to simplify and clarify the design process. Readers should understand that the calculations and values presented here are not precise. However, given the errors inherent in field engineering, they should be more than accurate enough to provide the basis for good design and safe construction.

A simple *suspension crossing* consists of four basic parts: the load, stringers, the main cable, and fixtures. Figure 13-1 illustrates these components. The *load* is what the structure is supporting. This part is the aqueduct pipeline (also referred to as a tube line). The design of the suspension crossing depends on the design of the aqueduct. The *stringers* hang vertically from the main cable and support the load of the water-filled pipeline. The design of the stringers primarily requires determining the length of each stringer. The main cable supports the water-filled pipeline, the stringers, and itself. It is attached to both *fixtures* and supports the load over its length. Designing the main cable includes determining its length and strength.

Special dimensions have been defined in this chapter. They differ from traditional suspension bridge conventions but are used here for simplification because they are more easily measured and calculated in a field setting. The horizontal distance between the fixture points is referred to as the *span* of the crossing. The lowest point of the cable's arc is referred to as the *apex*. The *sag* is defined as the vertical distance between the highest points of the cable's arc (at the fixture points) and the lowest point (the apex). The dimensions of the cable are measured from fixture point to fixture point, regardless of where the actual pipeline enters the ground or its distance below the cable.

When a crossing has one cable fixture point lower than the other (illustrated in Fig. 13-2), an additional dimension is introduced. This is the *drop*, which is defined as the vertical distance between one fixture point and the other. In this chapter, the sag of an uneven crossing is always measured from the higher of the two fixture points. Once again, these dimensions are measured regardless of what path the pipeline takes. This chapter also uses the terms *sag ratio* and *drop ratio*. These are defined as the sag and the drop divided by the span, respectively. In a crossing where both fixture points are at the same height, the drop and the drop ratio are zero.

Lyle J. Stone contributed to this chapter.

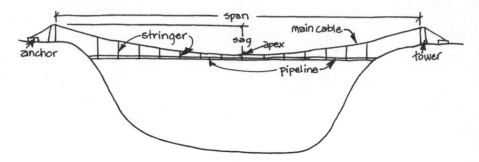

Figure 13-1. A Simple Suspension Crossing Showing the Four Basic Components: (1) the Load (Tube or Pipe), (2) Stringers, (3) Main Cable, and (4) Fixtures (Tower and Anchor).

Before the design is made, perform a detailed survey for the main conduction line. From this survey, roughly determine the cable and pipe sizes and ensure that materials are available. At this point, purchase the pipe and cable. The limiting factor may not be finding materials but getting the materials out to the job site. It is common for materials not to be available locally.

The final construction of a crossing typically waits for the conduction line to progress to the point where the crossing is located. The site of the pipeline crossing may change from its original design location because of rough ground, land disputes, and a variety of other reasons. Once the conduction line is constructed on both sides of the crossing, fix the final location and base the design of the anchors on the actual location and the materials that are available. After the anchors are constructed, measure the actual positions of the fixture points and fit the cable and pipes to these positions.

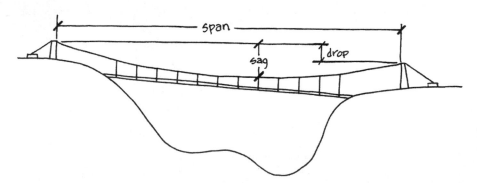

Figure 13-2. A Suspension Crossing Where One Cable Fixture Point Is Lower Than the Other.

Note: In this case, the vertical distance between the two fixture points must be determined. This vertical distance is called the *drop*.

))) 13.2 Planning and Surveying

During the planning and surveying phase of a water supply system, determine the best route. For a system that includes nothing but pipes, it is a simple undertaking. The most direct route is generally the easiest and least expensive. When considering obstacles such as canyons or rivers, however, adjust the pipeline to take advantage of more favorable terrain. If changing the proposed route of a pipeline will add to the length of the system but prevent a crossing from having to be constructed, compare the overall costs of the two options. Often preliminary decisions are made in the field as the survey is progressing.

The first step in planning is to determine the cost of materials. A preliminary visit to several suppliers is highly recommended. Obtain rough estimates, even though prices may change drastically. Even more importantly, a trip to several stores provides information on what materials are readily available. If a certain size of cable or pipe is only found at one

---

))) Box 13-1. Locating the Crossing

Imagine a small stream located roughly at elevation 0 is flowing from right to left on the topographic map. Which of the three crossings (labeled as A, B, and C) would you select?

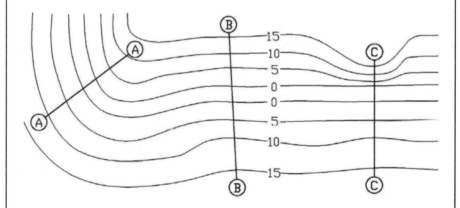

Location C has one of the two shortest spans. However, there is concern about the stability of the bank. The bank on one side (the upper side) is considerably steeper than the other bank. Inspect this bank. Perhaps it is bedrock, or perhaps it is made of the same soil and this section has not slumped off yet.

Location A has an easy gentle slope on both sides but is located at a bend in the canyon. The forces of floodwaters and small streams erode the outside corners of bends more. Avoid this location.

Location B is the best choice. The slopes are not any greater than the surrounding area and are therefore probably stable. Also, the stream is fairly straight in this section and will probably not erode more quickly than surrounding areas.

location, it may not be available when construction begins, or it may only be available at a premium.

The next step in planning is to identify what type of materials will be needed. The biggest variable cost of a crossing is the cable. Cable is specified by diameter, weave, and material type. *Diameter* is a nominal dimension roughly measured at the widest point of the cable. The *weave* is expressed by numbers such as 7 × 7, which stands for seven bundles of seven strands. The more strands a cable has, the more flexible it is. Cable comes in materials such as steel, stainless steel, and galvanized steel, and it sometimes has a plastic coating. Inspect any cable that has previously been used in a construction project in the area to see how well it is holding up to the local environment. This result can be compared to the design life of the new system.

Cable size is determined by the length of the crossing, the size of the pipe, and how much sag is allowed. Before the cable is purchased, the total tension that the cable will have to withstand should be determined. The quickest method to size the cable is to use Eqs. 13-1 to 13-3 (taken from Jordan 1980).

Equation 13-1 estimates the horizontal cable tension based on a parabolic model.

$$\text{Horizontal tension} = \frac{(\text{weight pipe per unit length}) \times (\text{length of span})^2}{8 \times \text{sag}} \quad (13\text{-}1)$$

Equation 13-2 estimates the angle of force at the fixture point based on a trapezoidal model.

$$\text{Angle of tension} = \arctan\left(\frac{4 \times \text{sag}}{\text{length of span}}\right) \quad (13\text{-}2)$$

Equation 13-3 estimates the total tension in the cable.

$$\text{Total tension} = \frac{\text{horizontal tension}}{\cos(\text{angle of tension})} \quad (13\text{-}3)$$

The span and sag are in units of length and are shown in Figs. 13-1 and 13-2. Appendix A-1 provides information on the weight of commonly used pipes that are full of flowing water. To estimate the length of span in Eq. 13-1 for a short crossing, throw a string across the crossing, pull it tight between the locations of the two anchors, and place a force on it to simulate the sag by adding a simple spring scale, pulley, or known weight. Then mark the string and take it to a flat area, stretch it, and reapply the same force to simulate the sag, and measure the horizontal distance.

The sag used in Eq. 13-1 can be estimated by visual inspection of the profile of the canyon, looking carefully at where the conduction line leaves and enters the ground and how tall the towers can be efficiently constructed.

A safety factor of 2.5 to 3 is applied to the value of total tension obtained from Eq. 13-3. This value is then used to select a cable size from the cable that is available. Materials often need to be purchased well in advance. With a greater safety factor, there is more leeway in the design if conditions change (Example 13-1).

Galvanized iron (GI) pipe is always recommended for crossings. It is far more durable than PVC pipe, which also degrades with exposure to sunlight. In an exposed condition in a rough area such as a canyon, the added durability is worth the initial cost. During the design of the pipeline, it is a good idea to check the effects of using smaller pipe diameters for the crossings. The added frictional losses can often be offset by using larger pipe diameters in other areas of the water system, where PVC pipes would be more appropriate. The reduced costs because of decreased GI pipe sizes and cable sizes can be significant.

Three-inch GI pipes and larger are not recommended for crossings. Although safe and secure crossings have been constructed using these pipes, their weight makes construction difficult and possibly dangerous. The weights of GI pipes are provided in Appendix A-1, along with other useful information.

))) Example 13-1. Preliminary Sizing of Steel Cable

A 140-ft span is selected as the best location for a crossing of a suspension bridge that supports 2-in. GI piping. It is determined that a sag of roughly 6 ft will be easy to construct and provide good clearance for surrounding topography. A safety factor of 3 will be used for the cable strength

From Appendix A-1, the weight of standard (40-Std-40S) 2-in. GI pipe is 3.65 lb/ft, and the weight of water in the 2-in. GI pipe is 0.121 lb/ft. The total weight is thus 3.77 lb/ft.

Equations 13-1 through 13-3 can be used to estimate the total tension.

$$\text{Horizontal tension} = \frac{(3.77\,\text{lb/ft}) \times (140\,\text{ft})^2}{8 \times 6\,\text{ft}} = 1{,}539\,\text{lb}$$

$$\text{Angle of tension} = \arctan\left(\frac{4 \times 6\,\text{ft}}{140\,\text{ft}}\right)$$

Therefore, the angle of tension = 9.7°.

$$\text{Total tension} = \frac{\text{horizontal tension}}{\cos(\text{angle of tension})} = \frac{1{,}539\,\text{lb}}{\cos(9.7°)} = 1{,}562\,\text{lb}$$

Multiplying by the safety factor of 3 results in a value of 4,686 lb.

From this estimate, a cable size can be specified and added to the total weight. From Appendix A-2, it can be seen that 7×7 $^7/_{32}$-in. steel cable has a breaking strength of 4,800 lb. Therefore, this cable size would be used in the design if it were available.

However, assume that only two cable types are available locally: 7×7 $^3/_{16}$-in. steel cable, which has a breaking strength of 3,700 lb and a weight of 0.062 lb/ft; and 7×7 $^1/_4$-in. steel cable, which has a breaking strength of 6,100 lb and a weight of 0.106 lb/ft.

The $^3/_{16}$-in. cable is too small, and the $^1/_4$-in. cable is too large and thus uneconomical. Either of these cables can be used, however, if the parameters are optimized. The locally available cable can be used by adjusting the design for the sag.

To optimize to either the ³⁄₁₆-in. cable size or the ¼-in. cable size, we need to determine the safe applicable load by dividing the minimum breaking strength (listed in Appendix A-2) by the safety factor of 3.

$$\text{³⁄₁₆-in. cable: safe load} = 3{,}700/3 = 1{,}233 \text{ lb}$$

$$\text{¼-in. cable: safe load} = 6{,}100/3 = 2{,}033 \text{ lb}$$

Next, a geometric design is determined that will better fit the cable strengths by solving for a more appropriate design sag. The calculations will be made first for the ³⁄₁₆-in. cable. Use Eq. 13-1 and solving for the sag:

$$\text{Sag} = \frac{(\text{weight of pipe per unit length}) \times (\text{length of span})^2}{8 \times \text{horizontal tension}}$$

The horizontal tension equals the "safe load" in this case:

$$\text{Sag} = \frac{(3.77 \text{ lb/ft}) \times (140 \text{ ft})^2}{8 \times 1{,}233 \text{ lb}} = 7.5 \text{ ft}$$

A similar calculation can be performed for the ¼-in. steel cable. The resulting sag (given the "safe load" of 2,033 lb) is determined to be 4.5 feet.

The diameter of the cable is inversely proportional to the length of the sag. For a ³⁄₁₆-in. 7×7 steel cable, a 7.5-ft design sag is a better choice. For the ¼-in. 7×7 steel cable, a 4.5-ft design sag makes more sense.   )))

))) 13.3 Masonry Column and Anchor Design

Once the rough design of the crossing and conduction line is completed, the detailed design of the anchors can begin. From the site survey, determine the locations of the anchors and select the type of anchor that is best suited to the location.

This section presents two types of anchor design. A *column and anchor* design uses a column to raise the cable, along with an anchor, or pair of anchors, to resist the force. An alternative is the *integrated tower and anchor*, which can be constructed from rubble masonry.

The advantage of using a column and anchor is that it has the ability to lift the cable fixture points higher. Higher fixture points allow for greater sag (thereby resulting in a smaller required cable diameter). Columns may be constructed from steel pipe, masonry units, or precast concrete. The anchors are usually constructed from rubble masonry.

Failure modes that should be considered are material failure (crushing or breaking of the concrete, masonry, or steel) and buckling failure. Whether a column will buckle or not is a function of the ratio of height to width, the force applied, and the material of the column. Roughly defined, *buckling* is the condition when a column fails from bending when only an axial force is applied.

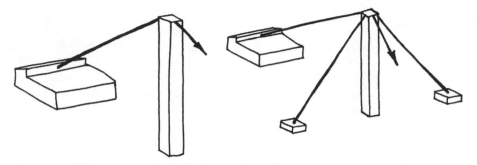

Figure 13-3. Column Without Guide Cables (Left) and Column Braced with Guide Cables (Right).

Note: The effective height of the column on the left is twice the actual column height because of lack of side bracing. The effective height of the braced column on the right is 0.7 of the actual column height.

To reduce the possibility of buckling, it is recommended that columns be braced laterally with guide cables, as shown in Fig. 13-3. Bracing by this method increases the maximum load by a factor of 8.2 over that of a nonbraced column. Lateral bracing can be achieved either by placing guide cables at right angles to the main cable or by using two anchors offset from each other and the main cable by 120° (Fig. 13-4). This angle reduces the required number of anchor blocks.

13.3.1 Steel Pipe Column Design

A steel pipe column has the advantage of being strong and easy to construct. Because it is premade, its strength is not as dependent on the quality of local construction. The most critical failure mode is buckling failure.

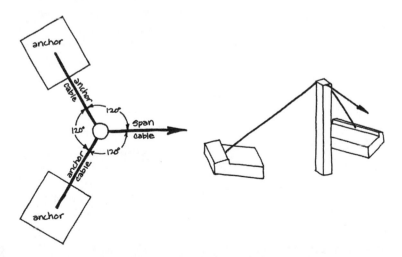

Figure 13-4. Column Braced with Split Anchors at 120°.

The design method of the *American Institute of Steel Construction LRFD Design Manual* (American Institute of Steel Construction 1998) is reduced here to Eq. 13-4. The values of 0.55 and 0.60 are safety factors incorporated into the equation. The critical axial load (in pounds) can be found in Appendix A-3, and critical bending moments (in in.-lb) can be found in Appendix A-1. To use Appendix A-3, locate the size pipe selected, and then check the critical axial load (also referred to in the equations as critical axial force) listed for the effective height.

$$\frac{(\text{Applied Axial Force})}{(0.55) \times (\text{Critical Axial Force})} + \frac{(\text{Applied Bending Moment})}{(0.60) \times (\text{Critical Bending Moment})} \leq 1 \quad (13\text{-}4)$$

Example 13-2 provides an example for how to use this equation in design of steel columns.

))) **Example 13-2.** Design of a Steel Pipe Column

Consider a 7-ft tall column. A 1,026-lb axial force is applied at the top, and there is no external bracing or applied bending moment. As stated in the caption of Fig. 13-3, a column without bracing has an effective height that is twice the actual column height. Thus, the effective height is 14 ft. The standard strength of steel is assumed to be 36,000 lb/in.^{2}. Determine the size of pipe that should be used.

Solution
Using Eq. 13-4:

$$\frac{(1,026\,\text{lb})}{(0.55) \times (\text{Critical Axial Force})} + \frac{0}{(0.60) \times (\text{Critical Bending Moment})} \leq 1$$

The critical axial force that is solved for equals 1,866 lb because the bending moment is zero. Assume that the steel pipe available at the local hardware store is 40-Std-40S with a 2-in. diameter. From Appendix A-3, we can see that this pipe has a critical axial force at an effective height of 14 ft of 2,945 lb. Thus, it will be acceptable in this situation.

Consider whether the same 2-in. steel pipe column will be acceptable in a situation with different forces applied to it. In this case, the axial force is 849 lb, and a bending moment of 238 ft-lb (2,856 in.-lb) is applied. The critical bending moment is based on the pipe size and material and is found from Appendix A-1 (value of 20,183 in.-lb).

$$\frac{(\text{Applied Axial Force})}{(0.55) \times (\text{Critical Axial Force})} + \frac{(\text{Applied Bending Moment})}{(0.60) \times (\text{Critical Bending Moment})} \leq 1$$

$$\frac{(849\,\text{lb})}{(0.55) \times (2,945\,\text{lb})} + \frac{(2,856\,\text{in-lb})}{(0.60) \times (20,183\,\text{in-lb})} = 0.75 \leq 1$$

In this case, the steel column would be stable.   )))

13.3.2 Concrete and Masonry Column Design

A column can be cast in place, prefabricated, or built up by masonry (Example 13-3). The concrete and the reinforcing steel carry the forces. This section represents a simplified design method based on the Uniform Building Code (UBC) (International Conference of Building Officials 1997). To simplify our discussion, only columns with a square cross section are used. Figure 13-5 depicts a concrete column and anchor set-up.

Figure 13-6 shows a cross section of the concrete column with the location of rebar. The effective width is considered to be the concrete toward the interior of the column starting at the inside edge of the steel bracing bar. The concrete outside the steel is thin and may break off or crack during extreme loadings. Its main function is to cover the steel.

The effective height of a column with regard to buckling relates the actual height of the column and how well it is braced (refer back to Fig. 13-3). A column that is not braced out of plane (in a direction other than the direction the load is applied) will have an effective height of the actual height times two. A column that is braced at the top with cables with a split anchor layout, as shown in Fig. 13-3, will have an effective length of 0.7 times the actual height.

The UBC applies safety factors to the materials and not to the force applied to the structure. The safety factor applied to masonry is 0.25 in Eqs. 13-5 and 13-6 because it is a nonhomogeneous material subject to the skill of the mason. The safety factor applied

Figure 13-5. A Square Steel-Reinforced Concrete Column and Anchor with Cable Attached.

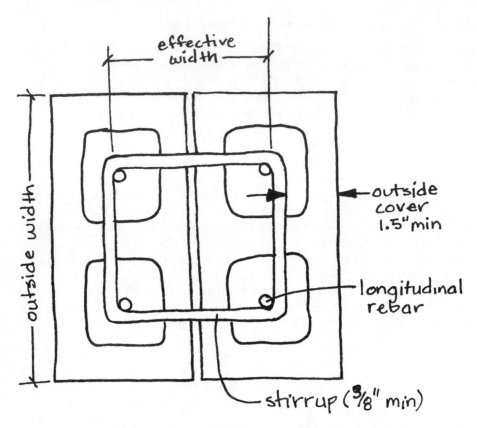

Figure 13-6. Cross Section of Standard Masonry Block Column That Consists of Two Masonry Blocks. The Outside and Effective Widths Differ.

to the steel is 0.65 in Eqs. 13-5 and 13-6 because it is a homogeneous material produced under controlled factory conditions.

Equation 13-5 can be used to calculate the allowable axial force for a non-slender (height/width < 29) reinforced masonry column, and Eq. 13-6 can be used for slender (height/width > 29) reinforced masonry column. Equation 13-7 is used for calculating the allowable bending moment on a reinforced masonry column. If an additional bending moment is applied to the column, more rebar can be added to account for this bending moment.

$$\text{Allowable axial force} = \begin{pmatrix} [0.25 \times (\text{masonry strength}) \times (\text{effective masonry area})] \\ + [0.65 \times (\text{steel strength}) \times (\text{steel area})] \end{pmatrix}$$
$$\times \left(1 - \frac{\text{height}}{40 \times \text{width}}\right)^2$$

(13-5)

$$\text{Allowable axial force} = \left(\begin{array}{l} [0.25 \times (\text{masonry strength}) \times (\text{effective masonry area})] \\ \quad + [0.65 \times (\text{steel strength})(\text{steel area})] \end{array}\right)$$

$$\times \left(\frac{20 \times \text{width}}{\text{height}}\right)^2 \qquad (13\text{-}6)$$

$$\text{Allowable bending moment} = 0.25 \times (\text{masonry strength}) \times (\text{width})^3 \qquad (13\text{-}7)$$

The column will be able to handle both a compression axial load and a bending load simultaneously, provided the combined load does not overstress the column. Equation 13-8 can be used to check the combined load. If the sum of the ratio of applied loads to allowable loads is ≤1, the column will be stable.

$$\frac{(\text{Applied Axial Force})}{(\text{Allowable Axial Force})} + \frac{(\text{Applied Bending Moment})}{(\text{Allowable Bending Moment})} \le 1 \qquad (13\text{-}8)$$

⟫⟩ Example 13-3. Design of a Masonry Column

Consider a 7-ft tall column. A 1,026-lb axial force is applied at the top and there is no external bracing. Assume a masonry column 8 in. × 8 in. After allowing for 1.5-in. cover and 0.375 in. for the bracing stirrups on each side, the effective width will be 4.25 in. Remember from Fig. 13-3 that a column without bracing has an effective height that is twice the column height. Thus, the effective height is 14 ft (168 in.). Assume masonry strength of 250 lb/in. and steel strength of 36,000 lb/in.^{2}. Four pieces of 3/8-in. rebar will be used, each with a cross-sectional area of 0.11 in.^{2}.

Solution

The height/width ratio is 168/4.25 = 39, so we will use Eq. 13-6.

$$\text{Allowable axial force} = 0.25 \times (250 \text{ lb/in.}^2) \times (4.25 \text{ in.} \times 4.25 \text{ in.})$$

$$+ 0.65 \times (36,000 \text{ lb/in.}^2)(4 \times 0.11 \text{ in.}^2)$$

$$\times \left(\frac{20 \times 4.25 \text{ in.}}{168 \text{ in.}}\right)^2$$

$$= 2,925 \text{ lb}$$

Because the allowable axial force (2,925 lb) is greater than the applied axial force (1,026 lb), the column is stable.

Consider the same column, but with an axial force of 849 lb and a bending moment of 238 ft-lb (2,856 in.-lb). The allowable bending moment is first determined from Eq. 13-7:

$$\text{Allowable bending moment} = 0.25 + (250 \text{ lb/in.}^2) \times (4.25 \text{ in.})^3 = 4,798 \text{ in.-lb}$$

These calculated values are used in Eq. 13-8 as follows:

$$\frac{(849\ \text{lb})}{(2{,}925\ \text{lb})} + \frac{(2{,}856\ \text{in.-lb})}{(4{,}798\ \text{in.-lb})} = 0.89$$

Because this value (0.89) is ≤1, the combination of axial and bending force does not over-stress the column. Therefore, the column is stable.

》》》

13.3.3 Masonry Anchor Design

A *masonry anchor* is nothing more than an efficiently designed mass that is heavy enough to withstand the forces applied and tall enough to supply the required sag (Example 13-4). These are the simplest form of anchors. The main failure modes of a masonry anchor are overturning or sliding. Overturning occurs when the force of the cable causes the back of the anchor to lift off the ground. Sliding occurs when the force of the cable causes the entire anchor to shift forward.

The anchor must be heavy enough to keep it from sliding under the force of the cable and from being lifted. The forces acting on the anchor are the vertical and horizontal forces of the cable that are countered by weight of the anchor and the friction with the soil. The friction angles for various soils are provided in Table 13-1. When designing the anchor on soil, safety factors of 3 or even up to 4 are not excessive because soil behavior is difficult to predict, and the cost of the added safety is often limited to a small amount of additional concrete. Reduced weights must be used, however, if the anchor will be submerged because the anchor will want to float because of buoyant forces.

The tower-and-slab type anchor, as illustrated in Fig. 13-7, fits best in canyons with sudden drop-offs and mostly flat land surfaces. The cable passes up over the tower and then attaches at the back to a rebar loop.

The stability of an anchor is analyzed by balancing all the forces. To check for overturning, balance the moments around the front bottom edge. To check for sliding, balance the forces in the horizontal direction. A safety factor of 2 to 4 is recommended, depending on the construction. An anchor design that uses rubble masonry will typically have a larger safety factor applied to it because it tends to be less uniform and often has large voids. New concrete is, however, more uniform.

Table 13-1. Friction Angle of Various Soils, ϕ

Type of Soil	Friction Angle, ϕ
Sand and gravel mixture	33–36
Well-graded sand	32–35
Fine to medium sand	29–32
Silty sand	27–32
Silt	26–30

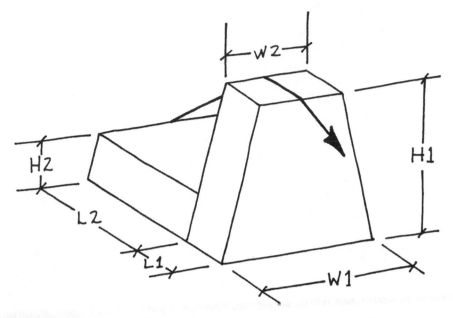

Figure 13-7. Tower-and-Slab Anchor Mass with Dimensions Used in the Design Example.

⟩⟩⟩ Example 13-4. Design of Mass Block Anchor

In this example, we will design a simple block anchor mass. It differs from the tower and slab anchor mass depicted in Fig. 13-7 because it is just a rectangular mass with dimensions of $H \times L \times W$.

An anchor is required for a 2.75-ft tower. The anchor is subject to a cable load with a vertical component force of 268.3 lb and a horizontal component force of 1,984.5 lb. Soil conditions suggest setting the base of the anchor at least 1.5 ft below the surface on a solid sand layer. Provide the dimensions of the anchor. The width (W) is set at 3 ft based on experience. Use a safety factor (SF) of 3, and assume the density of the masonry mass is 150 lb/ft^3.

Solution
Because the anchor will be buried 1.5 ft, the anchor height (H) is 4.25 ft.

First, check the system to ensure that it will not overturn. Sum the moments around the front bottom edge so they equal zero as follows:

$$\Sigma M = 0 = \left[(\text{horizontal force}) \times (H) \times SF \right]$$

$$- \left[L \times W \times H \times (\text{weight of mass per ft}^3) \times \frac{L}{2} \right]$$

$$\Sigma M = 0 = \left[1{,}985 \text{ lb} \times 4.25 \text{ ft} \times 3\right] - \left[L \times 3 \text{ ft} \times 4.25 \text{ ft} \times (150 \text{ lb/ft}^3) \times \frac{L}{2}\right]$$

Solving for $L = 5.14$ ft, round up to 5.25 ft.

Now check for sliding. Here, balance the horizontal forces and determine if the safety factor is within our allowable range.

$$\Sigma F_x = 0 = \left[SF \times (\text{horizontal force})\right]$$
$$- \left[L \times W \times H \times (\text{weight per ft}^3) \times (\tan \varphi)\right]$$
$$- \left[\text{vertical force}\right]$$

$$\Sigma F_x = 0 = \left[SF \times 1{,}985 \text{ lb}\right]$$
$$- \left[5.25 \text{ ft} \times 3 \text{ ft} \times 4.25 \text{ ft} \times 150 \text{ lb/ft}^3 \times (\tan 29°)\right] - \left[268 \text{ lb}\right]$$

Solve for the safety factor, $SF = 2.9$. Because the SF of 2.9 is much greater than 1 and close to the recommended value of 3, the block is considered stable.

Redesign as a Tower-and-Slab Anchor

The anchor can be redesigned as a tower-and-slab type anchor (shown in Fig. 13-7) by determining the dimension $L2$. Using the following design dimensions and layout depicted in Fig. 13-7, $H1 = 4.25$ ft; $H2 = 2$ ft; $W1 = 3$ ft; $W2 = 1.5$ ft; and $L1 = 1.5$ ft.

$$\Sigma M = 0 = \left[(\text{horizontal force}) \times (H1) \times SF\right] - \left[\frac{W1 + W2}{2}\right]$$
$$\times \left[L1 \times H1 \times (\text{weight of mass per ft}^3) \times \frac{L1}{2}\right]$$
$$- \left[W1 \times L2 \times H2 \times (\text{weight of mass per ft}^3) \times \left(L1 + \frac{L2}{2}\right)\right]$$

$$\Sigma M = 0 = \left[1{,}985 \text{ lb} \times 4.25 \text{ ft} \times 3\right] - \left[\frac{3 \text{ ft} + 1.5 \text{ ft}}{2}\right]$$
$$\times \left[1.5 \text{ ft} \times 4.25 \text{ ft} \times 150 \text{ lb/ft}^3 \times \frac{1.5}{2}\right]$$
$$- \left[3 \text{ ft} \times L2 \times 2 \text{ ft} \times (150 \text{ lb/ft}^3) \times \left(1.5 \text{ ft} + \frac{L2}{2}\right)\right]$$

Solve for $L2 = 5.9$ ft. Now check this design for sliding.

$$\Sigma F_x = 0 = \left[SF \times \text{(horizontal force)} \right]$$

$$- \left[\left(\frac{W1 + W2}{2} \times L1 \times H1 \right) \times (W1 \times L2 \times H2) \times \text{(weight per ft}^3) \times (\tan \varphi) \right]$$

$$- \left[\text{vertical force} \right]$$

$$\Sigma F_x = 0 = \left[SF + (1{,}985 \text{ lb}) \right]$$

$$- \left[\left(\frac{3 \text{ ft} + 1.5 \text{ ft}}{2} \times 1.5 \text{ ft} \times 4.25 \text{ ft} \right) + (3 \text{ ft} \times 6 \text{ ft} \times 2 \text{ ft}) \times (150 \text{ lb/ft}^3) \times (\tan 29°) \right]$$

$$- \left[268 \text{ lb} \right]$$

Solving for the safety factor (SF), $SF = 2.2$. The mass is stable because all resisting forces are greater than the applied forces with an adequate safety factor.

Comparing the two designs, both masses withstand the same forces with comparable results. However, through efficient placement of material, the tower-and-slab design uses only 51 ft^3 of material, whereas the simple mass used 67 ft^3.   )))

))) 13.4 Construction and Design Details

13.4.1 Layout

A string line is the easiest method for laying out the construction. If the original survey stakes remain, use them as guides. If not, stake and measure two new survey benchmarks with string. The staked-out strings keep all the parts in line. It is crucial that both anchors be in a line. If they are crooked, it will cause out-of-plane horizontal forces that have not been accounted for in the design.

13.4.2 Anchors

Column and Anchor

In this type of anchor, set-up is vital for the structural integrity of the crossing. The column, no matter what type of construction, will be more difficult to line up. It is preferable to place the column and then lay out the anchor based on the true alignment of the column and not the planned alignment.

A suggested rebar detail for the anchor unit is shown in Fig. 13-8. The size of the pad can be reduced if the pad is covered with something, such as rock, that can provide the required dead weight. However, this rock must never be removed. A minimum pad thickness of 350 cm (1 ft) is required. The rock placed on top of the concrete pad can be

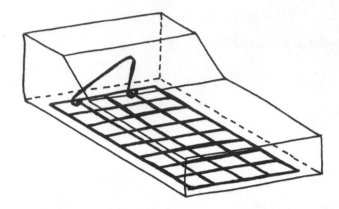

Figure 13-8. Rebar Detail for an Anchor Mass.

used to cover some of the cable hardware, which will make it less vulnerable to malicious intent and the hands of curious children.

Concrete columns can be built up with masonry units, cast in place, or cast on the ground and then tilted up into place. An in-place pour is not recommended, because the tall, thin column would be problematic to construct. The formwork would be complicated, and it would be difficult to maintain a consistent mix throughout the column if concrete were dropped in from the top. A better solution is to construct the column on the ground and then tip it into place after it has cured.

Pipe Column

If a different tension is to be put on the anchor cable than on the span cable, use a turnbuckle to adjust the tension. The point of adjusting the tension in the anchor cable is to eliminate the bending force. If the bending moment is eliminated, the column will not bend. If the column is vertical when it was installed and vertical after the cable forces are applied, then the bending moment has been effectively eliminated. It is also best to keep the column as vertical as possible during construction. During installation of a pipe column, often one person is assigned with the specific task of maintaining the true vertical level of the pipe by holding a level or plumb line against the column while other workers fill the excavation with concrete.

The easiest method to get the correct tension on the cables is to use a level on the column. A small level can be used and attached to the column where it can be seen. Because the span cable is loaded with the pipes and will later be filled with water, the turnbuckle on the anchor cable may need to be adjusted to keep the column from bending.

If the column is not plumb, it is highly recommended that additional guide wires be added to the design. These additional anchors and cables will also be adjusted to replumb the post in the same manner.

Masonry Anchor

Using stone masonry for an integrated column and anchor can be an economical solution. The stones reduce the amount of concrete used and eliminate the need for forms. Large, flat masses can be built by practically anyone. Taller masses require more skill. Stones will fall off each other before the mortar cures, and making tall stacks can cause disastrous failures during construction. One solution is to create a rebar cage that can be filled with mortar and stones. The rebar will hold everything in place while the mortar cures.

Figure 13-9 shows an example rebar layout. All the rebar is ⅜-in. thick, with the exception of the ¼-in. rebar stirrups that wrap the column. The rebar cable attachment and the rebar in the bottom is the only rebar that is structurally significant. The rebar provides a great construction guide; the builder only has to reach through the rebar cage, align the stones within the rebar, and fill in the voids with mortar.

The rebar on the outside of the column must be covered to prevent corrosion. Even though the strength of the column does not depend on this rebar, corroding rebar expands and can destroy concrete. Wrapping chicken wire or some other form of wire fabric around the cage will help thicker layers of mortar to stick. The thinnest, cheapest chicken wire that is available is fine and will help to provide the structure with a solid cover.

13.4.3 Final Cable Adjustment

The placement of the cable is important for the integrity of the crossing. Therefore, remeasure the distance between the anchors after construction to account for any construction irregularities. Use string to measure directly from one fixture point to another. Often, a crossing will use a turnbuckle to make the final fine adjustment to the cable. Even if this method is used, the cable length will still need to be precise enough that the turnbuckle will be able to account for any small errors.

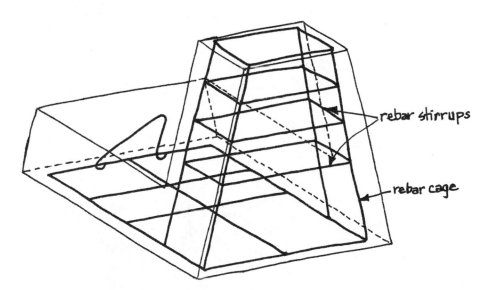

Figure 13-9. Typical Rebar Cage Layout for an Anchor and Column Unit Built with Stone Masonry.

13.4.4 *Stringer Length When Anchors Are at the Same Elevation*

The stringers tie the pipe to the main cable. Stringers usually consist of a thin (³⁄₁₆-in.-diameter or smaller) steel cable, doubled over at the top with two cable clamps around a cable eye, and a double loop at the bottom around the pipe secured with two more cable clamps (Fig. 13-10).

Stainless steel cable and high-strand flexible cables are more expensive, but for the small diameters and quantities used in the stringers, the added ease of construction may be worth the extra expense. It is thus advisable to devote some additional funds from the

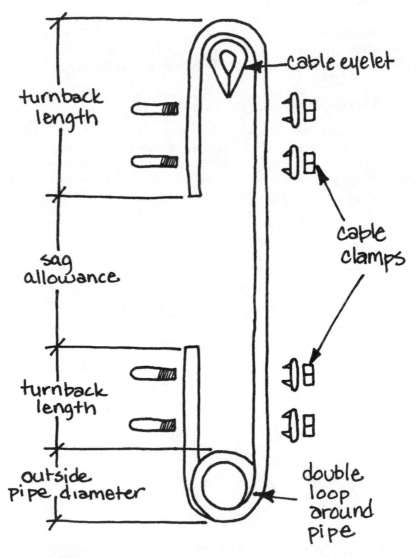

Figure 13-10. A Typical Stringer.

overall budget to the stringer cables. Cables with more flexible weaves cost a little more, and stainless steel cable is easier to work with than galvanized or plastic-coated cable. One-eighth-in. stainless steel looks thin but is strong enough to handle many size pipes. Three-sixteenths-in. stainless steel is usually not much more expensive and may be advisable for 2-in. GI piping and higher.

The stringers are most often cut and prepped the afternoon before the main cable and pipes are installed. The length of each stringer consists of
- the length needed to compensate for the sag that is specific to each stringer,
- enough cable to attach the stringer to the main cable, and
- enough cable to wrap the stringer twice around the pipe.

Figure 13-1 depicts the location of the stringers relative to other crossing components. The turnback lengths (Fig. 13-10) are determined by manufacturers' recommendations based on cable diameters or by what is considered practical in the field. The sag allowance that is specific to each stringer (also depicted in Fig. 13-10) can be calculated by first determining C. This calculation is done by either simultaneously solving Eqs. 13-9 (the catenary equation) and 13-10 (the parabolic equation) or using the sag ratio to look up the value of C (Appendix A-4).

The C constant is what gives the curve its shape, and it is different for every sag ratio. When determining the value of C using the following equations, the distance from the apex to the tower is x (or one-half the span length when the anchors are at the same elevation). When the sag allowance associated with individual stringers is determined later, x will be defined as the distance from the apex to the location of the stringer.

$$y = C \cosh\left(\frac{x}{C}\right) - C \qquad (13\text{-}9)$$

$$y = \left(\frac{C}{2}\right)\left(\frac{x}{C}\right)^2 \qquad (13\text{-}10)$$

When determining C by solving Eqs. 13-9 and 13-10 simultaneously, x can be set to half of the span and y to the sag. C can also be determined using design tables, as presented in Example 13-5. Determining the C value for a cable hung between two anchor points at different elevations (Fig. 13-2) is more difficult and is discussed elsewhere (Stone 2006).

))) Example 13-5. Determining Sag Using Tabulated C Values

A 140-ft span has an actual sag of 11.75 ft (sag ratio is 0.084). The Appendix A-4 value for C using this sag ratio on a crossing with the two anchor points at the same elevation produces a value of 1.501890 for a unit length of span. Scaling the crossing up by multiplying the unit C value by the length produces an overall C for the span of 210.2646.

$$y = 210.2646 \cosh\left(\frac{70\text{ ft}}{210.2646}\right) - 210.2646 = 11.76\text{ ft}$$

$$y = \left(\frac{210.2646}{2} \right) \left(\frac{70 \text{ ft}}{210.2646} \right)^2 = 11.65 \text{ ft}$$

The two calculated values of the sag (y) are close to 11.75 ft. Although this result is not exact, this difference over a 140-ft span is more than acceptable for our purposes. ⟫⟫

⟫⟫ **Example 13-6.** Determining the Length of the Stringers

Assume a preliminary design that had a 140-ft span with a sag of 7.5 ft. When the anchors were completed and remeasured, it was discovered that the span was actually 142.5 ft from fixture point to fixture point. Using the new span length and original sag, the new sag ratio (sag divided by span) is calculated to be 0.0526.

Appendix A-4 provides values of C as a function of the sag ratio. To use Appendix A-4, the C value is lowered to its closest value in the table, 0.052. This amount corresponds to a sag of 7.41 ft (142.5 ft × 0.052) and a tabulated C value per unit length of the span of 2.412463. Multiplying by the length yields a C value for the span of 343.776.

For the design, we plan to place the stringers at 5, 15, 25, 35, 45, and 55 ft from the apex.

Use Eq. 13-9 to determine the sag at the location of the first stringer. This usage is different than our definition of sag, which was defined as the location of the apex.

$$y = C \cosh\left(\frac{x}{C} \right) - C = 343.776 \cosh\left(\frac{5 \text{ ft}}{343.776} \right) - 343.776 = 0.0364 \text{ ft}$$

Use this calculation to determine the sag allowance at each stringer location.

There is no stringer at the center because it would have a zero sag allowance. We recommend that readers refer back to Fig. 13-1 for a visual representation for why the stringer has zero sag allowance at this point. The lowest value is thus set to zero (which is the stringer closest to the apex), and the other calculated values are reduced by the same amount (set as the normalized value in the table below). Subtracting this amount from every stringer raises the pipe.

Cut the stringers to a length that includes the sag allowance specific to each stringer, the turnback lengths, enough cable to wrap twice around the pipe, and enough to go around the eyelet (refer to Fig. 13-10). For a ⅛-in. steel cable, the recommended turnback length is 4 in. Allow for two lengths at the top and two at the bottom. An additional 4 in. is added to go around the eyelet, and wrapping the cable around two pipe circumferences adds an additional 15.71 in. This amount totals 35.71 in. in extra cable for connections.

Distance (ft)	Sag Allowance (ft)	Normalized (in.)	Normalized (in.)	With connections
5	0.0364	0	0	35¹¹/₁₆
15	0.3273	0.2909	3⁸/₁₆	39³/₁₆
25	0.9094	0.873	10⁸/₁₆	46³/₁₆

Distance (ft)	Sag Allowance (ft)	Normalized (in.)	Normalized (in.)	With connections
35	1.7832	1.7468	$20^{15}/_{16}$	$56^{11}/_{16}$
45	2.9494	2.913	$34^{15}/_{16}$	$70^{11}/_{16}$
55	4.4091	4.3727	$52^{8}/_{16}$	$88^{3}/_{16}$

$$\rangle\rangle\rangle$$

13.4.5 Calculating the Final Cable Length

Cable stretches under tension. The design calls for a certain sag over a certain distance. A cable marked to length on the ground without tension and then placed under tension will elongate. The longer cable will also have a greater sag. If not accounted for, the stringers may be too short near the banks of the ravine. This mistake could cause unnecessary and unaccounted forces on the pipes and may change the geometry of the crossing. All of this can be avoided by marking the cable on the ground at a length that, when placed under the weight of the fully loaded pipes, will stretch to the correct length for the given span and sag.

To determine the length of a cable that is fully loaded, first determine the average force in the cable. The force changes from a purely horizontal force at the apex to a combination of vertical and horizontal forces at the fixture point. Although it is not technically correct, it is well within the acceptable margin of error to take the average of these two forces.

What makes steel cable more challenging to design with is that it stretches at two rates. It stretches more at the beginning as the individual strands tighten around each other, and then at a lesser rate as the strands themselves stretch. These two rates depend on the weave of the cable. The first rate occurs at approximately 0–20% of the failure load, and the second occurs between 20% and 65% of the failure load. Accordingly, to determine the length of the cable between the two fixture points (e.g., the towers) first use Eq. 13-11 to calculate the arc length from the apex to a distance x. Then apply Eqs. 13-12 and 13-13 to determine the final cable length after it is stretched and after it is loaded.

$$y = C\sinh\left(\frac{x}{C}\right) \tag{13-11}$$

$$\text{Proportional change in length} = \left[\frac{20\% \text{ of Minimum breaking strength}}{\text{Steel area} \times 13{,}500{,}000 \text{ lb/in}^2}\right]$$
$$+ \left[\frac{\text{Total load} - 20\% \text{ of Minimum breaking strength}}{\text{Steel area} \times 15{,}000{,}000 \text{ lb/in}^2}\right] \tag{13-12}$$

$$\text{Starting length} = \frac{\text{Final length}}{1 + \text{Proportional change in length}} \tag{13-13}$$

⟫⟫ **Example 13-7.** Determining the Required Length of the Main Cable

Assume a 142.5-ft span with a sag of 7.41 ft and a *C* value of 343.776. Also assume that 7×7 $\frac{3}{16}$-in. steel cable will support 2-in. galvanized pipe over the crossing.

Use Eq. 13-11 to determine the arc length from the apex to a distance *x*. For a level crossing, the length from the apex to the fixture point is half the total length.

$$y = C \sinh\left(\frac{x}{C}\right)$$

Here, *y* is the arc length from the apex to the distance *x*, which is located at a fixture point, such as a tower.

$$y = 343.776 \sinh\left[\frac{(142.5 \text{ ft}/2)}{343.776}\right] = 71.8$$

The total length along the arc between fixture points is thus 143.52 ft.

Eqs. 13-3 and 13-1 are used, respectively, to estimate the total tension (the tension in the cable at the fixture point) and horizontal tension (the tension at the apex). The resulting horizontal tension equals 1,312.6 lb, and the resulting total tension equals 1,340.0 lb. The average of these two values is 1,326.3 lb. The minimum breaking strength of the cable is 3,700 lb (value obtained from Appendix A-2).

Equation 13-12 can be solved for the proportional change in length:

$$\text{Proportional change in length} = \left[\frac{0.20 \times 3{,}700 \text{ lb}}{0.0166 \text{ in}^2 \times 13{,}500{,}000 \text{ lb/in}^2}\right]$$

$$+ \left[\frac{1{,}312.6 \text{ lb} - 0.20 \times 3{,}700 \text{ lb}}{0.0166 \text{ in}^2 \times 15{,}000{,}000 \text{ lb/in}^2}\right]$$

$$= 0.005602$$

The proportional change in length thus equals 0.005602.

Then use Eq. 13-13 to determine the starting length that must be installed to account for stretch in the cable after it is loaded:

$$\text{Starting length} = \frac{143.52}{1 + (0.005602)} = 142.72 \text{ ft}$$

Note how the cable stretches a small amount after the load is placed on it. The cable is thus marked at 142.72 ft and then attached so that the marks are placed exactly on both fixture points. The sag at this time will be less than the final sag, but after the pipes are placed and filled with water, it will stretch to the final position. Note in Figs. 13-1 and 13-2 how additional cable is still required to connect to the anchors.   ⟫⟫

⟩⟩⟩ 13.5 Attaching the Cable and Pipes

After the lengths are calculated for the stringers and the main cable (Example 13-7), attach the cables and pipes. This should take place all in one day so that the crossing is secured as soon as possible. Do not leave the cables and the pipes hanging loose overnight. Before the final assembly begins, prepare the stringers, measure the cable, perform all calculations, and transport heavy materials such as the pipes to the site.

The preparation for each stringer that allows for the quickest assembly on the day of construction is to set all cable clamps but one, as described below. Figure 13-11 shows the correct placement of cable clamps.

First set the top two cable clamps that hold the eyelet in at the correct turnback length. This procedure is easiest if you first set and tighten the bottom clamp and then place the upper clamp directly above it but loosely. Form the eyelet, slide the second cable clamp up to hold it, and tighten it.

Then set the bottom of the stringer for the correct sag allowance. Set the top cable clamp tightly and place the second loosely. During construction, slide the double loop around the tube loosely, slide the second cable clamp tightly, and secure it. This way, only one cable clamp needs to be tightened in the field, and all the distances are measured and set in advance.

When attaching the cable clamps, orientation is important. The cable clamp has two main parts: the *staple* (the horseshoe-shaped bolt, or U-bolt, to which the nuts attach) and the *saddle* (the cast steel part with the two holes in it). A cable that has been turned back and clamped to itself also has two parts: the cable that supports the load, known as the *live cable*, and the cable that was turned back, known as the *dead cable*. The saddle grips the cable smoothly while the staple tends to pinch or bend the cable to which it's attached. Therefore, the saddle always goes on the live cable (Fig. 13-11). A way to remember this is the phrase "Never saddle a dead horse!"

The Wire Rope Technical Board (2005) suggests using two cable clamps for all cable smaller than 0.5 in. and three cable clamps for 0.5-in. to 0.75-in. cable. In this book, we recommend that three cable clamps be used for 0.25-in. cable and higher and that an extra cable clamp be used anywhere a connection can be reached by curious hands.

The most effective method for placing the cables is to attach the main cable on both sides of the ravine and then slide the pipes out one by one from one side of the ravine (Fig. 13-12). Thread the main cable through the stringers before securing it. Then attach the first stringer to the first pipe and push the pipe out into the canyon. Screw in the next pipe and secured it to the stringers. The pipes continue to be pushed into the opening until the final pipe is in place and the bridge spans the distance.

Figure 13-11. Correct Placement of Cable Clamps.

Figure 13-12. Pipes Are Connected While the Crew Waits to Push Out the Next Length.

Note: The stringers are hanging on the cable next to the tower.

For long spans with thin pipes, pull the pipes from the far side with a rope in addition to pushing them. This procedure will help keep the pipes from bending or buckling under the force of gravity.

Pipes can be prepared in advance: attach a coupling to one end and check the threads to ensure that they are clean, straight, and will screw in smoothly. Mark the pipes with the correct cable placement, usually every 10 ft (directly in the middle of the pipe), with either a scratch of a file or a strip of tape. Tape is preferred because it is easier to see.

Before pushing out the first pipe, cover the opening and secure it with a cloth. This step is done because the pipe may bang into the ground when it reaches the other side. A pipe with soil in it will be difficult to clean out once the other pipes are connected.

))) 13.6 Finishing Touches

Plastic pipes are vulnerable to breakage and leakage where they connect to steel pipes. This change in materials occurs at crossings. Steel pipes are stronger and may not be damaged by impact, but they are not as flexible as plastic pipe. A strong jolt to a steel pipe can crack the plastic pipe at the adaptor, which is often the weakest link. To avoid this problem, cover the steel pipe with large stones where it enters the ground and attaches to the plastic pipe.

Crossings often occur at low points in the system. Because it is also a place where the pipeline is above the ground, it provides an excellent location to place a sediment clean-out valve. The critical issue here is that the clean-out valve needs to be easily reached and that it also will not cause excessive erosion under the anchor when it is opened. Of course, if the crossing is located at a localized high point, then an air-release valve can be installed as well.

⟫⟫ References

American Institute of Steel Construction. (1998). *American Institute of Steel Construction LRFD design manual.* American Institute of Steel Construction, Chicago, Ill.

International Conference of Building Officials. (1997). *Uniform Building Code,* Vol. 2. Whittier, Calif.

Jordan, T. D., Jr. (1980). *A handbook of gravity-flow water systems,* Intermediate Technology Publications, London.

Stone, L. J. (2006). *Suspension canyon and gully crossings for small scale community aqueducts: A design guide based on experience and observations from Peace Corps, Dominican Republic,* <http://cee. eng.usf.edu/peacecorps/Resources.htm> (Jan. 24, 2009).

Wire Rope Technical Board. (2005) *Wire rope users manual,* fourth ed., Alexandria, Va.

14

Water Storage Tanks

))) 14.1 Rationale for a Storage Tank

Storage tanks are incorporated into gravity-fed water system designs to buffer peaks in demand (i.e., when peak demand exceeds source production) and to ensure reserve when supply is intermittent or interrupted. They are also incorporated into rainwater harvesting and spring box systems. This chapter focuses primarily on gravity-fed water systems. Chapter 17 provides guidance on sizing tanks for rainwater harvesting systems. Chapter 15 discusses sizing a spring box reservoir to hold all the water that exits a spring after the last user in the evening and before the first user in the morning.

Storage tanks may reduce the overall cost of water systems by eliminating the need for the large-diameter pipes that would be needed to meet peak demand if the cost of the increased supply line pipe is more than the cost of the storage tank. They can also improve water quality by decreasing turbidity because during storage, some particles settle to the bottom of the tank. The tank also protects water quality by maintaining positive pressure in the downstream distribution network, thus preventing contaminants from seeping into the pipes. The factors that affect the type and size of storage tank most appropriate for a given water system design include the following:

- source supply,
- water demand,
- desired critical reserve,
- economics, and
- engineering materials and local construction knowledge.

))) 14.2. Factors That Influence Sizing

14.2.1 Source Supply

The design parameter related to source supply is the minimum source flow rate ($Q_{\text{min-source}}$). This value, measured in volume/time, is obtained through measurements taken at a water source during the lowest source flow of the year. $Q_{\text{min-source}}$ is the design flow rate used to size pipes from the source to the storage tank (here referred to as the *supply line*). Q_{peak}, related to

---

Daniel Nover, Ryan W. Schweitzer, Matthew A. Niskanen, Erlande Omisca, and Maya A. Trotz contributed to this chapter.

demand (see below), is the design flow rate within the distribution system. Because $Q_{min\text{-}source}$ is often smaller than Q_{peak}, the required pipe diameter for a similar head loss is smaller.

14.2.2 Water Demand

Water demand is a function of the design population, minimum personal water requirements, and factors such as seasonal activities and infrastructural demands (e.g., from schools, clinics). *Design population* (P_N) is the projected population in the last year of the design life. It is becoming a more difficult value to predict because of less predictable trends of population growth, trade, urbanization, and migration. It can be calculated as follows:

$$\text{for populations} \leq 2{,}000 \qquad P_N = P_O\left(1 + \frac{r \times N}{100}\right) \qquad (14\text{-}1)$$

or

$$\text{for populations} > 2{,}000 \qquad P_N = P_O\left(1 + \frac{r}{100}\right)^N \qquad (14\text{-}2)$$

In Eqs. 14-1 and 14-2, r is the percentage projected growth rate, P_0 is the initial population, and N is the number of years in the design life.

Water demand is characterized by the following:

$$\text{Average daily usage} \qquad V_{ave} = [P_N \times \omega_R] + \text{infrastructure} \qquad (14\text{-}3)$$

$$\text{Maximum daily usage} \qquad V_{max} = V_{ave} \times \text{safety factor (1.25 to 1.5)} \qquad (14\text{-}4)$$

In Eq. 14-3, V_{ave} is the daily average volume of water required by the population, P_N is the population at the end of the design life of the tank, ω_R is the volume of water required per capita per day, and infrastructure is the daily volume of water required to support nonhousehold infrastructure in the community (e.g., hospital, hotel, fire protection). Per capita water consumption usually ranges between 25 and 100 L/capita-day.

In Eq. 14-4, V_{max} is the daily maximum volume of water needed for all uses. The safety factor values of 1.25 to 1.5 are derived from Instituto Nacional de Aguas Potables y Alcantarillado data (PCDR 2000). Although V_{ave} and V_{max} are sometimes thought of as the volume of water required on a daily basis, they are in fact flow rates. Later, in Example 14-1, we refer to demand flow rates (Q_{max} and Q_{ave}). The values are the same as V_{max} and V_{ave}, respectively, except that units for demand flow rate are typically converted to liters per second.

Peak demand flow (Q_{peak}) is calculated once the daily water demand pattern is identified. Typical water usage patterns have two peaks, with one occurring in the morning (5:00 A.M.–11:00 A.M.) and one in the late afternoon (4:00 P.M.–7:00 P.M.). Identifying the exact water demand pattern of a population can be a complicated, time consuming, and often unnecessary process. It is usually sufficient to estimate the time of peak collection by counting the number of people collecting water at a community well or by interviewing a representative sample of households. If it is logistically difficult to estimate the time of peak collection, it is acceptable to assume 6 h of peak demand. From there, Q_{peak} can be calculated, assuming that the demand occurs during a peak period.

14.2.3 Economics

Increasing the size of a storage tank can be a way to increase food security by making water more regularly available for irrigation of gardens. Often, compared to the overall cost of a water project, the cost of increasing the size of a tank is low. It is possible to work backward and design the largest tank a budget will allow. Large tanks have the disadvantage of higher cost, but the extra storage can be used in the rainy season when the flow from the source is larger than the design value, $Q_{min\text{-}source}$.

14.2.4 Building Materials and Practices

The availability of materials and local construction knowledge affect tank size. Table 14-1 provides guidelines on maximum tank size based on the type of construction material.

⟫⟫ 14.3 Methods for Calculating Tank Size

There are several methods for determining the appropriate water storage tank size:
- sizing by comparison of supply and demand,
- sizing by ability to meet the peak demand,
- sizing by critical reserve, and
- sizing to store all of the available water during hours of no use (covered in Chapter 15).

We recommend that an analysis of the four methods be conducted before choosing a specific size.

14.3.1 Sizing by Comparison of Supply and Demand

A supply versus demand comparison can be used to estimate the required storage capacity. In this case, the ratio of the minimum source production ($Q_{min\text{-}source}$) to maximum daily demand (Q_{max}) is determined to identify the appropriate storage capacity, as related in Table 14-2.

Table 14-1. Practical Size Limitations for Water Storage Tanks

Tank Type	Practical Size Limitations (gal)
Ferrocement	700–3,000
Stone	5,000–100,000
Cement block	Unlimited
Poured concrete	Unlimited
Plastic	125–3,000

Note: Local construction knowledge may decrease or increase these size limitations.

Source: Curtis 2000.

Table 14-2. Method to Determine Tank Size by Comparing Supply and Demand

$Q_{min\text{-}source}/Q_{max}$	Tank Size
<1	Source is probably not feasible.
1–2	$1/2\ V_{max}$
2–3	$1/4\ V_{max}$
3–4	$1/8\ V_{max}$
>4	Open system (no tank) might be more appropriate.

Source: PCDR 2000.

This method simplifies demand to a constant flow when in fact demand fluctuates over the day from peak demand to a value as low as zero when all taps are closed at night. Although tank sizing by this estimate can meet the needs of the community over the course of a day, perhaps by forcing the individuals to change their water consumption habits, the system will not provide water on demand. To overcome this issue of demand over the day, community members may store their water, causing recontamination of water and insect breeding.

14.3.2 Sizing by Ability to Meet the Peak Demand

This method is based on the ability to meet the peak demand of a community and is explained in Example 14-1.

))) Example 14-1. Sizing Water Storage Tanks to Meet Peak Demand

Calculate the tank volume to satisfy peak demand provided the following information:
- a current population of 402 individuals;
- a projected growth rate of 1.4%;
- a project design life of 20 years;
- per capita water consumption of 50 L/capita-day;
- a dry season flow at the source of 1 L/s;
- a maximum daily usage safety factor of 1.25; and
- a peaking factor of 4.

To characterize the water demand:

Design population

$$P_N = P_O\left(1 \times \frac{r \times N}{100}\right)$$

$$P_{20\ yr} = 402 \times \left(1 + \frac{1.4 \times 20}{100}\right)$$

$$= 515\ \text{people}$$

Average daily usage
$$V_{ave} = P_N \times \omega_R + \text{infrastructure}$$
$$= 515 \text{ people} \times 50 \text{ L/capita-day} + 0$$
$$= 25,750 \text{ L/day}$$

Maximum daily usage
$$V_{max} = V_{ave} \times \text{safety factor}$$
$$= 25,750 \text{ L/day} \times 1.25$$
$$= 32,200 \text{ L/day}$$

Demand flow rates
$$Q_{ave} = 25,750 \text{ L/day} \times (1 \text{ day}/86,400 \text{ s})$$
$$= 0.30 \text{ L/s}$$
$$Q_{max} = 32,200 \text{ L/day} \times (1 \text{ day}/86,400 \text{ s})$$
$$= 0.37 \text{ L/s}$$

To determine the maximum and minimum design flow rates for a treatment facility or piping network, a peaking factor is applied to the average daily flow rate. A *peaking factor (PF)* is a multiplier that is used to adjust the average flow rate to design or size components in a water or wastewater treatment plant or components of a water distribution or wastewater collection system (e.g., pipes, pumps, storage tanks) (Mihelcic and Zimmerman 2009). In this example, the peaking factor is a scaling factor that tells us how much greater the maximum demand is within the distribution system (Q_{peak}) in the short-term demand compared to the maximum demand in the overall system on a daily or long-term basis.

$$Q_{peak} = Q_{max} \times 4$$
$$Q_{peak} = 1.5 \text{ L/s}$$

The situation requiring the largest volume of storage would be when the entire maximum daily demand volume is withdrawn at the peak demand flow. The time that it would take to draw V_{max} at Q_{peak} is termed τ_{peak}. It is calculated as

$$\tau_{peak} = V_{max}/Q_{peak} = 32,200 \text{ L}/1.5 \text{ L/s} = 21,500 \text{ s} = 6 \text{ h}$$

By our definitions, $\tau_{peak} \times$ peaking factor equals 24 h.

To calculate the storage volume required in this scenario, the rate at which water can be provided by the source is subtracted from the peak demand:

$$V_{tank} = (Q_{peak} - Q_{min-source}) \times \tau_{peak} = (1.5 \text{ L/s} - 1 \text{ L/s}) \times 21,500 \text{ s} = 10,700 \text{ L}$$

This is the most conservative case. Under actual conditions, the maximum daily water volume is not withdrawn all at once (i.e., under peak flow conditions).

One method to double-check this tank size is to determine the amount of water available per person if service to the tank is interrupted. In this case, the storage tank would provide approximately 21 L per person for the design population if service was disrupted for 24 h. This amount is close to the World Health Organization's definition of *reasonable access,* which provides for the availability of at least 20 L per person per d (L/capita-day) from a source within 1 km of the user's dwelling. 》》》

14.3.3 Sizing by Critical Reserve

Another method for determining tank volume is based on the desired amount of critical reserve that should be available if service is interrupted for a given time (Table 14-3). This method requires the user to determine if the critical time to provide water is 1 day or 1 week, and whether the flow is based on supply by the source or demand throughout the system.

》》》 14.4 Construction

As mentioned previously, the type of materials and construction techniques are typically based on local conditions. The decision to build a circular tank versus a rectangular tank depends on the method of construction. Typically, block masonry tanks are rectangular, whereas brick, stone, and ferrocement tanks are circular. Tank roofs are usually constructed as elevated concrete slabs. However, tanks could be made from ferrocement, precast concrete, plastic, or metal.

14.4.1 Tank Location

Locate storage tanks at a point higher than all of the existing (and planned) houses in the community. As much reinforced concrete tank as possible may be constructed below ground level because the surrounding earth will keep the stored water cool and will provide additional exterior support to the tank.

14.4.2 Foundations and Walls

Tank floors are usually reinforced concrete, but depending on tank size, could be brick or rock and concrete composite slab construction. Chapter 7 describes foundations, slabs on grade, and wall construction for various materials. Chapter 6 provides information on slab thickness and reinforcement, as well as concrete mix designs. After clearing and excavating the site, a 10–15-cm layer of gravel is placed and compacted in the excavated

Table 14-3. Critical Reserve Method Used to Determine Tank Volume

Reserve Period	Calculation Method to Find Tank Volume, V_{tank}
1 day of system-wide demand	V_{ave} or V_{max}
1 day of source flow	Q_{source} $(L/s) \times 86,400$ (s/day)
1 week of system-wide demand	7 days $\times V_{ave}$ (L/day)
1 week of source flow	Q_{source} $(L/s) \times 86,400$ (s/day) $\times 7$ (days/wk)

area to accommodate runoff during the rainy season and thus prevent erosion under the tank foundation (Fig. 14-1). As also shown in Fig. 14-1, the walls of the tank are connected to the foundation by rebar bent in an L shape and tied to foundation rebar before the foundation concrete is poured.

14.4.3 Tank Piping

Piping consists of inlet, outlet, and overflow piping. Figure 14-2 shows how piping may differ between a community gravity-fed water storage tank and a rainwater storage tank. For gravity-fed systems, the inlet piping includes a gate valve and is made of galvanized

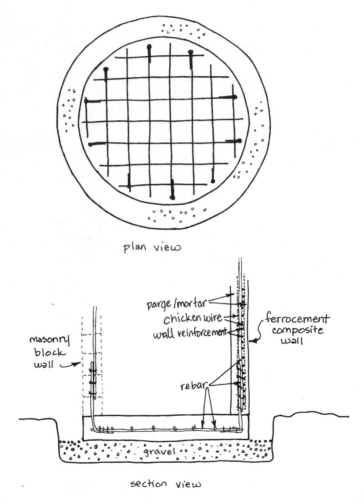

Figure 14-1. The Plan and Cross-Sectional Views of a Water Storage Tank, Showing Foundation and Wall Reinforcement.

The cross-sectional view shows a masonry wall (left) and a ferrocement wall (right) that are tied to the foundation slab.

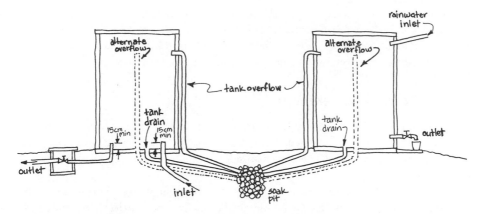

Figure 14-2. Examples of Piping for a Gravity-Fed Water Storage Tank (Left) and for a Rainwater Storage Tank (Right).

Note: The inlet and outlet pipe for the gravity-fed system (left) are a minimum of 6 in. above the tank floor to prevent sediment from entering the pipes.

iron (GI) the same size as the supply line. The outlet piping is also made of GI; it is located opposite the entrance and is the same size as the distribution line. Locate the outlet pipe 15–20 cm above the floor of the tank to keep sediment out of the line, and include a control or shutoff valve. Locate the shutoff valve outside the tank and protect it by a valve box made from a suitable material. Size the valve box so that a hand and tools can easily fit in the box to perform maintenance or replace the valve. Place a cover on the valve box that can be locked to prevent unauthorized water release from the tank. Rainwater harvesting tanks typically have a faucet on the side of a tank wall that acts as a shutoff valve.

For all tanks, install an overflow pipe so that tank overflow will not lead to erosion of the tank foundation. Install the overflow pipe about 20 cm below the top of the tank. Direct the overflow to an outlet situated away from the tank, a stream, or a location protected from erosion by rocks. If desired, install a clean-out pipe or drainpipe in the center of the tank foundation to facilitate cleaning of the tank because sediment may collect on the tank bottom. If a gravity-fed tank is not sized for rainy season flow, then the excess water will need to be routed away without causing erosion or ponding that could lead to mosquito breeding.

Table 14-4 provides information on typical sizes of clean-out and overflow piping. Though not depicted in Fig. 14-2, ventilation piping is located on the roof of a tank

Table 14-4. Sizing Clean-Out and Overflow Piping in a Storage Tank

Volume (gal)	Diameter (in.)
5,000	2
10,000	3
15,000	3
20,000	4

receiving water via a spring or gravity-fed distribution systems. The ventilation pipe allows air to escape when water enters the tank. It is made of 1- to 2-in. GI pipe for tanks < 20,000 gallons and 3-in. pipe for tanks > 20,000 gallons. It is curved downward and screened at the end to prevent rain, insects, and animals from entering.

14.4.4 Other Considerations

Place an access hatch in the roof, at least 60 cm × 60 cm, with a secure and sanitary cover. Access hatches may be prefabricated or constructed locally. They may be galvanized iron, wood, or cast concrete (see Chapter 7). Consider location, environment, and vandalism in the decision.

Place a small tank (the *chlorinator*) above the storage tank to provide disinfection. Dissolve some form of hypochlorite solution in this smaller tank and drip it into the larger storage tank. The period of contact must be at least 30 min to ensure that the chlorine is mixed well in the storage tank. Adjust the drip of chlorine manually so that the chlorine residual in the distribution line is maintained at > 0.3 mg/L.

⟩⟩⟩ 14.5 Ferrocement Tanks

The expense of concrete construction is sometimes a major barrier. As an alternative, *ferrocement tanks,* or wire-reinforced cement–mortar tanks have been used successfully. Such tanks are preferred because they rely on commonly available materials and simple, easily transferable skills. The construction likewise requires simple equipment so that local people can easily contribute labor and local materials like sand instead of money. The framework may be prefabricated and used in the construction of many different tanks. Finally, the lifetime of a properly constructed ferrocement tank has been proven to be longer than that of a masonry tank. Although these tanks may be built in many shapes, they are most commonly built as cylinders or jar shapes (this shape is aesthetically pleasing, and the curved walls add strength). Ferrocement tank construction was discussed in Chapter 7.

⟩⟩⟩ 14.6 Household Tanks

Throughout the world, households and businesses invest in water supply systems even when they are connected to and pay for water services from a private or government provider. These supply systems usually include storage tanks, pumps, pipes, and a structure to elevate storage tanks above the house. The systems provide water throughout the day even when the supply from the water main is low or zero. Generally, bottom tanks are connected to the main supply lines and are filled when water pressures are high. Water from the bottom tank is then pumped to the top tank at a higher elevation, where it is then connected to all of the house pipes and it is used for drinking, cooking, washing, and flushing toilets. Wooden, aluminum, and plastic tanks are commonly used. In addition to paying the water company, individuals and businesses incur additional costs associated with their own water tank systems. Although these individual systems help to guarantee a more constant supply of water, they may affect water quality when it does reach the household tap.

In some locations, the tanks may become breeding grounds for mosquitoes responsible for the spread of diseases like dengue fever. The type of tank, the manufacturing process for the plastic, and the source of the water may also affect the concentrations of bacteria (Tokajian and Hashwa 2004), heavy metals, and organics in the water.

))) References

Curtis, J., Contributing Ed. (2000). *Rural aqueducts and community development in the Dominican Republic.* U.S. Peace Corps, Santo Domingo, Dominican Republic.

Mihelcic, J. R., and Zimmerman, J. B. (2009). *Environmental engineering: Fundamentals, sustainability, design,* John Wiley and Sons, New York.

Peace Corps Dominican Republic (PCDR). (2000). *Training information packet.* PCDR, Santo Domingo.

Tokajian, S., and Hashwa, F. (2004). "Microbiological quality and genotypic speciation of heterotrophic bacteria isolated from potable water stored in household tanks." *Water Qual. Res. J. Canada,* 39(1), 64–73.

))) Further Reading

Jordan, T. D. (2000). *A handbook of gravity-flow water systems.* Intermediate Technology Publications, London.

Lifewater International, Designing a ground level storage tank, Technical Note No. RWS. 5.D.2, Water for the World, Development Information Center, Agency for International Development, Washington, D.C. <http://www.lifewater.org/resources/rws5/rws5d2.pdf> (Jan. 23, 2009).

Reents, N. (2002). *Designing water supply systems in rural Honduras,* <http://cee.eng.usf.edu/peacecorps/Resources.htm> (Jan. 23, 2009).

Watt, S. B. (1978). *Ferrocement water tanks and their construction,* Intermediate Technology Publications, London.

15

Development of Natural Springs

⟫⟫ 15.1 Capturing and Protecting Natural Springwater to Reduce Water-Related Illnesses

In the developing world, many communities rely on natural springs as a source of drinking water. Often, these springs are unprotected, as shown in Fig. 15-1. Springwater can generally be considered safe because it has been filtered through the ground, provided that there are no nearby groundwater pollution sources (e.g., latrines or livestock pens). However, if the spring is open, the spring pool can be contaminated by children wading as they draw water, people bathing in or near the spring, water users dipping dirty buckets into the spring, or nonpoint source runoff from agricultural activities. Springs can also act as potential breeding grounds for mosquitoes, creating unnecessary opportunities for people drawing water to be infected by malaria.

Figure 15-2 shows a typical arrangement of an unimproved spring. Figure 15-3 shows several engineered methods for developing natural springs. In this chapter, the construction of the spring box that is shown in Fig. 15-3c is presented in detail. An alternative to spring box construction is a spring collection system with filter material and perforated pipes buried to capture water in groundwater veins (Fig. 15-3a). Such spring collection systems may be more appropriate in settings where the spring does not conform to the classic pond-type spring shown in Fig. 15-1. Huisman et al. (1981), the Natural Resource Conservation Service's *Engineering Field Handbook* (NRCS 1983), and Pickford (1991) provide details of other designs for spring development.

Spring improvement can be accomplished incrementally. For example, a community may begin by establishing rules for watershed protection and use of agricultural chemicals near the source, protecting the spring with a fence, using only clean buckets for drawing water, and washing clothes downgradient of the source. Then, once the community establishes these practices, it may move to a construction project of either a spring box or another type of spring collection system.

As summarized in Table 15-1, water-related diseases include four classifications, and each classification requires different prevention strategies. Table 15-1 also shows where a spring improvement project can intervene in disease prevention strategies.

Table 15-1 shows that spring boxes are most effective at preventing water-borne diseases. Instead of being allowed to pool in an open pond, water is protected on leaving the ground. Water in unprotected springs is typically turbid and contains leaves and other

Figure 15-1. An Unprotected Spring.

Note: Before construction of a spring box, the spring water is allowed to collect in this open pool. Water stagnates, people dip dirty buckets into the pool, and the spring is a mosquito breeding area.

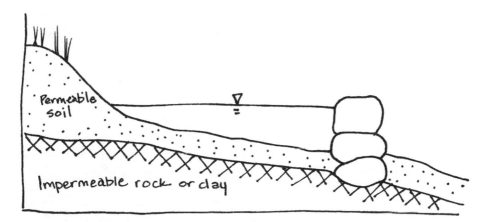

Figure 15-2. A Typical Arrangement of an Unimproved Spring.

Note: To prepare for spring box construction, remove the permeable layer of sand or soil at the spring pool.

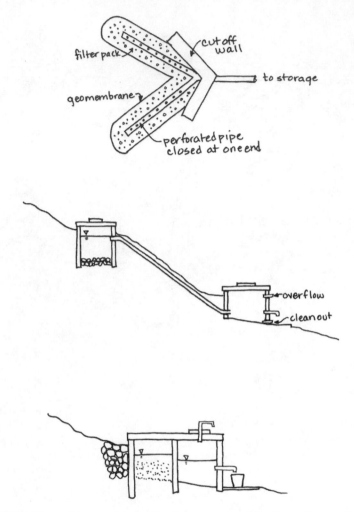

Figure 15-3. Three Alternative Spring Improvement Designs (Top to Bottom).

Note: (a) Spring collectors using perforated pipes, filter pack, and geomembranes; (b) spring box with reservoir downgradient of collection box; and (c) spring box with collection, filter, and reservoir at the source.

objects (e.g., frogs and aquatic plants). A noticeable improvement in water quality can usually be observed after the construction of a spring box.

Spring boxes have only a minimal effect on water-washed diseases. Because these diseases result from insufficient bathing, their prevention depends on improving the accessibility and quantity of water (see Chapter 9). Although the quantity of water cannot be augmented, the reliability is improved by allowing water to be stored in a reservoir. Spring boxes generally fail to improve the accessibility of water. However, if the spring is situated higher than the community, a gravity-fed water distribution system (see Chapter 11) can be installed to improve accessibility. Improved hygiene is achieved through education and behavior change; construction of a spring box alone is not enough.

Table 15-1. Classification of Water-Borne Illnesses, Strategies for Their Prevention, and Potential Effects of Spring Improvement Projects

Classification	Definition	Prevention Strategies	Can Spring Improvement Help?
Water-borne	The pathogen is in water ingested by a person or animal, who then becomes infected.	Improve drinking water quality.	Yes
		Prevent casual use of unprotected sources of water.	Yes
Water-washed	Results from inadequate bathing. Depends on quantity of water available.	Increase water quantity.	No
		Improve accessibility.	No
		Improve reliability.	Yes (store water)
		Improve hygiene.	No
Water-based	The pathogen spends part of its life cycle in a water snail or other aquatic animal.	Reduce need for contact with infected water.	Yes
		Control snail populations.	Yes
		Reduce contamination of surface waters.	Yes
Insect–vector	The spread of the disease relies on insects breeding or biting in or near water.	Improve surface water management.	No
		Destroy breeding sites.	Yes
		Reduce the need to visit breeding sites.	No
		Use mosquito netting.	No

Source: Adapted from Cairncross and Feachem 1993.

A spring box can have an effect on water-based diseases. Before construction, springs are often home to a number of aquatic animals, including snails, frogs, and small fish. When the pond is drained and water is stocked in a protected reservoir, these animals can no longer exist in the spring. Spring box construction may also have an effect on diseases transmitted by insect vectors, such as malaria. By eliminating stagnant water, mosquitoes lose their breeding ground. However, improved surface water management, overall reduction of breeding sites, and promotion of mosquito netting must also be integrated into the project.

Because spring box construction projects are relatively inexpensive and a large proportion of the materials are usually available at the community level (e.g., sand, gravel, and rocks), spring box projects are generally feasible for villages. A spring box requires minimal maintenance because there are no mechanical parts, and it is not easily damaged by children, who are often responsible for drawing water. If the springs are already well used by the communities, few behavior changes are necessary.

⟫⟫ 15.2 Spring Box Design

15.2.1 Choice of Spring

If the community has several springs from which to choose, evaluate each source. Choose the spring based on surveying community habits, output of the spring, quality of the water (if no water quality testing kits are available, this can be based on a sanitary survey, turbidity testing using a turbidity tube, visual appearance, odor, and opinions of villagers on taste), and accessibility. Also, choose a spring from which water drains readily. If the spring appears to be simply a pool of water and water does not flow away from the spring, it will not be an appropriate location for a spring box. Some springs are difficult to access because of rough terrain. In this case, it may be difficult for women and children to carry buckets of water up the steep slope. More information on the hydrology of springs and selecting a spring for a community water source can be found elsewhere (e.g., Huisman et al. 1981).

Reasons that villagers are in the habit of using a specific spring instead of others may include the following:

- the water looks cleaner and tastes better,
- it is closer to more people,
- the flow is constant and sufficient, and
- the location has been historically used by a community's ancestors.

Use the preferred spring of a community if possible. Water collection habits are difficult to change.

15.2.2 Design and Principles

A basic spring box design is shown in Figs. 15-4 and 15-5. Because this is a general design and each spring is different, design changes are made on a site-specific basis.

Example 15-1 provides design calculations for sizing each component of the spring box. The output of the spring is required for design calculations and can be measured using a weir in the drainage channel (Mann and Williamson 1973), as shown in Fig. 15-6. Another method of measurement is to dam the spring temporarily and record the time needed to fill a bucket of a known volume with the overflow from the dam. The output can then be used to determine the size of the reservoir.

⟫⟫ Example 15-1. Example Calculations for a Spring Box Design

Primary Filter and Capture Area

Sizing the primary filter and capture area depends on the placement of the veins. Therefore, no calculations are required. To find the veins, open the spring by digging a channel that allows the spring pool to drain. Once the pool is empty, the veins will be visible. Excavate to the impermeable layer of rock or clay (shown in Fig. 15-2), and then embed the walls into the surrounding earth to capture water coming into the spring from the veins.

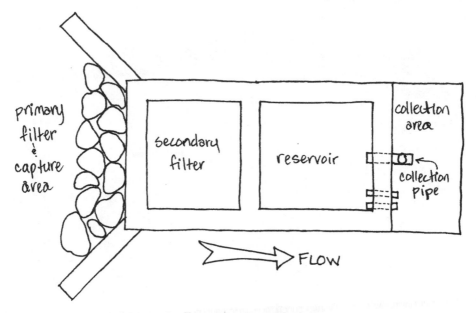

Figure 15 4. A Plan View of a General Spring Box Design.

Note: Dimensions are specific to the output of the spring and the storage needs. The filter may be omitted from the design.

Source. Adapted from work of Lauren Fry with M. E. Ekoe from April 2002 to April 2004.

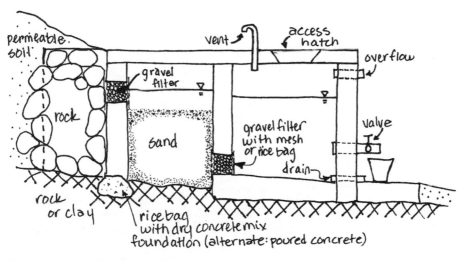

Figure 15-5. A Side View of a General Spring Box Design.

Note: The filter may be omitted from the design.

Source: Adapted from work of Lauren Fry with M. E. Ekoe from April 2002 to April 2004.

Figure 15-6. Measurement of Spring Output Using a Weir.

Sizing the Slow Sand Filter

For a spring with a measured output of 6 L/min during the peak of the rainy season (6 L/min = 0.36 m³/h) and a desired filtration rate (v) of 0.1 m³/m²/h (see Chapter 18 for guidance on desired filtration rates):

$$v = \frac{Q}{A}$$

Solving for the area, A:

$$A = \frac{Q}{v} = \frac{0.36 \text{ m}^3/\text{h}}{0.1 \text{ m}^3/\text{m}^2/\text{h}} = 3.6 \text{ m}^2$$

In some cases, where space or materials are not available to build a slow sand filter, gravel can be substituted, assuming that water is sufficiently filtered by the ground before it enters the spring.

Reservoir

Assuming that no one collects water between 6:30 P.M. and 5:30 A.M., the reservoir can be sized to hold the amount of water that can be stored after 11 h of filling. For 11 h of filling during the peak of the rainy season, the maximum volume, V, that can be stored overnight can be determined using this equation:

$$V = Q \times t$$

$$V = 0.36 \text{ m}^3/\text{h} \times 11 \text{ h} = 3.96 \text{ m}^3$$

The volume of water actually available to the community is that located above the level of the water tap. The location of the water tap is usually located close to the ground; thus, there is not much difference between the design volume and the volume of water actually available to a user.

(Source: Adapted from Fry et al. 2006).   》》》

15.2.3 Primary Filtration and Capture Area

Spring boxes are a simple design. As shown in Fig. 15-4, the primary filtration and capture area is built directly over the spring pool, after digging to the impermeable layer of bedrock or clay.

Water leaving the ground enters the primary filtration and capture area, built with walls angling into the surrounding ground to capture as much water as possible. Rocks fill the primary filtration and capture area to the level of the entrance to the secondary filter. The entrance to the secondary filter is a layer of gravel 5–10 cm thick cast in a wall, about 0.75 m above the bottom of the wall.

The size of the primary filtration and capture area depends on the location of the veins augmenting the spring. If the veins are widely spaced, the capture walls must also be spread out. Because the springwater pool is usually located in a low spot just below a steeper edge of soil, build walls into the surrounding earth and extend them approximately 30–50 cm into the soil to provide stabilization and ensure that water is guided into the spring box. The natural impermeable layer of bedrock or clay acts as the floor of the primary filtration and capture area, so no poured concrete floor is required.

15.2.4 Secondary Filter

When water leaves the primary filtration and capture area, it enters the secondary filter. Because protected groundwater is essentially clean as it leaves the spring, the filter only removes sediment that may be caught by the water as it leaves the spring and moves through the primary filtration and capture area. In some cases, if the groundwater exiting the spring has greater potential for pollution, design the secondary filter as a slow sand filter (see Chapter 18). If the spring box is located far from a potential pollution source, such as a latrine, the relatively large surface area required for slow sand filtration may an unnecessary expense. In fact, a filter is omitted in many spring box designs, with the understanding that natural subsurface processes have already filtered the springwater.

Instead of constructing slow sand filters, the filters may be filled with coarse sand or small gravel so that filtration is rapid. In this case, design the filter to be approximately 0.5 m, 1.5 m, and 1 m for length, width, and height, respectively, with a filter media height of 0.4 m. Water enters the filter through the gravel entrance built into the wall and travels down through the filter. At the bottom, the wall between the filter and the reservoir contains another gravel filter entrance.

15.2.5 Reservoir

Size the reservoir based on the output of the source to maximize the amount of water that is stored overnight (Example 15-1). For instance, as shown in Example 15-1, if a spring has a maximum output during the rainy season of 6 L/min, assuming that no one draws water between 6:30 P.M. and 5:30 A.M. (11 h of filling), the reservoir size required for 1 day of storage would be calculated as 3.96 m³:

$$\frac{6L}{\min} \times \frac{60\,\min}{h} \times \frac{11\,h}{day} \times \frac{m^3}{1,000L} = 3.96\ m^3 \tag{15-1}$$

However, the size may be restricted by the means of the community to provide not only local sand and gravel but also budgeted funds for items such as cement, rebar, and *sikalite* (a waterproofing admixture used when placing concrete in wet conditions). Slope the floor to the drain (note the drain in Fig. 15-5). In addition, parge the walls and floor inside the reservoir to make them watertight and easy to clean (see Chapter 7).

Locate a concrete slab at the exit of the reservoir to provide a surface on which to set buckets. Dig drainage canals to channel overflow and spillage away from the site, to avoid stagnant water. Pipes that extend out of the spring box should be short to prevent breakage. Breakage can be caused by children standing or swinging on the pipe, or large animals rubbing against it. Additionally, at the community's request, construct a stand for washing clothes a few meters downstream of the reservoir.

Some adjustments may be necessary depending on the situation of the spring. For example, the design may include stairs to the collection area, retaining walls, or extra drainage.

Chapter 4 provided a list of items to include in a budget estimate for a typical spring box project. The materials are usually readily available, if not in the village itself, then in regional centers. Spring box maintenance and repair costs are minimal. For example, items required include cement and sand for mortar to reseal the reservoir lid after cleaning (less than half a sack per cleaning) and replacement faucets and pipes in case of breakage. The system is durable to withstand everyday use by children and has few parts that break. After construction is completed, maintenance is the responsibility of the management committee. A maintenance schedule is provided in Table 15-2.

))) 15.3 Spring Box Construction

Once funding is received, construction can begin immediately and can take as little as 3 weeks, depending on the motivation and participation of the community. A typical spring box construction schedule was provided in Chapter 4. The construction phase is also an opportunity to educate community members about operation and maintenance, as well as general issues of public health.

Table 15-2. Typical Maintenance Schedule for a Completed Spring Box Project

Schedule	Tasks
Monthly or more often	Cleaning around the site (i.e., raking leaves, clearing the drainage canal, clearing brush from the access trail)
Every 6 months or when water does not look clean when exiting the reservoir	Open the reservoir and filter. Clean reservoir walls and floor, washing with bleach. Rinse or remove the top layer of the filter material and clear debris from gravel layer entrances. Covers can be secured with a cement mortar.
Unexpected maintenance	Replace broken pipes, repair cracks and leaks.

Note: Educating and training the community and management committee on the importance of maintenance is crucial to ensure the long-term sustainability of the project.

Source: Adapted from Fry 2004.

15.3.1 Preparation of the Spring

Construction takes place in five phases. To begin, drain the spring pool completely by removing barriers that had been used to dam the water. This process involves digging a channel down from the spring and clearing leaves, branches, and vegetation from the immediate area. Once the spring is drained, the veins augmenting the spring become visible. This step shows where the primary filtration and capture area need to be built and determines the size of the capture area.

15.3.2 Construction of Primary Filtration and Capture Area, Filter, Reservoir, and Finishing Touches

After the source is drained and cleared of debris, begin construction on the three walls of the primary filtration and capture area. First, excavate to the impermeable layer of rock or clay just below the pool (see Fig. 15-2). It is important not to break through the impermeable layer, which could cause permanent loss of the spring. Foundations may be poured reinforced concrete or dry concrete mix secured in rice bags. The floor and slabs may be poured concrete or other composite slab on grade construction options (see Chapter 7).

Then construct walls in about 2 days. Reinforced concrete, reinforced masonry block, and rock masonry covered with mortar are viable wall construction options. Take special care when constructing and embedding the gravel filters in the filter and reservoir walls. After the capture area is in place, construct the secondary filter. Filter construction generally takes 2 days and is followed by the construction of the reservoir walls. Construction of the reservoir takes 3–4 days: 2 days for the walls and another day or 2 to place the sloped floor and parge or seal the interior reservoir walls. After the reservoir is complete, construct covers and access hatches for the filter and reservoir, place rocks in the capture area, and fill the filter with sand or gravel. Final improvements and reinforcement of surrounding earth can take up to 4 days. Box 15-1 describes potential problems that may be encountered during the construction of a spring box.

>>> **Box 15-1. Potential Problems Encountered During Spring Box Construction**

Every spring box construction project encounters difficulties. Suggestions to avoid problems during construction are listed below:

- Avoid accidentally digging through the impermeable layer because this accident may cause water to be lost from the spring.
- When filling the primary filtration and capture area, do not block the veins augmenting the spring. Be sure that the veins are not obstructed by mud.
- Covering the primary filtration and capture area with earth after it is filled with rocks may be appropriate in some places, but where soil is fine, the earth could filter through the rocks and block the gravel layer entrance. In this case, fill the primary filtration and capture area with rocks and cover them with a plastic tarp, rice bags, or other impermeable material before covering them with earth.

>>> **References**

Cairncross, S., and Feachem, R. G. (1993). *Environmental health engineering in the tropics: An introductory text,* Second ed. JohnWiley and Sons, Chichester, U.K.

Fry, L. M. (2004). *Spring improvement as a tool for prevention of water-related illness in four villages of the Center Province of Cameroon,* <http://cee.eng.usf.edu/peacecorps/Resources.htm> (Jan. 18, 2009).

Fry, L. M., Mihelcic, J. R., and Watkins, D. W. (2006). "Improving public health by improving water supply: Results from springbox projects in Cameroon." *J. Engineering for Sustainable Development: Energy, Environment, and Health,* 1(1), 33–42.

Huisman L., De Azevedo Netto, J. M., Sundaresan, B. B., Lanoix, J. N., and Hofkes, E. H. (1981). *Small community water supplies: Technology of small water supply systems in developing countries,* IRC International Water and Sanitation Centre, The Hague, Netherlands.

Mann, H. T., and Williamson, D. (1973). *Water treatment and sanitation: Simple methods for rural areas.* The Russel Press, Nottingham, U.K.

National Resource Conservation Service (NRCS). (1983). "Springs and wells." *Engineering field handbook,* National Resource Conservation Service, Washington, D.C. <http://directives.sc.egov.usda.gov/17550.wba> (Jan. 18, 2009).

Pickford, J. (1991). *The worth of water: Technical briefs on health, water and sanitation.* IT Publications, London.

Schweitzer, R. N. (2009). *Community managed rural water supply systems in the Dominican Republic: Assessment of sustainability of systems built by the National Institute of Potable Water and Peace Corps, Dominican Republic,* <http://cee.eng.usf.edu/peacecorps/Resources.htm> (May 21, 2009).

16

Manually Constructed and Operated Water Wells

))) 16.1 Producing Water in a Sustainable Fashion

Water wells have been used for water supplies since the beginning of recorded history. The first wells were certainly dug by hand, and manual drilling methods for water wells are thought to have been first used in China as early as 1000 B.C. (Brantly 1961).

Digging is the most labor-intensive and least equipment-intensive method for installing a well and continues to be used in many parts of the world. Digging becomes more difficult with depth because of the increase in compaction, greater lifting and shoring requirements, the occurrence of bedrock, and the presence of groundwater. Water well installation technologies and methods have evolved to help overcome these difficulties. Compact, portable, manual or motor-powered drills are available for drilling wells to diameters approaching 15 cm and depths of tens of meters.

In favorable conditions (plentiful and regular recharge conditions, shallow (<5 m) groundwater table, permeable aquifers, and acceptable water quality), groundwater supplies for small populations can be developed without technical training because wells are able to produce water at adequate rates in a sustainable fashion. Even for small hand-dug and hand-drilled wells, however, it is advantageous to use technical considerations to maximize the well yield (production) and provide for future enhancements to the pumping systems.

The more water a well is expected to produce, the more engineering analysis is required to optimize the hydraulic behavior of the well and minimize the pumping power requirements. Technical expertise is needed to overcome challenges related to supplying water for large populations and from deep aquifers in hard rock or fine-textured sediments where the water is not replenished regularly or rapidly (i.e., long dry seasons or extended droughts). Safety and hygiene considerations are equally important when planning a well project.

))) 16.2 Subsurface Geological Conditions and Groundwater

A typical geologic profile near the surface is demonstrated in Fig. 16-1. *Unconsolidated* materials (loose or compacted, broken particles) form from physical and chemical

---

This chapter was written by John S. Gierke.

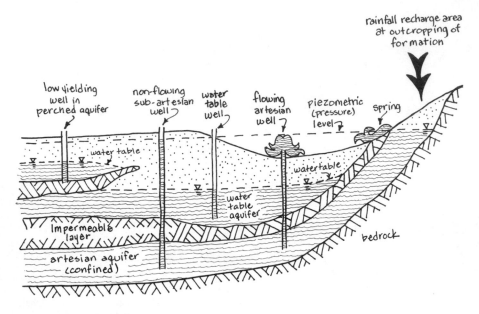

Figure 16-1. General Geological Profile Depicting the Occurrence of Groundwater for Water Supply Development.

Source: After Watt and Wood 1979.

weathering of rocks (*consolidated* materials). Unconsolidated materials can originate either from the underlying bedrock or from rocks elsewhere that have been deposited by wind, water, and glaciers. This kind of general geological profile exists everywhere, differing in types and depths of soils and rock.

The most important feature of the soil and rock profile for water supplies is that the bedrock is typically less permeable than the overlying sediments and soil. This impermeability tends to inhibit downward movement of infiltrating water and allow accumulation of water in the sediment and soil pore spaces, which can then be extracted with wells. Fractures in the underlying rock tend to diminish in number, size, and extent with depth, so the uppermost part of the bedrock is the most amenable to manual well installation methods, and the most likely to yield water at sufficient rates for a well. Situations do occur in bedrock aquifers where wells drilled to the same depth in close proximity (only meters apart) differ in their ability to produce water. This difference probably occurs because the different wells encountered different fractures.

The groundwater table depth varies depending on the geology, climate, surface hydrology, and land use. It tends to mimic the surface topography, albeit in a subtler manner (i.e., less relief). Thus, the water table is typically deeper (from the ground surface) in higher terrain and shallower in lower terrain. In the absence of human disruptions of the water cycle, the groundwater table fluctuates seasonally in an oscillatory fashion, trending toward peak elevations during wet seasons and deepest levels when it is dry.

⟫⟫ 16.3 Siting a Well

The needs and wishes of a community should be combined with knowledge about groundwater (both local knowledge and available outside technical knowledge) when determining the placement of a new well. This placement can become a highly political issue. Because women are often responsible for obtaining water, involve them in the determination of the new site. If the well is meant to serve an entire community, take care to ensure that the least influential families are also considered when the site is determined.

The location and depth of existing wells is one important source of information. However, an additional source of information that should not be overlooked is the location of failed wells (MacDonald et al. 2005). Where have people tried but failed to reach groundwater? At what depth did digging or drilling stop? Where have wells been installed but had yields so low that they were not usable?

Additional clues are available in the form of site topography:

- the tendency of groundwater to discharge toward low areas and permanent surface water bodies;
- groundwater seeps and springs;
- vibrant vegetation, especially succulent plants that require a shallow water table; and
- linear topographic features suggesting subsurface fractures and faults.

Dowsing is considered by some communities to be an important way to locate a well. Dowsers are often residents who, through experience, have developed an inherent knowledge of local groundwater conditions that can contribute to their ability to find water. Accepting the recommendation that a dowser provides may be an important social or cultural factor influencing community participation in a project. Unfortunately, doing so without tempering this recommendation with engineering judgment can often lead to illogical locations for wells. To prevent this situation from occurring, a community and an engineer can work together to predetermine the area in which dowsing will take place. Then, the dowser will already be working in an area with features that suggest the proximity of groundwater to the surface.

States in the United States promulgate setback distances for locating water wells relative to contamination sources, such as septic systems. For example, a minimum setback distance of 15 m (50 ft) is common to many states. The setback refers to the lateral distance from the potential contamination, such as a soil absorption system and the well. However, these guidelines ignore the underlying soil, rock, and aquifer conditions; infiltration capacity and recharge rates; depth to the groundwater; and radius of influence of the pumping well.

Because of the variety of natural complexities and community influences that surround locating a well, it is beyond the scope of this chapter to discuss all of the facets related to siting a well in relation to potential contamination sources. Rather than focus on a setback distance, it is more prudent to consider locating wells as far from upgradient (typically upslope) sources of contamination as is practically possible, avoiding the direct downgradient position. The location of the well relative to latrines and soil absorption systems is most important when the water table is shallow (less than about 7 m deep), especially in areas where the soils readily allow infiltration and recharge rates are high. In arid regions or where the water table is deep (more than 20 m), latrine and septic contamination of wells is less likely.

))) 16.4 Well Depth and Size: Hydraulic Considerations

Wells should be dug or drilled during the latest part of the dry season to ensure that they are deep enough to endure the seasonal drop in the water table. Dug wells are also less likely to collapse during construction at this time. However, keep in mind that withdrawing water from the aquifer may cause dry-season water levels to drop even further.

Drawing down the water level in a well causes groundwater to flow toward it, as shown in Fig. 16-2. The larger the difference between the water level in the well and the groundwater table near the well, the greater the driving force (*hydraulic gradient*, which is the slope of the groundwater table) causing flow (Q). Water flow rates into a well are proportional to the transmissivity (T) of the portion of the aquifer spanned by the well ($Q \propto T$). As the water level in the well recovers from being drawn down because of

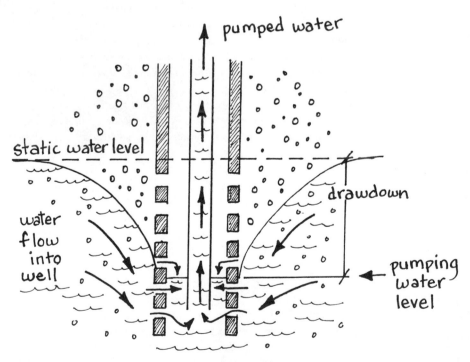

Figure 16-2. Drawdown in a Well Because of Pumping.

Source: After Watt and Wood 1979.

pumping, the hydraulic gradient decreases and the recovery rate decreases, as shown in Fig. 16-3.

When the groundwater table in an unconfined aquifer drops, whether naturally during the dry season or because of excessive abstraction, the aquifer transmissivity decreases because the total depth of the aquifer decreases. The transmissivity of confined

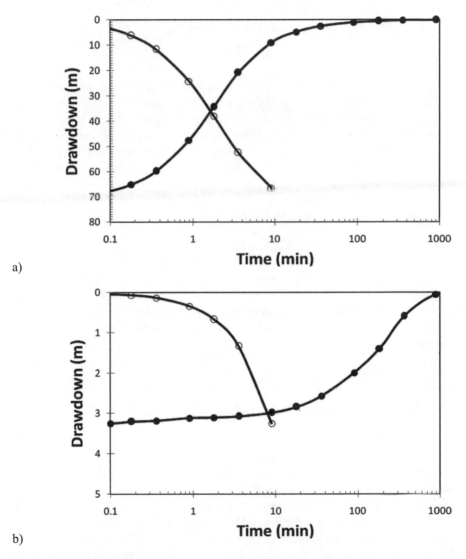

a)

b)

Figure 16-3. Examples of Drawdown (Open Circles) and Recovery (Dark Circles) in (a) a Small-Diameter Well and (b) a Large-Diameter Well.

Note: The time scales are logarithmic, and recovery takes significantly longer than the pumping duration. Curves were calculated using a solution from Papadopolous and Cooper (1967).

aquifers remains relatively constant until the potentiometric surface, as represented by the static water levels in wells, drops below the bottom of the confining unit.

From the perspective of well yield (Q, water volume produced per unit time), it is beneficial to quantify the replenishment or recuperation rate of the well. An approximation for estimating specific well capacity (Q/s, where Q is the flow and s is the corresponding drawdown in the well) is described in Driscoll (1986):

$$\frac{Q \text{ m}^3/\text{day}}{s \text{ m}} = 0.72 \times T \text{ m}^2/\text{day} \quad \text{or} \quad \frac{Q \text{ gal/min}}{s \text{ ft}} = \frac{T \text{ gal/day/ft}}{2,000} \qquad (16\text{-}1)$$

Equation 16-1 estimates the amount of sustainable flow that is available from a well in a confined aquifer per unit drawdown in the well. The result is based on the geological formation's limitations, represented by T. The rate from an unconfined aquifer is about 33% greater (Driscoll 1986). Equation 16-1 is useful when geological information is known. For example, a well completed in an aquifer with $T = 1$ m^2/day can continuously provide 720 L (1,000 L = 1 m^3) of water per meter of drawdown. If geological information is not available, an alternative is to determine the storage capacity of a well by estimating the recovery rate of the well, as in Box 16-1. If the available drawdown decreases, such as during the dry season, then specific well capacity will likewise diminish.

Water supplies should be adequate as long as the average daily rate at which the well recovers can sustain the average daily water demand. At a minimum, the well should provide enough storage capacity (horizontal area multiplied by the depth interval available for drawdown each day) to supply daily water demand. However, household wells in aquifers with a transmissivity (T) greater than 10 m^2/day (900 gal/day/ft) are not restricted to a minimum storage capacity because the potential replenishment rate will likely be sufficient.

⟫⟫ Box 16-1. Demand and the Storage Capacity of a Well

If a well is to be installed for a family, and their daily demand for water is 360 L (95 gal), a dug well that is 1.0 m (39 in.) in diameter would have to be deep enough to allow for the water level to be drawn down at least 0.46 m (1.5 ft) every day. The well can sustain this demand if it replenishes faster than the daily demand of 360 L/day. The rate of recovery is determined by timing how long it takes for the water level in the well to return to the static (preremoval) level. A drilled well that is 0.1 m (4 in.) in diameter would require at least 46 m (150 ft) of submerged depth (and the daily replenishment rate of the well likewise must satisfy the amount pumped). If a well cannot supply sufficient water to meet demand, additional wells or greater storage capacity of the well will be required. Water needs could also be met by drawing on additional sources (e.g., rainwater, greywater) or providing education that promotes water conservation.

For dug wells, depth and diameter can be increased to encompass more aquifer volume to intercept more groundwater flow and to allow for water storage in the well. The size of the well is determined primarily during the process of the digging, adapting for the rock, soil, and water conditions that are encountered. A well encountering sufficient water yields in unconsolidated materials above bedrock might be larger in diameter to take advantage of the easier digging conditions. Hard-rock aquifers are extremely difficult to dig manually, so some people keep the diameter to a minimum, digging instead to greater depths to gain storage capacity. Drilled wells are primarily deepened to achieve more storage, and their diameter is a function of the size of the available drilling equipment.

))) 16.5 Dug Wells

16.5.1 Configuration and Components

Figure 16-4 shows the components of a dug well that allow water to enter the well (intake), maintain the well opening (shaft), and protect people and animals from falling into the well and the well from contamination (wellhead).

In wells dug or drilled into consolidated rock, either porous or fractured, the strength of the rock allows the hole to remain open without an intake structure. In loose, highly fractured rock and unconsolidated geological materials (soils and sediments), the intake structure can be kept open at the bottom using some sort of porous lining material. This step is typically accomplished using the same structures that are used for lining the shaft, modified to have openings to allow for water inflow. Possibilities include a windowed concrete ring (Egboka et al. 1988), a dry wall of bricks placed without mortar (Shaw 1999), and caissons made of porous pavement or made with holes (Watt and Wood 1979).

Inflow can be enhanced by increasing the open area (increasing the number of holes) or the hydraulic gradient. However, the increased flow into the well is not directly proportional to the increase in the open area. The openings in the sidewalls experience less hydraulic gradient because of their position above the well bottom. In loose materials where water flows readily into the well, fine-grained sediments (silts and clays) can be mobilized and suspended because of the inflowing water, in which case the bottom of the well would require a porous plug or layers of sand and gravel to filter out the fines (Watt and Wood 1979).

If the hole is dug larger than the shaft lining, the annulus between the lining and the hole sidewalls must be backfilled, not only to shore up the lining, but also to lessen the opportunity for surface contaminants to migrate vertically to the well intake. The excavated materials can be used for this purpose after removing rocks large enough to prevent the sand and finer materials from being properly compacted into place. Concrete and clay, especially the swelling kinds, like bentonite, provide the best seal from surface contamination and are used to fill the space above the water table.

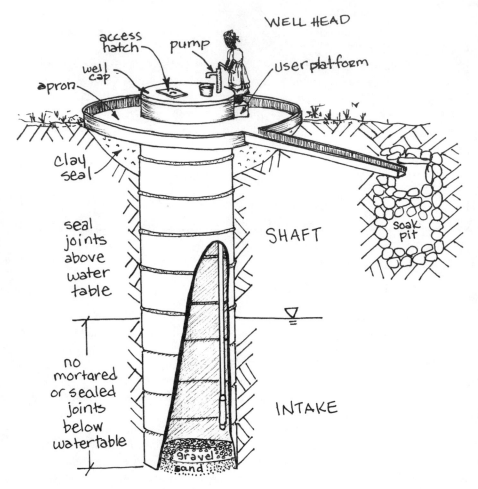

Figure 16-4. Dug Well Components.

Note: In this case, the well is lined with concrete caissons to support the surrounding unconsolidated material, but the joints below the water table are not mortared to allow for water flow.

16.5.2 The Excavation Process

The nature of the digging during the first meter of depth, where shovels and pickaxes can be used from the surface, changes drastically as progress leads to a depth that entails one person digging inside the well, with the excavated material brought to the surface via a bucket or scow. In addition, deeper soils and sediments are under greater confining stress and are often harder to dig. Once you encounter bedrock, the rate of digging is controlled primarily by the rate at which you can break the rock. Digging is difficult without the proper tools, including a digging hoe and a pick, both with shortened handles (Fig. 16-5). When you encounter the groundwater table, the collapse strength of the unconsolidated

Figure 16-5. Short-Handled Tools Needed for Digging in the Confined Space of a Dug Well.

materials is reduced to the point that digging becomes dangerous. Proper and frequent shoring is important.

16.5.3 Lining Options

Although steel culverts could be used as a shaft liner, concrete and bricks are more economical and available in areas where dug wells are commonly constructed. It is likely that you will have to make caissons on site with fabricated molds, or that bricks, stones, and mortar will be used to line the shaft. You can line the shaft section with precast caissons, masonry, and cast-in-place concrete (Fig. 16-6). Install sections of caissons as the digging proceeds.

Shore up the sides of the portion of the hole, spanning loose material (soil, sediments, and highly weathered bedrock) to prevent collapse. Shoring is unnecessary when the integrity of the materials is sufficient to withstand collapse. Some caving in is inevitable in unconsolidated settings, unless the well is lined. Wells can also be dug without lining, but the risks for collapse are great in all but hard-rock systems.

Precast Concrete Caissons and Bricks

Precast caissons can be made on site with reusable forms. For example, Pickford (1991) illustrates the reinforcing configuration using 16 m of 6-mm-diameter rebar and a concrete mixture of 45 L of sand, 90 L of gravel (smaller than 15-mm size), and 1 bag of cement to make one 0.8-m-high concrete ring with a 1-m diameter and a 5-cm wall thickness. A section of caisson this size would weigh approximately 320 kg (700 lb).

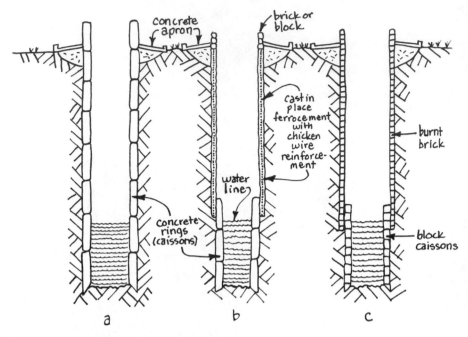

Figure 16-6. Different Methods for Lining a Well. (a) Concrete Caissons, (b) Concrete Caissons Below the Waterline and Reinforced Concrete Cast in Place Above the Waterline, and (c) Bricks Above the Water Line and Concrete Block Caissons Below the Water Table.

Source: WaterAid 2006.

The lowest caisson is typically beveled inward along the bottom to provide a cutting edge in contact with the bottom of the well as it is being dug. The effectiveness of the bevel for "cutting" is debatable (Box 16-2) because the downward pressure exerted by the weights of the stacked caisson lifts is minimal when the well is dug just large enough for the caissons to fit (see example pressure and force calculations below). To avoid having the caissons wedge themselves in the well shaft, the dig the bottom of the well as level as is practical. Figure 16-7 shows the pattern in which soil should be removed from inside the caisson to enable it to sink (the numbers in this figure show the order of excavating the soil). The soil type will determine the extent to which soil needs to be removed from underneath the edges of the caisson.

Concrete Cast In Situ

One excavation approach uses a temporary lining at the ground surface to inhibit collapse of the loose surface soils and subsequent excavation for about 5 m. (Detailed procedures for excavation and lining are provided by Watt and Wood (1979).) The shaft lining is then cast in place up to the bottom of the temporary lining. In addition to the tools for

))) Box 16-2. Earth and Downhole Pressures

A section of precast concrete casing nominally 1-m in diameter, 0.8-m in height, with 5-cm wall thickness exerts a downward pressure of almost 3 lb/in.^{2}, with each successive section adding in turn to the pressure at the bottom. Static earth pressures increase at most about 0.7 lb/in.^{2}/ft of depth. In an excavation, these pressures can translate to lateral pressures against the wall of the caissons of 0.25–0.33 lb/in.^{2}/ft (assuming a coefficient of earth pressure at rest between 0.35–0.47 lb/in.^{2}/ft, which is a typical range for cohesionless soils).

A typical coefficient of friction between concrete and sand is 0.35 lb/in.^{2}/ft, so the sidewall frictional force resisting the downward movement of the caisson might be as high as 250 lb/ft of depth if the caisson is in direct contact with the sidewalls of the hole. The force from the weight of the caisson is 270 lb/ft.

Accordingly, except in loose soils or where the flow of groundwater into the hole is causing some soil liquefaction, the net downward force on the caissons is negligible (<20 lb) if the casing is in contact with the walls of the hole. This lack of force is why installation typically requires some manual removal of soil underneath the casing, as shown in Fig. 16-7.

excavation (pickax, short-handled shovels, buckets), use a basic well-bore caliper to guide the excavation and make the sides as vertical and the diameter as uniform as possible. Construct a simple well-bore caliper by using two sticks of equal length tied together at the centers to make a cross. Use removable cylindrical forms for the inside of the well, and the borehole wall serves as the outside form.

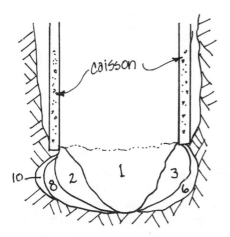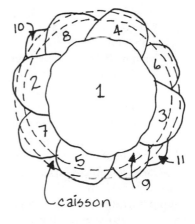

Figure 16-7. Procedure for Digging Within a Caisson. Soil Conditions Determine the Number of Steps Through Which Digging Must Proceed Before the Caisson Lowers Itself.

Source: Redrawn from Pickford (1991) with permission.

Telescoping

When lining a well with a porous caisson that will be used as the intake, install it after completing the lining of the shaft by telescoping the intake sections inside and below the shaft caissons. Instead of porous caissions, the intake can also rely on the open bottom of the caisson (if necessary, covered with a layer of sand and then gravel). Porous caissions are the bottom one or two caissons that are fabricated with windows (Egboka et al. 1988) or holes (Watt and Wood 1979). The juncture between the highest intake section and the lowest shaft caisson can be sealed with mortar if needed. It is appropriate to leave the joints between caissons below the water table open (not mortared) because sealing them is not necessary and the openings will provide another inlet for water. The intake sections must be held in place if there is uplift of the soil at the bottom, which occurs in loose soil conditions, but otherwise their weight should keep them in place.

A precast porous intake must be a smaller diameter to be lowered through the well shaft lining (including sufficient tolerance to clear the inside diameter of imperfectly constructed and eccentricity of the shaft lining lifts). Take care during the installation of telescoped porous sections to avoid dropping them or wedging them in the shaft. The greatest compression forces are exerted on the lowest section, and because openings in concrete reduce caisson strength and the edges of openings are subject to stress concentrations, it is prudent to install the shaft lining first and then telescope the porous intake into its position beneath the water table.

16.5.4 Wellhead Protection

Wellheads are ground-level construction to protect the well and prevent collapse. Basic wellheads are constructed from logs placed flush with the ground around the well opening. Figure 16-8 shows that incremental improvements can be made to wellheads to improve safety and water quality. For example, Fig. 16-9 shows an improved well that has a protruding shaft lining and a cover. This version may be an incremental step toward installing a pump on the well.

An impervious concrete apron surrounding and sloping away from the well shaft for at least 1 m prevents contaminants from migrating into the well. This structure helps prevent the entrance of contaminants, runoff, and debris directly into the well, while reducing the number of animals and mosquitoes that would gather in stagnant pools of water. Also, a curb at the apron perimeter channels spilled water into a soak pit (Chapter 21). A shaft lining that protrudes above the apron provides a barrier to runoff and debris entering the well and prevents animals and people from accidentally falling in.

Wells constructed at a basic level might be covered with scrap wood or sheet metal. Obviously, the more secure the cover, the better it will serve to protect water quality. If a pump is used, a removable cover (or secure access port) is required to allow for maintenance (repairing pump equipment, disinfection, and removing sediments). This cover can also be used as an emergency access hatch for buckets if the pump breaks down and is not able to be repaired immediately.

Do not install hand pumps if their ongoing maintenance is unlikely because of financial constraints, lack of coordination, or inability to obtain replacement parts. However, using a windlass and a dedicated bucket maintains water quality.

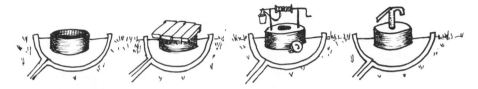

Figure 16-8. Protection of the Wellhead Is Important Both in Water Quality and in the Safety of the User.

Note: Wellheads can be improved incrementally(moving left to right), starting with an unprotected well (top left) and moving toward a sealed well with an apron, a raised lining, and a hand pump (bottom right), as resources allow.

Figure 16-9. Improved Well.

Note: This well has been incrementally improved by including a cover and a protruding shaft lining. Further improvements could include a pump and a soak pit.

People may be accustomed to washing dishes or clothes near the well. If this is the case, it is prudent to provide a convenient nearby wash area (Chapter 21) that is away (and downhill) from the well. This situation will help to prevent the infiltration of wash water (greywater) into the drinking supply.

))) 16.6 Drilled Wells

Drilling a well has the same steps as hand digging a well: breaking up the regolith (soil) into cuttings and removing the cuttings. Whereas in a dug well the breaking and removal is done manually, a well is drilled with systematic breaking and removal steps. Soil and rocks are broken by hitting with a chisel or grinding with a bit or auger.

The percussion method depicted in Fig. 16-10 uses a chisel bit that is moved in a reciprocating fashion. Raise the drill bit and drop it into the hole. The weight of the tool

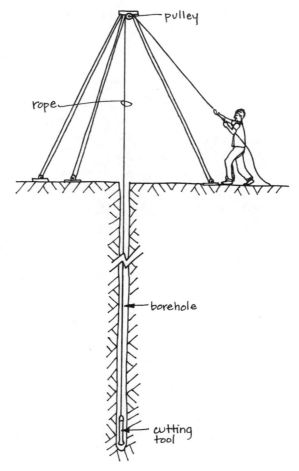

Figure 16-10. Percussion Drilling Process.

Note: Source: Redrawn from Shaw (1999) with permission.

and its design cause it to break up the soil and rock. Commonly, the bit is designed to rotate slightly (less than a half turn) to impart a sideways motion along with the downward hammering. The broken loose materials accumulate at the bottom of the hole and eventually inhibit further breaking. Then remove the drill bit and either flush or bail out the materials. Alternatively, a manual method called the Baptist well drilling method is a combination of sludging and percussion that has been developed to simultaneously percussion drill and flush out cuttings (Water for All International (n.d.)). Sludging is a drilling method where a pipe (bamboo, PVC, or metal) is moved up and down in a water filled bore hole, thereby loosening up soil.

Rotary and auger drills (Fig. 16-11) primarily use rotational motion to grind rock and move soil. These methods move the cuttings to the surface with either drilling fluids (such as water, water–clay mixtures, or air) or along auger flights. Drilling fluids can be injected through a rotary drill bit. As the fluid flows toward the surface within the annulus between the drill rod and the borehole casing, it carries with it those cuttings that settle more slowly than the water velocity. This process can also be powered by gas or diesel

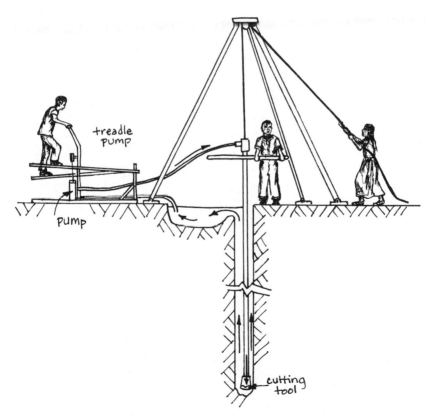

Figure 16-11. Rotary Drilling Coupled with Flushing of the Cuttings.

Note: Source: Redrawn from Shaw (1999) with permission.

engines. For example, the LS-100 (Lifewater Canada 2004) is a portable rotary drilling rig often used in developing countries to drill wells up to 15 cm in diameter.

Flushing is fast but requires significant quantities of water. Using conventional Stoke's law settling considerations, one can justify the rules of thumb for minimum uphole velocities of 1–1.5 m/min (e.g., Roscoe Moss Company 1990) for drilling fluids to flush out sand-size and smaller particles. More viscous and denser drilling fluids work with slower velocities. Minimum flow rates for drilling fluids need to be on the order of tens of gallons per minute to achieve these velocities. Alternatively, cuttings can remain in the hole until the drilling process becomes slowed by their presence, after which the drill stem can be removed and the cuttings either bailed from the well or flushed.

⟫⟫ 16.7 Driven Wells

Wellpoints can be driven into unconsolidated formations, with either a sledge hammer or a fence-post driver (Fig. 16-12). The depth achievable for driving a wellpoint depends on the nature of the lithology. In sandy soils where large cobbles or boulders are absent, a 32-mm-diameter, 1-m-long point can be driven to depths of 8–12 m. Hard-packed till or rocky soils can inhibit driving a point at almost any depth.

Conventional points are available in lengths of 0.9–1.2 m and nominal diameters of 32 or 38 mm (1.25 or 1.5 in.). The drive point itself is made from a length of steel pipe with perforations surrounded by a brass or stainless steel mesh. The bottom end has a hardened steel point attached, and the top end a female pipe thread accepts the pipe lengths that are added as the point is driven deeper.

Add pipe lengths (<2 m) as the point is driven deeper into the soil. Use a special coupling, called a *drive coupling* (Fig. 16-13), to connect lengths of point. These couplings allow the pipe to meet end to end in the center of the coupling, in contrast to conventional water-pipe couplings, where the ends do not meet and thus are weaker. Take care to drive the point vertically (plumb) and ensure that the threads on the end being driven remain intact. Use a drive coupling and a short length of sacrificial pipe with a drive cap to maintain the integrity of the pipe threads. Because of their small diameter (<4 cm), driven wells are usually equipped with suction lift pumps, like a pitcher pump (see Section 16.9).

⟫⟫ 16.8 Well Development and Disinfection

The final steps in preparing a well for use are well development and disinfection (MacDonald et al. 2005). Well development is performed by pumping water out of the well rapidly. This procedure is difficult for hand-dug wells unless you can use powered pumps, such as a sump pump. The purpose of the development is to remove fine-grained sediments inside the well left from intake installation and the fine sediments in the soil formation nearest the intake that eventually would infiltrate the well. To remove these sediments, use a pumping rate during development that is higher than the pumping rate during normal use (Box 16-3). Advantages of developing a dug well include achieving higher effectiveness of disinfection (suspended sediments reduce the effectiveness of disinfectants) and lessening the rate at which the well will fill in with sediments.

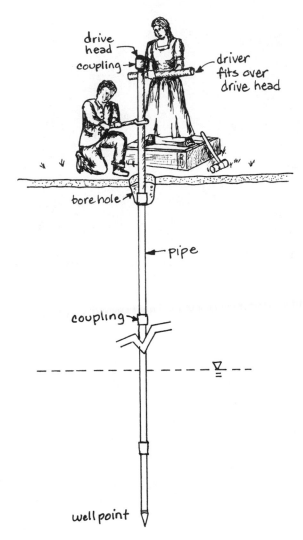

Figure 16-12. Components of a Driven Well.

Source: After Alth et al. 1992.

Disinfect the finished well with chlorine. Thoroughly mix the disinfectant in the well bore, resulting in a concentration of chlorine of at least 200 mg/L as Cl, and allow it to remain in contact for several hours (MacDonald et al. 2005). (See Chapter 18 for chlorination methods.) Remove the chlorinated water before drinking, which may be a tedious task in dug wells because several well volumes need to be removed. Disinfect after well development because suspended sediments will lessen the effectiveness of the disinfectant. It is also prudent to disinfect the pump periodically. A pump manager or the water committee should be charged with regular disinfection, as well as preventive maintenance.

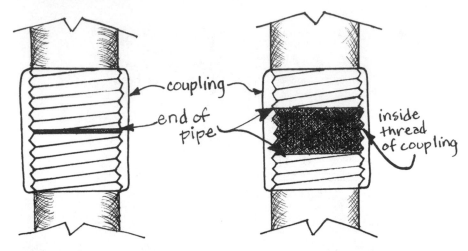

Figure 16-13. The Difference Between a Standard Coupling (Right) and a Drive Coupling (Left) Used for Wells.

Source: After Alth et al. 1992.

))) Box 16-3. Well Development Pumping Rates and Volumes

Typical withdrawal rates that can be achieved with buckets are on the order of a few gallons per minute (gal/min), and rates two to three times higher are achievable with hand pumps. Sump pumps, such as those commonly used to pump typical U.S. home basements, can achieve rates as high as 25 gal/min. A typical volume of water in a hand-dug well is approximately 1 m^3 (270 gal) (assuming a 0.9-m-diameter inside intake and a 1.6-m depth). To remove one well-water volume requires at least 60 (5-gal) buckets (at least a couple hours of work) or 11 min of pumping with a sump pump. Depending on the amount of fines in the formation, the gradation of the sediments, and the pumping rate during development, the well might have to be developed for 10 or more volumes.

))) 16.9 Pumping Systems

The basic purpose of a pump is to use externally supplied energy (e.g., from humans, wind, solar, electricity, diesel) to move water at a particular rate from some depth to an elevation at which it can be used. The energy required includes the amount of *static lift* (the depth to the static water level), the *back pressure* exerted by the system (i.e., the water elevation in a storage tank), the drawdown in the well during pumping, and the frictional losses in the piping and fittings as the water travels from the intake pipe to the storage tank. For hand-dug wells where buckets are used, the energy or work required is the height related to the lift. Some simple considerations for mechanical pumping systems are provided below.

Figure 16-14 depicts the primary factors that affect pumping requirements. Groundwater engineering texts refer to the energy required to pump water at a particular flow rate (Q) as the *total dynamic head* (*TDH*). The *TDH* is the sum of fixed (static lift plus back pressure, both independent of Q) and dynamic (drawdown and frictional losses at Q) heads. Physically speaking, *head* (h) is the water energy per unit weight, which is equivalent to water elevation above a specified point, such as the well bottom.

The water energy can be represented by pressure (P) by the following expression:

$$h = \frac{P}{\gamma_w} = \frac{P}{\rho_w g} \qquad (16\text{-}2)$$

where γ_w is the unit weight of water (9,810 N/m³ = 62.4 lb/ft³ = 0.433 lb/in.²/ft), ρ_w is the mass density of water (1,000 kg/m³), and g is gravitational acceleration (9.81 m/s² = 32.2 ft/s²)

The drawdown component (s_w) can be calculated from the specific capacity according to Eq. 16-1. The frictional losses can be estimated from nomographs or the Darcy–

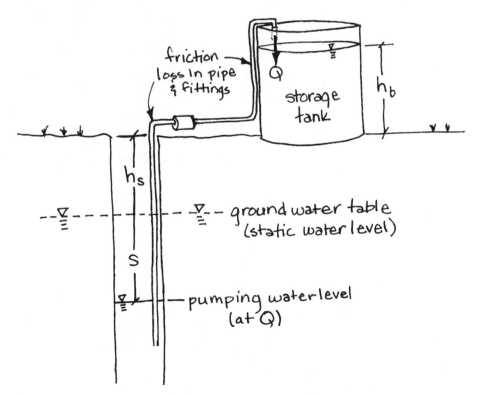

Figure 16-14. General Pumping Requirements for a Water Well.

Weissbach equation (see Chapter 11). The Hazen–Williams equation is a well-recognized formula for estimating the frictional losses in pipes flowing full:

$$h_f = \frac{10.43L}{d^{4.87}}\left(\frac{Q}{C}\right)^{1.85} \tag{16-3}$$

In Eq. 16-3, L is the length of pipe (ft) (the frictional losses caused by fittings, valves, inlets, and outlets can be treated as equivalent pipe lengths and added to the actual pipe lengths) with inside diameter d (in.), Q is the flow rate in gallons per minute, and C is a friction factor, typically 120 for galvanized iron and 150 for PVC. Equation 16-3 estimates friction losses in terms of head in feet. The factor 10.43 includes unit conversions for gallons, inches, and minutes in accordance to the powers in Eq. 16-3; alternative units for L, Q, and d can be used if the proper adjustments are made to the numeric factor.

The ability of a pump to move water depends on the *TDH*. As the *TDH* increases, the pumping rate decreases; the amount of the decrease depends on the type of pump. Mechanical pumps are classified as positive or dynamic displacement pumps. *Positive displacement pumps* can produce water at a relatively constant rate regardless of the *TDH*, but their capacity to produce water (i.e., flow rate) is somewhat limited. *Dynamic displacement pumps* use centrifugal action to cause water to flow and are capable of producing water at high rates. Their rates are influenced more by the *TDH*, and they tend to be less efficient and require motor power to achieve sufficient pump speeds to move water.

Pump types are also categorized in terms of whether they obtain water as suction lift or by pushing the water from downhole using a submersible pumping system. The maximum theoretical height that water can be lifted is the atmospheric pressure minus the vapor pressure of water (see Example 16-1). The theoretical height is reduced further because of practical influences, such as frictional losses in the suction pipe and the vacuum efficiency of the pump.

))) Example 16-1. Determining the Theoretical Height That Water Can Be Lifted

The atmospheric pressure at sea level is 1 atm (14.7 lb/in.^{2}). Imagine immersing the bottom end of a long, vertical tube in a bucket of freshwater (unit weight of 62.4 lb/ft^3). If the tube were evacuated completely, then the atmospheric pressure acting on the water in the bucket would push the water into the tube until the pressure of the water (because of the weight of the water suspended in the tube) at the end of the tube in the bucket is balanced with the atmosphere.

Equation 16-2 can be used to convert the given atmospheric pressure to the water energy represented by pressure (P):

$$h = \frac{14.7 \text{ lbs/in.}^2 \left(\frac{12 \text{ in.}}{\text{ft}}\right)^2}{62.4 \text{ lb/ft}^3} = 33.9 \text{ ft} = 10.3 \text{ m}$$

Remember, head (h) is the water energy per unit weight, which is equivalent to water elevation above a specified point, such as the well bottom. A perfect vacuum cannot be achieved because once the pressure is reduced to the vapor pressure of water, then the water boils and vaporizes and no further lifting occurs (the water level remains constant). At 20 °C, the vapor pressure of water is 17.54 mm Hg (0.33 lb/in.^{2}). Again, Eq. 16-2 can be used to convert the given vapor pressure to the water energy represented by pressure (P):

$$h = \frac{0.33 \text{ lbs/in.}^2 \left(\dfrac{12 \text{ in.}}{\text{ft}} \right)^2}{62.4 \text{ lb/ft}^3} = 0.76 \text{ ft} = 0.23 \text{ m}$$

The maximum theoretical height that water can be lifted is the atmospheric pressure minus the vapor pressure of water (10.3 m − 0.23 m = 10.1 m). ⟩⟩⟩

A classic example of a positive displacement well pump is the pitcher pump shown in Fig. 16-15. It is a type of reciprocating pump. Pushing down on the handle causes the piston to be drawn up, which creates a suction to lift the water. The weight of the water and the slight pressure created by the piston (which must slip on the downstroke) causes the check valve to close and prevent water from flowing back down the hole. The flowing water coming up on the upstroke causes the check valve to open. Figures 16-16 and 16-17 provide additional information about how reciprocating hand pumps work.

The piston is typically a leather seal that must be wetted, and the pump must be primed to get the process started. The maximum lift is a physical constraint (see Example 16-1). *Priming* means pouring water into the pump casing until the piston is covered. Having a pump that must be primed is problematic because water (possibly of questionable quality) must be on hand before the pump is operated.

Because of the physical limitation of suction-lift pumps, downhole positive displacement pumps are often used. The Tara (Fig. 16-18), the India Mk. II (Fig. 16-19), and the Afridev (Fig. 16-20) are reciprocating pumps, which operate much like the pitcher pump. In this case, though, the piston and check-valve system is installed downhole to push the water up, and thus is not subject to the maximum suction lift limitation.

The Mono pump (Fig. 16-21) is a *screw* pump, which augers the water to the surface. The Vergnet (not shown) is a diaphragm pump (like a heart), and the pumping action causes the diaphragm to inflate and expel the water. A rope pump is shown in Fig. 16-22. It operates without a check valve because the pistons (plungers) that move water to the surface are continually drawn toward the surface.

Downhole pumps are designed to be submerged, and thus are not subject to the same lift restrictions of positive displacement pumps. However, depending on the pump size and operating conditions, dynamic displacement pumps have a maximum capability to push water, called the *shutoff head*. Pumps run most efficiently at 50% or less of the shutoff head.

Pumping requirements are often summarized in terms of the *pump power* ($\dot{E}$) required to achieve a particular *TDH* at a desired *Q*. For freshwater, $\dot{E}$ (horsepower, hp) is given by

$$\dot{E} = \frac{Q \times TDH}{3,960 \, \eta} \tag{16-4}$$

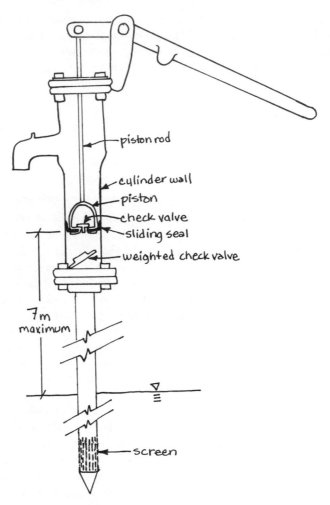

Figure 16-15. Schematic of a Conventional Pitcher Pump, Which Operates by Suction Lift.

Source: After WaterAid 2006.

where Q is the flow rate (gal/min); *TDH* is the total dynamic head described previously (ft); η is the pump efficiency (output power/input power); and the factor 3,960 includes unit conversions for gallons, minutes, and the unit weight of water. Equation 16-4 determines the power requirements in units of horsepower (hp). Typically, efficiencies of motorized pumps are above 60–70%.

The pump size, in terms of horsepower, is not often matched exactly to the pumping requirements. The *operating point*, as defined by the combination of *TDH* and corresponding Q, of a pump is where the energy supplied by the pump equals the energy required by the system. The energy required, represented by *TDH*, varies with Q, as described earlier. The relationship between the system *TDH* and Q is referred to as the

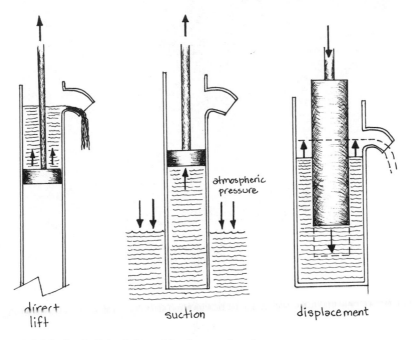

Figure 16-16. Basic Principles of Reciprocating Pumps.

Source: Redrawn from Fraenkel and Thake (2006) with permission.

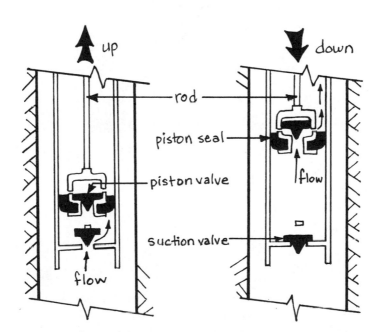

Figure 16-17. How a Reciprocating Hand Pump Works.

Source: Redrawn from Shaw (1999) with permission.

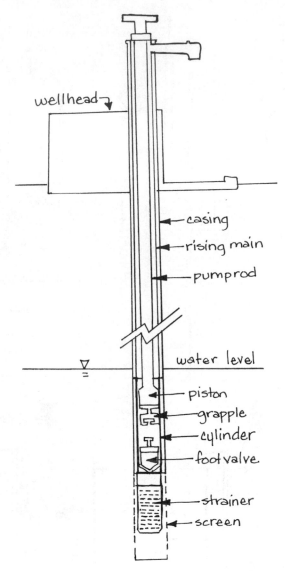

Figure 16-18. A Tara Hand Pump.

Source: Redrawn from Pickford (1991) with permission.

system head curve. Likewise, the pump produces *TDH* as a function of *Q*, and this relationship is referred to as the *pump characteristic curve.* Generalized relationships are shown in Fig. 16-23 between what are referred to as the system head curve and the pump characteristic curve. The operating point is the intersection of the system head and pump characteristic curves.

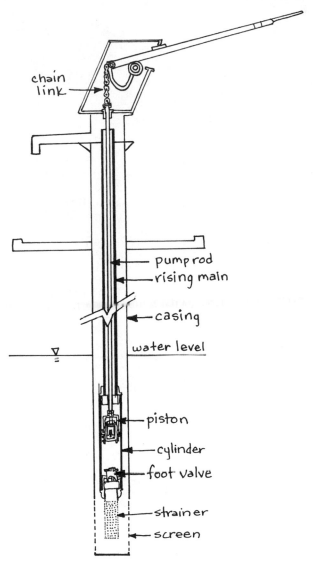

Figure 16-19. An India Mark II Hand Pump.

Source: Redrawn from Pickford (1991) with permission.

Pump characteristic curves (also called *pump performance curves*) are supplied by pump manufacturers. The system head curve is calculated from the fixed (static lift and back pressure) and dynamic (drawdown and frictional losses) heads, as described above. Example 16-2 illustrates the basic estimates that can be made using the approaches outlined earlier.

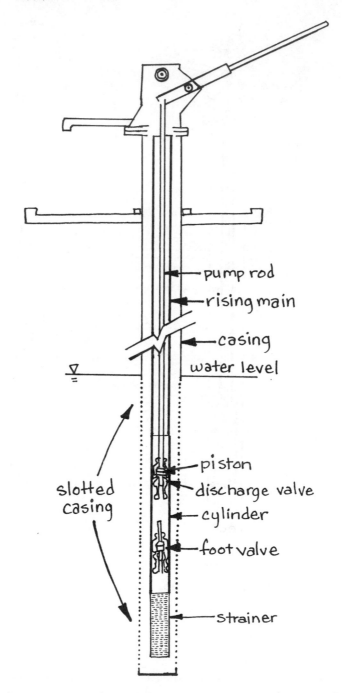

Figure 16-20. An Afridev Hand Pump.

Source: Redrawn from Pickford (1991) with permission.

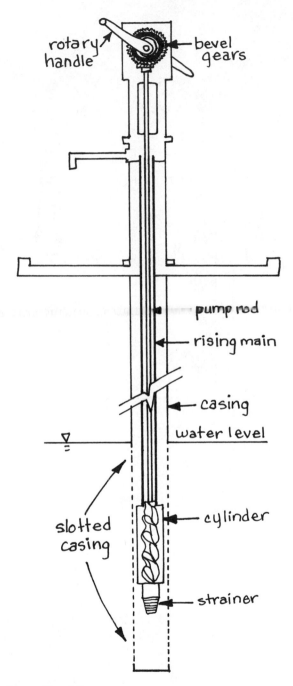

Figure 16-21. Mono Hand Pump.

Source: Redrawn from Pickford (1991) with permission.

Figure 16-22. A Rope Pump.

Source: Redrawn from Shaw (1999) with permission.

))) **Example 16-2.** Pump Selection

A farmer wants to know the size of a motorized pump needed to convert a rope pump system to an electrically powered pumping system that would fill a storage tank for irrigating crops. The static water level in the well is roughly 25 ft below ground level. The plan is to pump the water into a 600-gal storage tank at an assumed elevation of the water in a full tank of 20 ft above ground level. The tank would be about 1,200 ft from the well. The specific capacity of the well was high for a bedrock well, at 50 gal/min/ft. There is a range of pumping rates that would be suitable, but for illustration purposes a value of 6 gal/min is assumed.

The fixed head components that are provided (or can be calculated from the given information) are the following:

- $h_s = 25$ ft (static lift),
- $h_t = 20$ ft (back pressure caused by water level elevation in a full tank), and

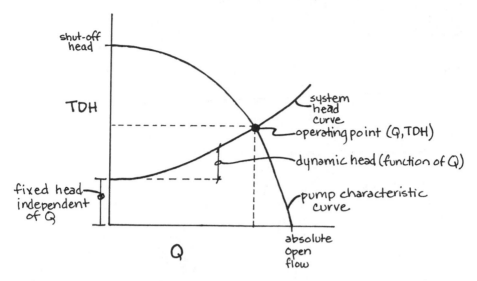

Figure 16-23. Generalized Relationship Between the System Head Curve and the Pump Characteristic Curve. The Operating Point Is the Intersection of the Two Curves.

- the drawdown because of pumping is obtained from the specific capacity and desired pumping rate: s_w = 6 gal/min ÷ 50 gal/min/ft = 0.12 ft.

We will see later when determining the TDH that this amount of drawdown is low enough to be negligible.

The friction loss in the pipe is determined using Eq. 16-3. PVC is a good choice for pipe material because it is smooth (friction factor, C = 150) and economical. However, exposed PVC is susceptible to UV oxidation.

Assume a diameter of 0.75 in. The total pipe length is the distance down to the pump (40 ft) plus the distance to the tank, which will be placed on a slight hill that is 1,200 ft from the well, to allow gravity to distribute the water.

$$h_f = \frac{10.43(1,240\ \text{ft})}{(0.75\ \text{in})^{4.87}}\left(\frac{6\ \text{gpm}}{150}\right)^{1.85} = 135\ \text{ft}$$

The total required energy (as head) at a pumping rate of 6 gal/min equals

$$TDH = 25\ \text{ft} + 20\ \text{ft} + 0.12\ \text{ft} + 135\ \text{ft} = 180\ \text{ft}$$

At the desired pumping rate of 6 gal/min and a typical efficiency of 70%, the pumping power required is determined from Eq. 16-4:

$$\dot{E} = \frac{(6\ \text{gpm})(180\ \text{ft})}{(3,960)(0.70)} = 0.39\ \text{hp}$$

Pumps typically are available in, nominally, 0.25-hp increments. In this case, a 0.5-hp pump would be appropriate. A spreadsheet program or graphing tabulated calculations is needed to determine the operating point based on the specified power. For these conditions, a 0.5-hp pump would operate at 6.6 gal/min, providing 209 ft of *TDH*.

The conditions for this example highlight the importance of pipe friction. If 0.5-in. pipe were used instead of 0.75-in. pipe, the pipe friction would be almost 10 times higher and the resulting pump size would have to be 3 times larger just to overcome the additional frictional losses. Likewise, if steel pipe were used instead of PVC, the higher roughness of steel would incur almost twice as much friction loss and require a pump almost twice as powerful.   **))))**

The approach outlined in Example 16-2 is useful for estimating power requirements for any type of mechanical pump. For example, a rope pump could be adapted for wind or motorized power. A large storage tank and sustainable pumping rate can provide a wide range of flexibility in matching water demand for domestic and agricultural use to well production and power supplies (e.g., wind, solar). Analyses can be performed with spreadsheets to explore alternative pumping rates and storage tank volumes.

The ability to replace broken or worn parts is critical for many remote locations. VLOM (village-level operation and maintenance) pumps are often a good choice for these situations. Rope pumps are built entirely with relatively common materials and parts. However, other than replacing the rope or a handle, anything mechanical that breaks or seizes in the well requires the removal of the entire system, which may be 60 ft or more of pipe and ropes.

))) References

Alth, M., Alth, C., and Duncan, S. B. (1992). *Wells and septic systems*, 2nd ed., McGraw Hill, New York.

Brantly, J. E. (1961). "Percussion-drilling system." *History of petroleum engineering*, D. V. Carter, ed., American Petroleum Institute, Washington, D.C., 133–269.

Driscoll, F. G. (1986). *Groundwater and wells*, 2nd ed., Johnson Filtration Systems, Minneapolis, Minn.

Egboka, B. C. E., Mbanugoh, R. E., Nwogute, N. S., Uma, K. O., and Okpoko, E. I. (1988). "Positive implications of hand-dug wells in water resources planning and management in a developing economy such as in Nigeria." *Water International, 13*, 98–105.

Fraenkel, P., and Thake, J. (2006). *Water lifting devices*, 3rd ed., Food and Agriculture Organization (FAO) of the United Nations, Rome, Italy.

Lifewater Canada. (2004). *Drilling and well construction reference manual*, Thunder Bay, Ontario, <http://www.lifewater.ca/resources/drillingtutor.htm> (Jan. 24, 2009).

MacDonald, A., Davies, J., Calow, R., and Chilton, J. (2005). *Developing groundwater: A guide for rural water supply*, ITDG Publishing, Warwickshire, U.K.

Papadopolous, I. S., and Cooper, H. H. (1967). "Drawdown in a well of large diameter." *Water Resour. Res., 3*(1), 241–244.

Pickford, J. (1991). *The worth of water*, IT Publishing, London.

Roscoe Moss Company. (1990). *Handbook of groundwater development*, J. Wiley & Sons, New York.

Shaw, R. (1999). *Running water*, IT Publishing, London.

Water for All International. (n.d.). <http://waterforallinternational.org> (Jan. 24, 2009).

WaterAid. (2006). *Technology notes.* <http://www.wateraid.org/documents/technology_notes_low_res.pdf> (Jan. 24, 2009).

Watt, S. B., and Wood, W. E. (1979). *Hand dug wells and their construction,* ITDG Publishing, Rugby, U.K.

⟫⟫ Further Reading

Rope pumps. <www.ropepumps.org> (Jan. 24, 2009).

Stewart, E. (2003). *How to select the proper human-powered pump for potable water,* <http://cee.eng.usf.edu/peacecorps/Resources.htm> (Jan. 24, 2009).

17

Rainwater Harvesting

))) 17.1 Advantages of Rainwater Harvesting

Rainwater harvesting is the collection and subsequent storage of water from surfaces on which rain falls. The United Nations Environment Programme reports that rainwater harvesting is one of the most promising alternatives for supplying freshwater in the face of increasing scarcity and demand. It is also considered an improved water supply technology (see Chapter 9).

Rainwater harvesting has a documented history that stretches back to ninth or tenth century Asia. In many rural areas of the developing world, it continues to be an important source of domestic water. Figure 17-1 shows a simple rainwater harvesting set-up. Rainwater harvesting has an important place in the management of surface and groundwater resources by promoting groundwater recharge.

Surface impoundments can be dug in the ground to store surface runoff generated during the rainy season. This stored water can be used later to provide water to livestock and small-scale agriculture. The impoundment also improves groundwater recharge and reduces soil erosion. This particular use of rainwater harvesting is important, not only because groundwater tables are dropping in many parts of the world, but also because nonirrigated agriculture accounts for 60% of crop production in the developing world. One notable example is in India, where small check dams (called *johads*) are constructed from local materials and placed in strategic locations to facilitate groundwater recharge and pond formation after seasonal rain events (Fig. 17-2). Check dams work by slowing the velocity of the water and minimizing soil erosion.

As shown in Box 17-1, there are many advantages associated with rainwater harvesting. Successful rainwater harvesting projects are generally associated with communities that consider water supply a priority. However, cultural perceptions and religious views regarding the use of water, as well as traditional preferences for the taste, smell, and color of water, are as important as technical feasibility. Some communities simply prefer the taste of groundwater or surface water over rainwater, even when rainwater harvesting appears to be a technically feasible and healthier alternative.

In addition to providing drinking water, rainwater harvesting can also be used as a supplementary source of water for a household or community center (e.g., school, clinic). Thus, in communities where people prefer the taste of groundwater over rainwater

---

Joshua R. Cowden and Milagros JeanCharles contributed to this chapter.

Figure 17-1. Rainwater Can Be Collected from Roofs in a Variety of Manners.

Note: Here, a simple gutter provides collection, and a metal drum provides storage. Covering the drum after the rain event is important to protect the water quality and prevent mosquito breeding. Better-connected gutter systems can be combined with underground cisterns or above-ground ferrocement tanks up to 5,000 L in size.

as their source of drinking water, they may be amenable to using rainwater for washing hands, cleaning clothes, and other hygiene activities.

))) 17.2 The Domestic Rainwater Harvesting System

There are several questions to consider when evaluating the technical feasibility of installing a domestic rainwater harvesting system:

1. What are the sources of water?
2. What is the current demand for water?
3. Could rainwater harvesting be used as the sole source of water, or would it be a supplementary source?

Figure 17-2. Small Check Dams Were Historically Constructed in India to Slow the Movement of Surface Runoff and Thus Enhance Groundwater Recharge.

Note: This indigenous engineering was lost for a period of time, but is now being successfully reintroduced throughout the country. As one example, construction of johads was reintroduced in the Alwar district of Rajasthan—an area with an average annual rainfall of 50 cm. The project was initiated by a nongovernmental organization along with traditional village rainwater harvesters, who had no formal education or modern computational tools but had in-depth practical knowledge of hydrological cycles, site topography, regional aquifer flows, and design and construction of earthen dams. Data from the year 2000 showed a general rise of the groundwater level of almost 6 m and a 33% increase in the area's forest cover. For more information, see the Center for Science and the Environment (http://www.rainwater harvesting.org/Rural/Bhaonta-Kolyala.htm).

⟫⟫ Box 17-1. Rainwater Harvesting Advantages

- It relieves demand and reduces reliance on groundwater resources and springs.
- It is not subject to the wide variety of pollutants discharged into surface waters or the arsenic and fluoride contamination associated with groundwater in some parts of the world.
- It is relatively cost effective: It reduces water bills and operation costs are low.
- It is a simple yet flexible technology. Local people can be trained to build, operate, and maintain the system.
- It is decentralized and independent of topography and geology.
- It cannot be privatized like other water systems.
- It can supplement current water supply, providing water for increased hygiene.
- Water is delivered directly to the household, relieving women and children of the time and burden to collect it.

> ###))) Box 17-1. (continued)
>
> - It can be used for agricultural purposes.
> - It can be used to recharge groundwater.
> - Storage tank construction techniques are relatively simple because of their smaller size compared to community systems.

4. Of what are the roofs constructed and what other surfaces exist for capturing rain?
5. Are there any site restrictions (below or above ground) that limit the type and size of storage?
6. Is there sufficient rain in the area? DTU (1987) recommends that the minimum monthly rainfall be 50 mm for at least half the year, and UNEP (1997) recommends that there be annual precipitation of at least 400 mm.

Figure 17-3 shows a schematic of the domestic rainwater harvesting system. A roof (or other surface) collects the rainwater, and a gutter system is used to transmit the water to the storage tank.

17.2.1 Relating Roof Area and Precipitation to Water Collection

The volume of water that can potentially be collected over a period of time equals

$$P \times A \times C \tag{17-1}$$

where P is the volume of precipitation (mm) that is generated over a set time (e.g., month, year) and A is the guttered roof area (m^2). C is called the runoff coefficient and is the fraction of rain hitting the roof that reaches the storage system (unitless).

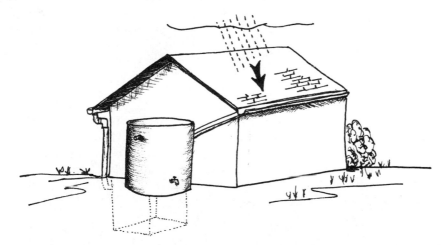

Figure 17-3. The Components of a Rainwater Harvesting System Are a Roof For Collecting Rain, a Gutter System to Transport the Water, and a Storage System.

Note: The storage system can be located below or above ground level, as shown in the figure.

Because most roofs are not sloped to a great degree, it is not critical that the roof area be calculated from the true area of the horizontal plane area. The *collection efficiency* (i.e., the runoff coefficient) is a function of the type of roofing material and the efficiency of the gutter and downspout system. However, collection efficiency is more strongly related to the efficiency of the gutters and downspouts because this is where most system losses occur (Gould and Nissen-Petersen 1999). Values of 0.8 to 0.85 are commonly used for *C*. However, this value may be as high as 0.9 or as low as 0.24.

A smooth, clean, impervious surface yields better water quality and greater quantity. Although the sloping of the roof is not critical for collection efficiency, sloped surfaces do provide better water quality by preventing the pooling of water on the roof and minimizing dirt and algae buildup.

Using a value of 300 mm for the precipitation generated over the rainy season and a roof area of 80 m², the volume of water that can be collected during the rainy season (assuming a runoff coefficient of 0.8) can be determined from Eq. 17-1 as follows:

$$300 \text{ mm} \times 80 \text{ m}^2 \times \frac{1 \text{ m}}{1,000 \text{ mm}} \times 0.8 = 19.2 \text{ m}^3 \text{ or } 19,200 \text{ L} \qquad (17\text{-}2)$$

The rainfall that would be required to meet a specific rate of water consumption can be related to the roof area by altering Eq. 17-1 as follows:

$$P_t = \frac{S_d t}{CA} \qquad (17\text{-}3)$$

In Eq. 17-3, P_t is the minimum rainfall (mm) required over time t to satisfy a desired consumption need. S_d is the desired consumption per unit t (units, for example, of liters per day), C is the runoff coefficient, and A is the guttered roof area.

Equation 17-3 can also be used to determine the required roof area to meet a consumption need (solve for A knowing P_t and S_d), and to determine under existing precipitation and roof constraints how much of the daily water consumption could be met (solve for S_d knowing P_t and A).

17.2.2 Roof Materials

The type of roofing material is constrained by finances, the availability of materials, and local building customs. In some communities, the school, clinic, church, or community center may have a more advanced roof structure than individual homes (e.g., galvanized steel roofing versus thatch). These community structures may also have a larger roof area, making them good candidates for rainwater harvesting. Table 17-1 provides examples of different roofing materials and considerations for each type.

17.2.3 Gutters

Most roofs require some type of gutter system (Figs. 17-1 and 17-3). Some roofs, however, are designed so that they can collect rainwater without gutters (Fig. 17-4).

Table 17-1. Types of Roof Materials

Type of Roof Material	Examples of Roof Material	Items to Consider
Organic	Straw, grass, palm leaves, bamboo, mud, clay, thatch	Attracts rodents and insects. Water treatment may be needed. Made of free, local materials. Consider using this water exclusively for sanitation and hygiene.
Wood	Shingles	Not feasible if chemically treated.
Metal	Galvanized steel, corrugated iron	Costly and not made locally. Can be noisy during the rainy season. Iron may leach into storage tank but is not a health problem.
Plastic	Plastic liner	Material will degrade in the presence of sunlight. May not hold up to extreme weather events. Creates mold issues for the underlying roof structure.
Asphalt		Asphalt will contain grit that should be removed by filtration or gravity settling before use. Material contains nonrenewable resources. No studies are available on how safe water is for drinking.

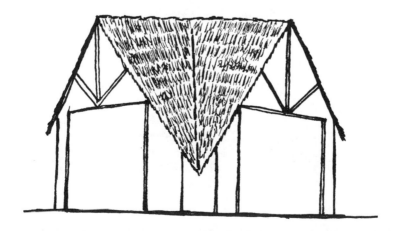

Figure 17-4. An Example of Roof Construction That Does Not Require Guttering.

As shown in Fig. 17-5, gutters can be manufactured from metal, PVC, or wood. Depending on the building construction, they can be attached to the fascia board, overhanging eaves, or wooden wall posts. Plastic and metal guttering systems can be obtained from local hardware stores. A piece of thin metal can be bent into a U-shaped, L-shaped, or V-shaped gutter. A wood gutter can be constructed by connecting two pieces of wood (approximately 120 mm wide and 20 mm thick) at a 90° angle. For the L-shaped gutter, one side will be slightly wider. Banana leaves and bamboo have been used historically in many parts of the world.

Figure 17-6 shows how a glide can be attached to a roof, eliminating the need to attach a gutter under the roof overhang. Glides can be constructed of metal or wood and are easily installed. They do, however, create a slight reduction in the capture area for rainfall.

17.2.4 Treatment of Rainwater and Flushing of Roof

During installation, ensure that the connection between any piping and the storage tank prevents access of rodents and other small animals. Place screens over the storage tank openings (which will also filter out large particulate matter, such as leaves).

One advantage of harvested rainwater is that it is typically free of pathogens, except for perhaps bird droppings. However, it can accumulate chemicals from the air, especially in urban areas and regions of industry or traffic (Mason et al. 1999). Because many hazardous chemicals (metals and organic pollutants) are associated with dust and other

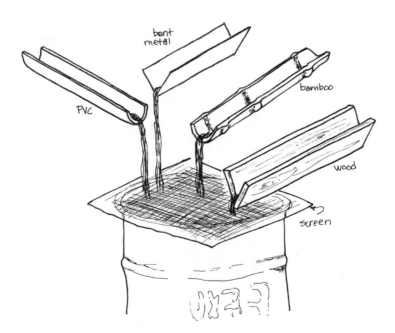

Figure 17-5. Simple Gutter Systems Include (Left to Right) Cut PVC Pipe, Bent Metal, Bamboo, and an L-Shaped Gutter Constructed from Two Pieces of Wood.

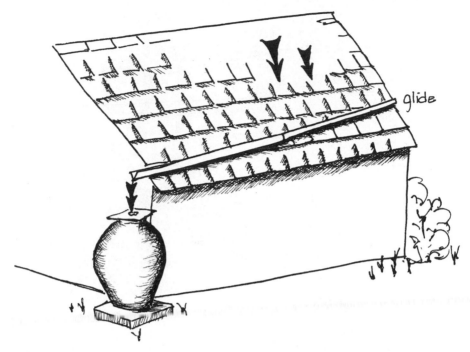

Figure 17-6. A Glide Constructed of Metal or Wood Can Be Attached to the Top Side of a Roof System as a Substitute for a Gutter That Would Be Placed Under the Roof Overhang.

particulate matter, collected rainwater may be treated with an appropriate water treatment technology. Also, a simple and inexpensive filtration method is to simply place a small fabric sack over the pipe at the point where it enters the storage tank, especially in areas where there is not extensive leaf litter.

At the beginning of a rain event, the roof may need to be flushed of dust and other particulate matter, which may include bird droppings. This quantity of water is referred to as the *first flush*. In many parts of the world, this quantity will correspond with the beginning of a rainy season. In this case, flushing does not have to occur every day. Roofs can also be swept by hand to remove accumulated particles.

A good rule of thumb is that approximately 40 L of water washes a 100-m² roof (this is approximately 10 gal of water to flush a 1,000-ft² roof). The roof can be cleaned manually by diverting the collection pipe at the beginning of the rain event to prevent the first flush from being transported to the storage tank (see Fig. 17-7). An even simpler manual method is to disconnect the section of PVC pipe attached to the gutter, which normally transmits water to a storage tank. In both cases, the roof can be visually observed to determine when it is wetted and producing runoff, which indicates that cleansing has occurred.

A simple automated cleaning system can be installed, as shown in Fig. 17-8. In this case, two pipes are connected to the gutter system. The one located at the end of the gutter is connected to the storage tank. The second is a roof washer pipe (the left pipe in

Figure 17-7. Manual System for Diverting the First Flush That Comes Off of a Roof.

Note: The tank is covered by a fine mesh screen to keep out debris.

Source: Redrawn from Pickford (1991) with permission.

Fig. 17-8) located before the pipe connected to the storage tank. After the roof washer pipe fills, the water then flows into the storage tank pipe. The roof washer pipe is sized to have a volume that corresponds to approximately 40 L per 100-m^2 roof area. The volume of the roof washer pipe (in liters) equals:

$$V = \text{pipe length} \times 3.14 \times (\text{pipe radius})^2 \times \left(\frac{\text{m}}{1{,}000 \text{ mm}}\right)^2 \times \frac{1{,}000 L}{\text{m}^3} \qquad (17\text{-}4)$$

where the pipe length is the height of the pipe (in meters) and the pipe radius is one half the inside diameter (in millimeters). The pipe can then be manually emptied by a clean-out valve.

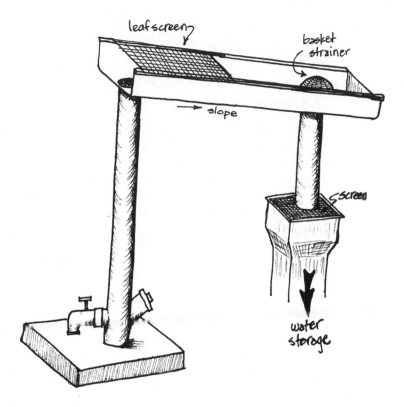

Figure 17-8. A Simple Automated System (on Left) That Collects the First Flush from a Roof. The First Flush Collected Can Then Be Drained Manually and Used as a Source of Water That Is Not Required to Be Pathogen Free.

❱❱❱ 17.3 Determining Storage Requirements

Many factors influence storage sizing. These factors include precipitation variability, site space constraints, costs, availability of local materials (e.g., clean 55-gal drums might be used as shown in Fig. 17-1), local construction knowledge, and the intended use of the collected rainwater. Equations 17-1 and 17-3 can be used to estimate the potential volume of harvested water. If 100% water coverage is the goal and the annual volume of harvested water does not meet the annual demand, either the roof catchment area needs to increase or demand needs to decrease. However, storage design should not be based on mean annual or monthly total volume availability.

Daily or monthly precipitation time series are needed to adequately assess a proposed tank size performance, with especially long records needed for arid locations or locations prone to drought (Example 17-1). Data sets can be obtained from local government agencies, regional airports or from several international data centers, such as the U.S. National Climatic Data Center, often at no cost. Several developing world locations, however, only have mean annual or monthly precipitation. These locations require disaggregation models or best estimates from local experience to properly design tanks for local climate variability. It may be possible to directly use or to interpolate from data from nearby gauges, although local climate factors (e.g., shadowing by mountains, proximity to large bodies of water) might promote error.

))) Example 17-1. Evaluating a Given Tank Size

Assume a roof area of 50 m² and an 85% capture efficiency ($C = 0.85$). A short monthly time series for precipitation obtained over a 3-year period is provided in Table 17-2. Monthly household demand is assumed constant at 6,000 L (6 m³/month). This corresponds to four people having access to 50 L/day. The tank size under consideration is 1,000 L (1 m³) because of site constraints. This size tank is perfect for ferrocement construction. Calculate the percentage of household demand met for each month.

Solution

Determine the volume of water that can be collected from the roof (i.e., the runoff) using Eq. 17-1. The volume of water in the storage tank at the end of each month (V_t) is calculated as

$$V_t = V_{t-1} + \text{runoff} - \text{demand} \qquad (17\text{-}5)$$

where V_{t-1} is equal to V_t from the previous month (Cunliffe 1998). Operating constraints on V_t include setting V_t to 0 if Eq. 17-5 produces a $V_t \leq 0$ and setting V_t to the tank volume if Eq. 17-5 produces a $V_t >$ tank volume. The percentage of the demand supplied by the rainwater harvesting system is the sum of runoff and V_{t-1} divided by the demand. These calculations can be easily done by hand or on a spreadsheet.

Table 17-3 shows that rainwater harvesting could typically supply 100% of the household demand from July through September and could provide a high level of water

Table 17- 2. Monthly Rainfall (mm) Assumed for Example 17-1 Calculations

Month	Year 1	Year 2	Year 3
Jan	2	4	1
Feb	3	14	4
Mar	17	29	9
Apr	77	60	24
May	90	104	56
Jun	131	156	126
Jul	165	208	180
Aug	240	289	217
Sep	214	222	125
Oct	89	121	61
Nov	17	27	14
Dec	1	2	8

Table 17- 3. The Percent of Monthly Demand That Can Be Met with a Specified Tank Size (Calculation Performed over a Three-Year Period)

Month	Rainfall (mm)	Runoff (m³)	V_t (m³)	V_{t-1} (m³)	% of Demand
Jan	2	0.07	0.00	0.00	1
Feb	3	0.13	0.00	0.00	2
Mar	17	0.74	0.00	0.00	12
Apr	77	3.26	0.00	0.00	54
May	90	3.83	0.00	0.00	64
Jun	131	5.58	0.00	0.00	93
Jul	165	7.02	1.00	0.00	100
Aug	240	10.21	1.00	1.00	100
Sep	214	9.10	1.00	1.00	100
Oct	89	3.79	0.00	1.00	80
Nov	17	0.72	0.00	0.00	12
Dec	1	0.05	0.00	0.00	1
Jan	4	0.17	0.00	0.00	3
Feb	14	0.61	0.00	0.00	10
Mar	29	1.24	0.00	0.00	21
Apr	60	2.55	0.00	0.00	43
May	104	4.42	0.00	0.00	74
Jun	156	6.62	0.62	0.00	100
Jul	208	8.83	1.00	0.62	100
Aug	289	12.30	1.00	1.00	100
Sep	222	9.42	1.00	1.00	100
Oct	121	5.14	0.14	1.00	100
Nov	27	1.14	0.00	0.14	21
Dec	2	0.07	0.00	0.00	1
Jan	1	0.04	0.00	0.00	1
Feb	4	0.16	0.00	0.00	3
Mar	9	0.38	0.00	0.00	6
Apr	24	1.00	0.00	0.00	17
May	56	2.36	0.00	0.00	39
Jun	126	5.37	0.00	0.00	90
Jul	180	7.66	1.00	0.00	100
Aug	217	9.20	1.00	1.00	100
Sep	125	5.31	0.31	1.00	100
Oct	61	2.59	0.00	0.31	48
Nov	14	0.58	0.00	0.00	10
Dec	8	0.33	0.00	0.00	5

supply enhancements during the 2 or 3 months before and after this time period. It is clear, however, that rainwater harvesting cannot be the sole source of water for this household, especially during the dryer months. This procedure can easily be modified to investigate other scenarios for different roof areas and tank sizes.)))

))) References

Cunliffe, D. A. (1998). *Guidance on the use of rainwater tanks.* National Environmental Health Forum Monographs, Water Series No. 3, National Environmental Health Forum, Rundle Mall, Australia.

Development Technology Unit (DTU). (1987). *Domestic roofwater harvesting research programme.* University of Warwick School of Engineering, <http://www.eng.warwick.ac.uk/DTU/rwh/index.html> (Jan. 20, 2008).

Gould, J., and Nissen-Petersen, E. (1999). *Rainwater catchment systems for domestic supply,* Intermediate Technology Publications, London.

Mason, Y., Ammann, A., Ulrich, A., and Sigg, L. (1999). "Behavior of heavy metals, nutrients, and major components during roof run-off infiltration." *Envir. Sci. and Technol.,* 39, 1588–1597.

Pickford, J. (1991). *The worth of water,* IT Publishing, London.

United National Environment Programme (UNEP). (1997). "Rainwater harvesting from rooftop catchments." *Sourcebook of alternative technologies for freshwater augmentation in Latin America and the Caribbean,* United National Environment Programme, Osaka, Japan <www.oas.org/usde/publications/unit/oea59e/ch10.htm> (Jan. 18, 2009).

))) Further Reading

Cowden, J. R., Watkins, D. W., and Mihelcic, J. R. (2008). "Stochastic rainfall modeling in West Africa: Parsimonious approaches for domestic rainwater harvesting assessment." *J. Hydrol.,* 361(1–2), 64–77.

Thomas, T. H., and Martinson, D. B. (2007) *Rainwater harvesting, a handbook for practitioners.* Technical Paper Series; no. 49, IRC International Water and Sanitation Centre, Delft, Netherlands.

18

Water Treatment

))) **18.1 The Need for Potable Water**

Environmental engineers entering developing communities are often confronted first by the need for potable water. Whether or not the fundamentals of disease transmission are understood, the importance of having a sufficient supply of high-quality and good-tasting water is obvious. Although chemical contamination should be a consideration for drinking water treatment, the majority of water-related health problems in developing countries are related to microbial contamination (WHO 2006). This chapter does, however, cover treatment of arsenic and fluoride, because these chemical constituents are important in many parts of the world.

))) **18.2 Drinking Water Guidelines**

According to the World Health Organization (WHO) (2006), *safe drinking water* is water that "does not represent any significant risk to health over a lifetime of consumption, including different sensitivities that may occur between life stages." The WHO views the risk–benefit approach to be more appropriate for setting individual nations' standards than setting international standards for drinking water. The *risk–benefit approach* involves analyzing the risks occurring throughout a water supply, including catchment, source, and point of use, and then identifying methods of managing these risks. Instead of publishing international standards, the WHO publishes guidelines for drinking water quality. The most recent guidelines are available online (WHO 2006). Where national standards exist, they should also be met.

The WHO recommends that at minimum, *Escherichia coli* (*E. coli*), thermotolerant (fecal) coliforms, and chlorine residuals (where there is chlorination) be monitored in community water supply systems. This minimum monitoring should be supplemented by monitoring of turbidity and by pH adjustment where the water is chlorinated.

Guidelines that are applicable to many developing world settings are summarized in Table 18-1. Arsenic, nitrate, fluoride, and turbidity can all be measured directly. Microbial water quality is typically measured by analyzing indicator microorganisms, such as *E. coli* or thermotolerant coliforms, or by assessing specific pathogen densities. The presence of *E. coli* in drinking water is conclusive evidence of recent fecal contamination. On the other hand, some viruses and protozoa are more resistant to disinfection,

---

Erlande Omisca, Maya A. Trotz, and Qiong Zhang contributed to this chapter.

Table 18-1. WHO Guidelines for Common Contaminants

Contaminant	Guideline
Arsenic	0.01 mg/L (P)
Fluoride	1.5 mg/L
Nitrate	50 mg/L
E. coli	0/100 mL[a]
Thermotolerant coliforms	0/100 mL[a]
Turbidity[b]	0.1 NTU
Arsenic	10 µg/L
Fluoride	1.5 mg/L

Note: P = provisional guideline value.

[a]In communities where the water supply fails to meet this guideline, the WHO suggests using a grading scheme based on the percentage of negative samples. See Table 5-2 in the WHO guidelines for an example grading scheme. Total coliforms is not considered an appropriate indicator for fecal contamination because many bacteria of no significance occur naturally, especially in tropical waters.

[b]There is currently no WHO guideline for turbidity, however the guidelines do suggest 0.1 nephelometric turbidity units (NTU) for effective disinfection.

Source: WHO 2006.

so a negative *E. coli* test does not necessarily mean that water is free of pathogens caused by fecal contamination.

))) 18.3 Water Treatment Processes

Water treatment can be accomplished by several different methods, and in conventional treatment, these methods are combined. The methods that are conventionally used in the industrialized world include mechanical separation, coagulation and flocculation, chemical purification, disinfection processes, biological processes, aeration, and membrane technologies. A schematic of the flow of water through a conventional treatment plant is shown in Fig. 18-1. As a last resort, boiling is used to treat disease-carrying water in homes, though it is estimated that hundreds of millions of people use this method. Table 18-2 defines each of these treatment methods and links them to the technologies discussed in this chapter. All of the processes used in conventional treatment systems can be scaled down for point-of-use treatment. The technologies described here may similarly be used in combination to increase effectiveness.

In addition to treatment, water storage at the household is an important component of a safe water supply plan. Proper water storage in the home protects treated water from contamination and can also allow sedimentation of untreated or poorly treated water.

))) 18.4 Point-of-Use Treatment vs. Treatment at the Source

In developed countries, drinking water treatment for communities is often centralized treatment at the source, with distribution to households. Therefore, much emphasis has

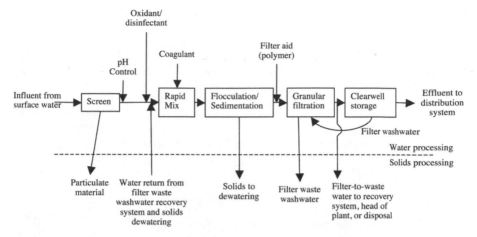

Figure 18-1. Schematic Flow Diagram of Conventional Treatment of Surface Water.

been placed on implementing community-scale drinking water treatment and distribution systems in developing communities. This method may be a viable option for many communities, but in some cases it may be necessary to consider implementing treatment options at the household level for several reasons, among them:

- the requirement for substantial organizational capacity for community-level systems,
- the potential for recontamination during the time between collection of treated water at the source and use in the household, and
- the personal responsibility and pride that may motivate people to maintain a household system more than contribute time or money to the maintenance of a community system.

Point-of-use treatment interventions have been shown to significantly reduce diarrhea morbidity. In fact, a review of water quality interventions found that point-of-use treatment was more effective at reducing diarrhea morbidity than improvement of water quality at the source, shown in Table 18-3 (Clasen et al. 2006). An additional consideration for determining whether to use point-of-use or source interventions would be the objective of the intervention. For example, if reduction of water-washed diseases such as trachoma was the main objective, then improvements at the source and improved access to water (e.g., via distribution) may be a preferred intervention. Community treatment and distribution systems require a great deal of organization. A positive side of this is that it may be an opportunity to bring a community together to build capacity for other development initiatives.

))) 18.5 Pretreatment (Sedimentation, Coagulation, and Flocculation)

Many water treatment processes are affected by turbidity, which is a measure of how cloudy water is. The higher the turbidity, the less light can pass through the water. Turbidity

Table 18-2. Methods Used During Conventional Treatment of Drinking Water

Treatment Method	Definition	Technologies Discussed in This Chapter
Mechanical separation	Removal of large particles through gravity (sedimentation), screening, or adhesion.	Sedimentation Clay pot filtration Slow sand filtration Filtration through cloth Guinea worm straws
Coagulation and flocculation	Coagulation involves adding a coagulant such as alum to water to destabilize suspended, colloidal, and dissolved matter. After coagulation, the destabilized particles and precipitation products aggregate through the process of flocculation. These larger "floc" particles are then removed through mechanical separation. Natural coagulants exist, such as seeds from the *Moringa olifera* tree.	Alum *Moringa olifera* seeds
Chemical purification	Softening, removal of iron, and acid neutralization.	
Disinfection processes	Inactivation of microbiological contaminants through the addition of disinfection agents such as free chlorine, combined chlorine, ozone, chlorine dioxide, UV light, or heat.	Chlorine disinfection Solar disinfection Heat
Biological processes	Oxidation of organic matter as it is consumed by organisms, or the death of undesirable organisms from the absence of food or being killed by other organisms.	Slow sand filtration
Aeration	The evaporation of carbonic acid and gases in solution, and supplying oxygen necessary for some chemical disinfection processes and water-purifying organisms.	Not discussed in this chapter, but aeration can be accomplished by nonmechanical methods.
Membrane technology	Physiochemical technique that uses the differences in permeability of water constituents as a separation method.	

Source: Adapted from Crittenden et al. 2005.

can reduce the effectiveness of water treatment in a number of ways. It can clog filters, prevent UV rays from reaching pathogens, and exert excessive oxygen demand during chlorination. For these reasons, many water treatment processes have pretreatment turbidity limits (see Table 18-4). Turbidity can easily be measured with the use of a turbidity tube (see Section 18.9). Treatment techniques that can reduce the turbidity of water before further treatment include settling and decanting, coagulation and flocculation, and roughing filters.

18.5.1 Sedimentation

Gravity is the oldest and most widely used process in water treatment and is effective at removing much of the suspended matter from water. Sedimentation can take place at

Table 18-3. Reduction in Diarrhea for All Age Groups Following Different Interventions to Improve Water Quality

Intervention Type (No. Trials)	Relative Risk (RR) Estimate (Random)	Percentage Reduction (1−RR)	95% CI of Relative Risk Estimates
Source (6)	0.73	27	0.53–1.01
Household total (32)	0.53	47	0.39–0.73
Household filtration (6)	0.37	63	0.28–0.49
Household chlorination (16)	0.63	37	0.52–0.75
Household solar sisinfection (2)	0.69	31	0.63–0.74
Household flocculation and disinfection (7)	0.48	52	0.20–1.16
Household improved storage (1)	0.79	21	0.61–1.03

Note: Although all interventions resulted in reduced diarrhea, point-of-use treatment interventions generally had a greater effect. There was significant variation in the relative risk estimates among studies of the same intervention type, which the authors suggest resulted from a variety of conditions that would require further research to understand. Additionally, these studies only considered reduction in diarrhea; results might have been significantly different if water-washed diseases such as trachoma had been considered.

Source: Clasen et al. 2006

both the community and household level. It can be a stand-alone treatment in situations where other treatment processes are unfeasible, or it can be a part of a larger treatment process, preceded by coagulation and flocculation and followed by filtration and disinfection. For a neighborhood or community, a simple rectangular sedimentation basin provides removal of much of the turbidity in surface water. At the household level, sedimentation can occur in the storage containers, and this process can be maximized by improving storage procedures.

Community Sedimentation Basin Design

In conventional systems, sedimentation often follows coagulation and flocculation and involves water moving through a sedimentation pond or tank, allowing time for particles

Table 18-4. Pretreatment Turbidity Limits for Various Water Treatment Processes

Treatment Type	Pretreatment Turbidity Limit (NTU)
Drinking water (general recommendation)	<5
Ceramic filtration	<15–20
Slow-sand filtration	<20
Chlorination	Ideally <1 Acceptable <5 In emergencies <20 for a very short period
UV disinfection	<30
Disinfection with heat	(No limit)

to settle by gravity. The speed at which the particles move downward—the *settling velocity*—determines the size of the sedimentation pond.

The method used to design a sedimentation tank depends on the treatment objective: removal of particles with a minimum diameter or particle removal efficiency. If the treatment objective is to remove particles of a certain diameter, then calculate the settling velocity to determine the dimensions of the tank.

If particles are assumed to act as discrete particles and settle without the interference of neighboring solids, the settling velocity is determined using Stokes' law:

$$v_s = \frac{g(\rho_p - \rho)d_p^2}{18\mu} \tag{18-1}$$

where v_s is the settling velocity of the particle, g is the acceleration because of gravity 9.81 m/s², ρ_p is the density of the particle (kg/m³), ρ is the density of water (kg/m³), d_p is the diameter of the particle (m), and μ is the dynamic viscosity of water (kg/m-s).

Equation 18-1 determines the settling velocity for laminar flow (low Reynold's numbers, Re). In turbulent water (high Reynold's numbers), the settling velocity is determined using Newton's law:

$$v_s = \sqrt{\frac{4g(\rho_p - \rho)d_p}{3C_D\rho}} \cong \sqrt{\frac{4g(sg_p - 1)d_p}{3C_D}} \tag{18-2}$$

where C_D is the unitless drag coefficient and sg_p is the unitless specific gravity of the particle. C_D is calculated using Eq. 18-3 for Re $\leq$ 1 or Eq. 18-4 for 1 < Re < 10,000.

$$C_D = \frac{24}{Re} \text{ for Re} \leq 1 \tag{18-3}$$

$$C_D = \frac{24}{Re} + \frac{3}{\sqrt{Re}} + 0.34 \text{ for } 1 < Re < 10,000 \tag{18-4}$$

The process of calculating the settling velocity is an iterative process. The general procedure for calculating a settling velocity is:

1. Calculate v_s using Stokes' law (Eq. 18-1).
2. Use the v_s calculated from step 1 to determine the Reynold's number.
3. If Re > 1, go to step 4. Otherwise, Stokes' law is valid.
4. Calculate C_D using Eq. 18-4 and the Reynold's number from step 2.
5. Calculate v_s from Eq. 18-2 using the drag coefficient from step 4.
6. Recalculate C_D using v_s.
7. Repeat steps 4–6 until C_D converges.

Particle sizes of pollutants and naturally occurring substances are shown in Fig. 18-2, along with treatment processes that correspond to a particular size range. Where

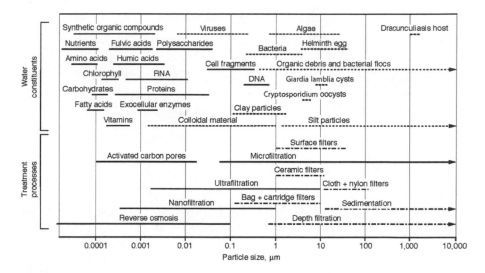

Figure 18-2. Characterization of Particulate Water Quality Constituents Found in Water by Type, Size, and Appropriate Treatment Methods

Note. Particles that cause health concern in the developing world and appropriate treatment methods are presented by dashed lines. Microbial constituents can be seen in the upper right corner. 1 μm is a millionth of a meter and equals 1/25,400 of an inch.

Source: Adapted from Mihelcic and Zimmerman (2009) with other data obtained from AWWA 1990; Colwell et al. 2003; and Ovanedel-Craver and Smith 2008.

coagulation and flocculation are used for pretreatment, typical settling velocities of floc types are provided in Table 18-5.

A rectangular sedimentation basin is a common treatment method used in community water treatment systems (Fig. 18-3). The design of a sedimentation basin is determined such that the detention time in the basin is sufficient for particles to settle at the calculated settling velocity. Figure 18-4 shows how particles settle as water moves through a rectangular sedimentation basin at fluid velocity, v_f.

In Fig. 18-4, it is clear that particles at the surface will be removed when the hydraulic retention time of fluid in the sedimentation basin is long enough for a particle to settle at its settling velocity, v_s, a distance of h_0. A particle (particle 2 in Fig. 18-4) that enters at the top of the basin and settles before it flows out of the basin is called a *critical particle*. Its settling velocity is defined as the *critical particle settling velocity* (v_c), which is determined as

$$v_c = \frac{h_0}{\tau}$$

(18-5)

where τ is the hydraulic retention time.

Table 18-5. Settling Velocities of Selected Types of Floc

Floc Type	Settling Velocity (m/s)
Small alum floc	2–4.5
Medium-sized alum floc	3–5
Large alum floc	4.0–5.5
Heavy lime softening floc	4.5–6.5
Iron floc	2–4

Source: Adapted from Crittenden et al. 2005.

Because the hydraulic retention time equals the volume (V) of the sedimentation basin divided by the process flow rate, Q, Eq. 18-5 can be written as

$$v_c = \frac{Q}{A} = OR \qquad (18\text{-}6)$$

where OR is defined as the *overflow rate* (m^3/m^2-h). Note that the OR is the hourly flow rate divided by the surface area of the sedimentation tank. In reality, its units are velocity (m/h). Tanks are designed such that if the settling velocity of individual particles is greater than (or equal to) the OR, then 100% of incoming particles are removed. Typical design criteria for horizontal-flow rectangular sedimentation basins are provided in Table 18-6. Example 18-1 shows how the settling velocity and flow rate are used to design a sedimentation basin that meets the criteria in Table 18-6.

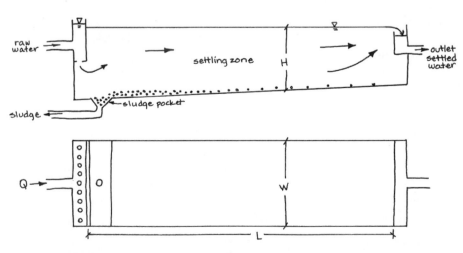

Figure 18-3. Basic Horizontal-Flow Rectangular Sedimentation Basin Design in Cross Section (Top) and Plan View (Bottom).

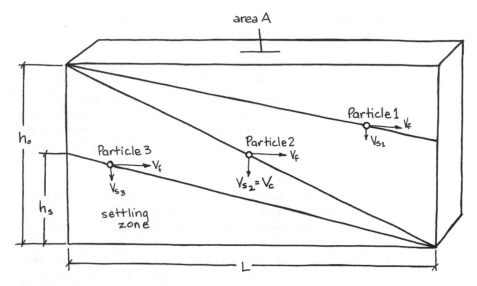

Figure 18-4. Trajectories of Three Discrete Particles Settling in a Rectangular Clarifier.

Note: Particles 2 and 3 will be removed in this particular situation.

Table 18-6. Design Criteria for Horizontal-Flow Rectangular Tanks

Design Parameter	Recommendation
Minimum number of tanks	2
Water depth	3–5 m
Length-to-depth ratio	15:1
Width-to-depth ratio	3:1 to 6:1
Length-to-width ratio	4:1 to 5:1
Surface loading rate (*OR*)	1.25–2.5 m/h
Horizontal mean-flow velocity at maximum daily flow	0.3–1.1 m/min
Detention time	1.5–4 h
Reynold's number	<20,000
Froude number	$>10^{-5}$
Bottom slope for manual sludge removal systems	1:300 m/m

Source: Crittenden et al. 2005.

》》》 **Example 18-1.** Sedimentation Tank Design

A water treatment plant with a maximum daily flow of 3 m³/s (assume this value is 1.5 times the average flow of 2 m³/s) is treating surface water. Particle settling velocity was determined to be 2.2 m/h at 10 °C. The dynamic viscosity of water at 10 °C is

0.00131 kg/m-s, and the density of water at 10 °C is 999.7 kg/m^3. Design a horizontal-flow rectangular sedimentation basin based on the maximum flow rate.

Solution

1. Determine the number of basins. Two basins would satisfy the minimum requirement for maintenance purposes. However, if one basin were off line, the entire plant flow would be directed through the remaining basin, possibly resulting in overloading of the basin. To minimize the risk of basin overloading, three basins will be selected, and overloading will be checked to verify that three basins are adequate.
2. Determine the size of each basin.
   a. Select the basin width and depth. The basin width will be governed by the standard size of sludge removal equipment. For example, in the United States, the standard maximum width of the chain-and-flight sludge collector is 6 m, so basin widths in increments of 6 m will be considered, starting at 18 m. Water depths from 3 to 5 m are appropriate, according to Table 18-6. Deeper basins are recommended over shallower basins, so a depth of 4 m will be selected.
   b. Determine the basin area. The settling velocity such that the particle is removed in the sludge zone just before the outlet, v_c, is given as 2.2 m/h at 10 °C. (This value is also equal to the overflow rate.) Use Eq. 18-6 to determine the basin surface area:

$$A = \frac{Q}{v_c} = \frac{3 \text{ m}^3/\text{sec}}{(2.2 \text{ m/hr}) \times (1 \text{ hr}/3{,}600 \text{ s})} = 4{,}909 \text{ m}^2$$

   c. Determine the length using the design guidelines in Table 18-6 for length-to-width ratios. For three tanks that are 18 m wide, the tank length and length-to-width ratio can be estimated:

$$L = \frac{4{,}909 \text{ m}^2}{3 \text{ basins} \times 18 \text{ m}} = 90.9 \text{ m} \qquad \frac{L}{W} = \frac{90.9}{18} = \frac{5.0}{1}$$

   The length-to-width ratio meets the recommendation of 4:1 to 5:1.
3. Check the various design parameters listed in Table 18-6.
   a. Check the detention times at Q_{max} and Q_{ave}.

$$\text{Detention time for } Q_{max} = \frac{(18 \times 90.9 \times 4) \text{ m}^3 \times 3 \text{ basins}}{(3 \text{ m}^3/\text{sec})(3{,}600 \text{ sec}/\text{h})} = 1.8 \text{ h}$$

$$\text{Detention time for } Q_{ave} = 1.5 \times 1.8 \text{ h} = 2.7 \text{ h}$$

   These detention times are within the acceptable range of 1.5 to 4 h.
   b. Check the length-to-depth ratio:

$$\frac{L}{D} = \frac{90.9}{4} = \frac{22.7}{1}$$

The basin length-to-depth ratio is 22.7:1, which is greater than the recommendation of 15:1.

c. Check the horizontal-flow velocity. The mean velocity is provided by the expression

$$v_f = \frac{Q}{A} = \frac{(3\,\text{m}^3/\text{sec})(60\,\text{sec/min})}{18\,\text{m} \times 4\,\text{m} \times 3\,\text{basins}} = 0.83\,\text{m/min}$$

The mean velocity is greater than 0.3 m/min and less than 1.1 m/min.

d. Check the Reynold's and Froude numbers:

$$\text{Re} = \frac{\rho v_f R_h}{\mu}$$

$$R_h = \frac{A_x}{P_w} = \frac{4\,\text{m} \times 18\,\text{m}}{18\,\text{m} + 2(4\,\text{m})} = 2.77\,\text{m}$$

$$v_f = \frac{0.833\,\text{m/min}}{60\,\text{sec/min}} = 0.014\,\text{m/sec}$$

$$\text{Re} = \frac{(999.7\,\text{kg/m}^3)(0.014\,\text{m/sec})(2.77\,\text{m})}{0.00131\,\text{kg/m}\cdot\text{sec}} = 29,594$$

The Reynold's number of 29,594 is higher than the recommended value of 20,000 for a horizontal sedimentation basin.

$$\text{Fr} = \frac{v^2}{gR_h} = \frac{[(0.014)^2\,\text{m}^2/\text{sec}^2]}{(9.81\,\text{m/sec}^2)(2.77\,\text{m})} = 7.2 \times 10^{-6}$$

The Froude number is lower than the recommended value for sedimentation tanks, so the tank design must be modified.

4. Consider the addition of two longitudinal baffles per basin and recompute the Reynold's and Froude numbers.

$$R_h = \frac{A_x}{P_w} = \frac{4\,\text{m} \times 6\,\text{m}}{6\,\text{m} + 2(4\,\text{m})} = 1.71\,\text{m}$$

$$\text{Re} = \frac{(997.7\,\text{kg/m}^3)(0.014\,\text{m/sec})(1.71\,\text{m})}{0.00131\,\text{kg/m}\cdot\text{sec}} = 18,162 < 20,000\,\text{OK}$$

$$\text{Fr} = \frac{[(0.014)^2\,\text{m}^2/\text{sec}^2]}{(9.81\,\text{m/sec}^2)(1.71\,\text{m})} = 1.17 \times 10^{-5} > 10^{-5}\,\text{OK}$$

Source: Crittenden et al. 2005.

If the treatment objective is to remove a percentage of total suspended solids (TSS), then the depth and detention time of the sedimentation tank can be determined using a column settling test, as described in Box 18-1. The column settling test measures the percent removal of TSS for a given sedimentation basin depth and hydraulic detention time. A design basin depth and detention time can be determined by adjusting the depth and hydraulic detention time of the settling test until the required percent removal is achieved.

Sedimentation in Household Storage Containers

Within a few hours, sand, silt, and large microbes settle to the bottom of a typical container used for household drinking water storage. After 1 or 2 days, helminth eggs, some parasites, certain types of algae, and large clay particles also settle out. Sedimentation cannot reliably remove viruses, bacteria, or small clay particles, although some of these are removed if they are attached to larger particles that do settle (Fig. 18-2).

After 1 day of storage, up to 50% of bacteria may die off (Shaw 1999). The transmission of schistosomiasis can also be prevented by storing water for more than 48 h, which is the limit of how long *cercariae* (the intermediate host) can live after leaving the snail and before reaching a human or animal host (Shaw 1999).

After enough time has been allowed for settling, the cleaner water (supernatant) at the top of the container can either be decanted or ladled out, being careful not to disturb the layer of sediment. At least two storage containers are required, one in which the larger particles settle, and one to store the supernatant. The first container should be cleaned after each use for small vessels, and as frequently as necessary for larger cisterns or tanks.

⟫⟩ Box 18-1 Simple Column Settling Test to Be Used for Determining Depth and Settling Time of a Sedimentation Basin

1. Measure the total suspended solids (TSS) of the water to be treated.
2. Select a depth and hydraulic detention time for the sedimentation basin.
3. Build a column of any diameter that is equal in height to the depth of the sedimentation basin.
4. Introduce a sample of the water to be treated so that it fills the column to the height of the sedimentation basin. Ensure uniform distribution of particle sizes as you introduce the water.
5. Allow settling to take place over a period of time. This time will be equal to the settling time of the proposed sedimentation basin.
6. Draw off the settled material at the bottom of the column, and mix the remaining water.
7. Measure the TSS of the remaining water in the column. The percent removal of TSS is simply the final TSS over the initial TSS. Adjust the height and settling time of the column test until the desired percent removal is achieved.

Source: Metcalf and Eddy 2003.

A three-pot treatment system (Fig. 18-5) has also been developed to promote sedimentation (Shaw 1999). The three required pots do not include the bucket or other vessel used for collecting water. Pot No. 1 should occasionally be sterilized using boiling water.

Every day, the following steps should be followed:

1. Take water for drinking and cooking from pot No. 3 (far right in Fig. 18-5). This water has already been stored for at least 2 days.
2. Pour water carefully from pot No. 2 to pot No. 3, and wash out pot No. 2 using some water from pot No. 3.
3. Pour water carefully from pot No. 1 to pot No. 2, and wash out pot No. 1 using some water from pot No. 3.
4. After collecting water from the source, pour it from the vessel used for collection into pot No. 1, with the possible use of a cloth to filter the water entering the pot.

Because of its low cost and low level of required skill, sedimentation can be an attractive way to pretreat water by reducing its turbidity, but it should not be relied on as the sole means of pathogen removal.

18.5.2 Coagulation and Flocculation

Because of their small size and charge repulsion, colloidal particles and smaller suspended particles cannot be removed by simple gravity sedimentation. Coagulation and flocculation are used to remove these particulate contaminants. *Coagulation* involves adding a chemical coagulant to water to destabilize suspended, colloidal, and dissolved matter. After coagulation, the destabilized particles and precipitation products aggregate through the process of flocculation. These larger "floc" particles are then removed through gravity settling or mechanical separation. Coagulants include alum, ferric chloride, ferric sulfate, and natural plant materials or synthetic polyelectrolytes. Table 18-5 provides typical settling velocities for selected types of floc.

Figure 18-5. A Three-Pot System Used for Household Treatment of Water Using Principles of Sedimentation in All Three Pots and Filtration in the First Pot.

Note: Cover the pots when water is not entering or leaving them.

Alum, Ferric Chloride, and Ferric Sulfate

A commonly used coagulant is alum, $Al_2(SO_4) \cdot xH_2O$, where x is usually 14. When added to water, alum dissociates and hydrolytic reactions proceed, producing aquo aluminum ions, mononuclear and polynuclear species, precipitate, and aluminate ions. Depending on the pH of the system, different products will dominate. The goal for coagulation is for the precipitate to dominate. For $Al(OH)_3(s)$ to dominate, the pH should be between about 5.5 and 7.7. This reaction occurs because in this pH range the solubility of $Al(OH)_3$ is near its minimum (so less alum needs to be added to form the precipitate) and the Al species formed have a primarily positively charged composition, which can neutralize the negative charge of most natural colloids.

The overall reactions for addition of alum, ferric sulfate, and ferric chloride, with the metal precipitates as the product, are shown, respectively, in Eqs. 18-7 to 18-9.

Alum:
$$Al_2(SO_4) \cdot 14H_2O + 3Ca(HCO_3)_2 (s) \rightarrow$$
$$2Al(OH)_3 (s) + 3CaSO_4 + 14H_2O + 6CO_2 \tag{18-7}$$

Ferric sulfate:
$$Fe_2(SO_4)_3 \cdot 9H_2O + 3Ca(HCO_3)_2 (s) \rightarrow$$
$$2Fe(OH)_3 (s) + 3CaSO_4 + 9H_2O + 6CO_2 \tag{18-8}$$

Ferric chloride:
$$FeCl_3 \cdot 6H_2O + 3Ca(HCO_3)_2 (s) \rightarrow$$
$$Fe(OH)_3 (s) + 3CaCl + 6H_2O + 3CO_2 \tag{18-9}$$

Because the interdependence of the characteristics of the coagulant, the concentration and type of particulates, the concentration and characteristics of natural organic matter, water temperature, and water quality is not yet known quantitatively, there is no quantitative prediction of the optimum coagulant combination for a particular sample of water. Instead, a test referred to as jar testing (see Mihelcic and Zimmerman 2009 for explanation) is used at water treatment plants to monitor coagulant addition. Table 18-7 provides typical dosages for alum, ferric sulfate, and ferric chloride that are used to treat typical surface waters.

In Eqs. 18-7 to 18-9, naturally occurring alkalinity (i.e., $Ca(HCO_3)_{2\,(s)}$) is consumed in the reaction. If the naturally occurring alkalinity is too low, it may be necessary to add alkalinity in the form of lime ($Ca(OH)_2$), caustic soda ($NaOH$), or soda ash (Na_2CO_3). The stoichiometry of the reactions for alum with caustic soda, soda ash, and lime are provided in Eqs. 18-10 to 18-12, respectively. Caustic soda is often the alkaline of choice because it is relatively easy to handle and it requires a relatively small dosage.

Caustic soda:
$$Al_2(SO_4)_3 \cdot 14H_2O + 6NaOH \rightarrow$$
$$2Al(OH)_3 (s) + 3Na_2SO_4 + 14H_2O \tag{18-10}$$

Soda ash:
$$Al_2(SO_4)_3 \cdot 14H_2O + 3Na_2CO_3 \rightarrow$$
$$2Al(OH)_3 (s) + 3Na_2SO_4 + 3CO_2 + 14H_2O \tag{18-11}$$

Lime:
$$Al_2(SO_4)_3 \cdot 14H_2O + 3Ca(OH)_2 \rightarrow$$
$$2Al(OH)_3 (s) + 3CaSO_4 + 14H_2O \tag{18-12}$$

Example 18-2 shows how to determine the amount of alkalinity that will be required using the stoichiometry of these equations.

Table 18-7. Typical Dosages for Coagulants Used in Traditional Water Treatment Processes

Coagulant	Typical Dosage
Alum [$Al_2(SO_4) \cdot 14H_2O$]	10–150 mg/L
Ferric sulfate [$Fe_2(SO_4)_3 \cdot 9H_2O$]	10–250 mg/L
Ferric chloride ($FeCl_3 \cdot 6H_2O$)	5–150 mg/L

))) Example 18-2. Example Calculation of the Required Amount of Alkalinity for Alum to Completely Precipitate During Coagulation

A dosage of 10 mg/L of alum is added to 1,000 L of water. What is the required dosage of alkalinity (expressed as grams of $CaCO_3$) for the alum to completely precipitate? (10 mg/L of alum in 1,000 L of water equals 10,000 mg alum, or 10 g alum.)

Solution

From Eq. 18-7, 6 moles of alkalinity (HCO_3^-) are consumed for every mole of alum. This reaction occurs because 3 moles of $Ca(HCO_3)_2$ are required and there are 2 moles of HCO_3^- in each mole. The reaction stoichiometry and the molecular weight of different species can be used to determine the amount of alkalinity consumed during the reaction:

$$\text{Alkalinity consumed} = 10 \text{ g } Al_2(SO_4) \cdot 14H_2O \times \left(\frac{1 \text{ mol } Al_2(SO_4)_3 \cdot 14H_2O}{594 \text{ gm } Al_2(SO_4)_3 \cdot 14H_2O} \right)$$

$$\times \left(\frac{6 \text{ mole } HCO_3^-}{1 \text{ mole } Al_2(SO_4)_3 \cdot 14H_2O} \right) \times \left(\frac{1 \text{ eqv. alkalinity}}{1 \text{ mole } HCO_3^-} \right)$$

$$\times \left(\frac{50 \text{ gm } CaCO_3}{1 \text{ eqv. alkalinity}} \right)$$

$$= 5 \text{ gm as } CaCO_3$$

)))

Moringa oleifera

The *Moringa oleifera* tree is commonly grown in the tropics. It is also known as the drumstick or horseradish tree. The seeds from these trees can be cultivated and used as a natural coagulant to remove turbidity. In Sudan, the seeds are collected, powdered, placed in a small cloth bag, and tied with a thread. The bag is then swirled in the turbid water to promote coagulation and flocculation, and the flocculated solids are allowed to settle before the water is consumed (Shaw 1999).

Moringa oleifera seeds can be used for coagulation in community water treatment systems with flow ≤10 m³/h. The steps for obtaining a solution from whole seeds for use in

))) **Box 18-2 Using *Moringa oleifera* Seeds to Treat Water**

Preparation of **Moringa oleifera** *Solution*
1. Allow the seeds to mature and dry naturally to a brown color while still on the tree.
2. Harvest the seed pods. Remove the seeds and shell them.
3. Crush the seeds and sieve with a 0.8-mm mesh. The traditional large mortar and pestle is an appropriate method for crushing the seeds, where there is no mechanical alternative. Seeds and seed powder can be stored dry.
4. Mix the seed powder with clean water to form a paste [2 g (2 teaspoons) of seed powder treats 20 L of water]. As a rule of thumb, the powder from one seed kernel can treat 1 L of very turbid or 2 L of somewhat turbid water. The paste should be freshly prepared for every treatment. Table 18-A provides guidelines for dosage when turbidity is known.

Table 18-A. Dose Requirements for *Moringa oleifera* According to Raw Water Turbidity

Raw Water Turbidity (NTU)	Dose Range (mg/L)
<50	10–50
50–150	30–100
>150	50–200

Source: Shaw 1999.

5. Dilute the paste in 1 cup of water in a sealed bottle, and shake the solution for 5 min to release the chemicals in the powder.
6. Filter the insoluble material using either a fine mesh screen or a muslin cloth. The remaining solution is ready for use.

Treatment of Drinking Water with **Moringa oleifera** *Solution*
1. Pour the solution into the container of water to be treated.
2. Stir rapidly for 2 min, then stir slowly for 10–15 min.
3. Let the bucket sit for 1–2 h. Do not disturb the bucket. This allows the solid floc to settle.
4. After the floc has settled to the bottom of the container, carefully pour the clean water off the top.

Source: Fuglie 2001.

water treatment are outlined in Box 18-2. Once the *Moringa oleifera* solution is prepared, the instructions in Box 18-2 can be used in the treatment of water.

The seed solution can be prepared using either the shelled whole seeds or the solid residue that remains after the extraction of seed oil. If the solid residue is being used, the press cake is ground to a fine powder and sieved through a 0.8-mm mesh. The solution is then prepared using steps 4–6 of the solution preparation process for whole seeds described in Box 18-2.

When *Moringa oleifera* seeds are used for treatment of a community water supply, thought should be given to the cultivation of the seeds. According to Shaw (1999), the average mature *Moringa oleifera* tree yields 3 kg of seed kernels per year. Trees should be planted with a spacing of 3 m. Once trees are planted from seeds or cuttings, they grow rapidly, up to 4 m in height. Flowers and fruits are produced after 12 months. In locations with favorable climates, two harvests per year may be possible. As a rule of thumb, expect 3–5 metric tons of seeds per hectare per year. Example 18-3 provides an example calculation for determining the land requirement for treatment of drinking water with *Moringa oleifera* seeds.

))) Example 18-3. Example Calculation of Land Requirement for Treatment of Drinking Water with *Moringa oleifera* Seeds

Calculate the required land area to treat water with a turbidity of 150 NTU with a process flow rate of 10 m^3/h that operates 8 h per day for 1 year:

$$\frac{100\ mg}{L}\left(\frac{kg}{1,000,000\ mg}\right)\left(\frac{10\ m^3}{hr}\right)\left(\frac{1,000\ L}{m^3}\right)\left(\frac{8\ hr}{day}\right)\left(\frac{365\ days}{year}\right) = 2,920\ kg\ seeds$$

$$2,920\ kg\left(\frac{1\ metric\ ton}{1,000\ kg}\right)\left(\frac{1\ hectare}{3\ metric\ tons}\right) = 0.97\ hectare$$

Thus, approximately 1 hectare of land is required to cultivate *Moringa oleifera* seeds for the treatment process.

Source: From Shaw 1999.   )))

))) 18.6 Filtration

Filtration can be used to remove turbidity from drinking water, as well as for biological disinfection. Several methods of filtration exist that can be used at either a community scale or for point-of-use treatment. Here we focus primarily on point-of-use treatment methods because community-scale treatment design information is more readily available in traditional environmental engineering textbooks.

18.6.1 Filtration through Ceramic Clay Pots

Two types of ceramic clay pot filters exist: the candle type, where water flows from an outer bucket into a partially submerged, smaller receptacle, and the type where the clay filter rests inside a larger bucket and water flows down through the filter into the bucket (Fig. 18-6). Clay pot filters need to be cleaned regularly, both to prevent biofilm accumulation and to

Figure 18-6. Example of a Ceramic Clay Pot Filter Used by Potters for Peace (n.d.).

Note: The water storage container is covered by the clay pot and also has a faucet for withdrawing water.

restore the original flow rate. Even a slight crack in this type of filter makes it unsuitable to remove pathogens and turbidity, so users must be prepared to replace the filter on a regular basis.

The ability of clay filters to remove pathogens varies dramatically with their materials and methods of production. Many clay pot filters produced in developed countries have pore sizes small enough to remove bacteria (Fig. 18-2). A recent study has found that 50% of the pore size diameters range from 0.02 to 15 μm (Ovanedel-Craver and Smith 2008). Clay pot filters can also remove viruses through adsorption. Virus removal decreases over time as these adsorption sites are occupied. Many commercially available filters are also impregnated with silver to help prevent the growth of a biofilm on the surface of the pot. Although the effectiveness of many of the ceramic filters produced in developing countries is uncertain, any of these filters (as long as they are not cracked) likely reduces turbidity and pathogens in water and is recommended over no treatment.

18.6.2 Slow Sand Filtration

In *slow sand filtration*, influent water moves downward through a nonuniform bed of sand, and particles are filtered out in the first few centimeters (called *surface filters* in Fig. 18-2). In addition to this physical process, a biological layer called the *Schmutzdecke* forms an additional filtration layer and biologically degrades some organic matter. Thus, the slow sand filter provides both a filtration process and a biological treatment process.

The slow sand filter consists of three layers: underdrainage, gravel, and sand. Slow sand filters can be designed at a large scale for treatment of water at the source, for example, integrated with a spring box (see Chapter 15). Similar slow sand filtration technologies can also be implemented at the household level.

For a slow sand filter to function, the sand should have an effective grain size of 0.15–0.30 mm and uniformity coefficient (D_{60}/D_{10}) < 5. The D_{60} is the sieve size that allows 60% of the sand to pass, and D_{10} is the sieve size that allows 10% to pass. The minimum height of the sand layer should be 0.5 m, and the rate of filtration should be no faster than 0.1–0.2 m^3/m^2-h (Visscher et al. 1987). For a filtration rate of 0.1 m/h through a clean filter, the minimum head over the filter is 5 cm, based on the Rose equation for head loss through granular porous medium (Metcalf and Eddy 2003). *Depth filtration* occurs deep in a sand filter. Slow sand filtration is more characteristic of surface filtration where the *Schmutzdecke* layer causes most of the particle removal, with 98% removal of particles of size 1–60 μm reported to be removed (AWWA 1990).

Slow sand filters are ideal where water is drawn from a high-quality source. Turbidity should not exceed 50 NTU, and there should be little or no colloidal clay. If the source water has higher turbidity, pretreatment may be necessary for slow sand filtration to be effective. At the community scale, this step can be accomplished with a sedimentation basin. At the household level, the storage system described in Section 18.5.1 can be used for pretreatment before passing water through the slow sand filter.

Slow Sand Filters for a Community System

Pickford (1991) and Huisman et al. (1981) provide detailed instructions for building a community slow sand filter (Fig. 18-7). The gravel layer should not fill the entire bottom of the filter box, and it should be no closer than 0.6 m to the walls, ensuring that even water that short-circuits the system along the filter walls passes through sand before entering the gravel and underdrainage (Pickford 1991).

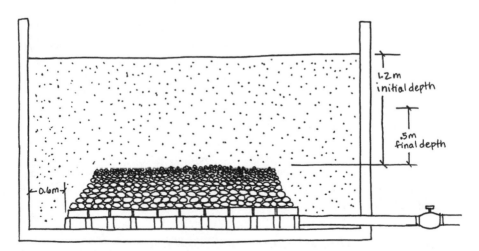

Figure 18-7. Potential Design of a Community Slow Sand Filter.

Source: Redrawn from Pickford (1991) with permission.

Filter walls can be either vertical or sloping. Walls should be roughened to prevent short-circuiting during treatment. The size of the tank depends on the flow rate and the filtration rate, as provided in Eq. 18-13.

$$A = \frac{Q}{v}$$ (18-13)

Here, A is the surface area of the tank, Q is the flow rate, and v is the filtration rate. For slow sand filtration, v should be 0.1–0.2 m^3/m^2-h. The initial depth should be 1.0–1.2 m (Pickford 1991). Application of Eq. 18-13 is provided for a spring box design in Chapter 15. For larger community systems, the design usually includes at least two units, so that one unit can run while the other is being cleaned.

Slow sand filters require a ripening period of several days for the formation of the *Schmutzdecke*, during which time effluent water quality should improve. Slow sand filters are run until the head loss reaches the available head of the system. At this point, the filter must be cleaned. The filter is cleaned by draining all water and removing the top 1 to 2 cm of sand. The sand is cleaned and stockpiled for reuse. When the minimum depth of sand is reached, the extracted sand is replaced and the process is started over again. After each scraping, the filter should be allowed to ripen, with the ripening time decreasing after several scrapings (Crittenden et al. 2005).

Slow Sand Filters for Point-of-Use Treatment

The slow sand filter in Fig. 18-8 can be constructed easily and functions under intermittent use. It consists of a drum or other large container (one that has not previously contained harmful chemicals), sand, gravel, and small amounts of pipe. If the container is metal, it may be necessary to line it with cement mortar to protect it from rust. Another option is to build the container out of concrete or ferrocement. If there is no faucet, it is important that the outlet pipe reach to 5–10 cm above the surface of the sand to allow 5–10 cm of water over the sand for development of the *Schmutzdecke* layer.

When water is first placed into the filter, allow 3 days for the water to sit before use, and ripen it for biological activity to begin. To use the filter, add the amount of water to be filtered and catch the exiting water from the outlet pipe. Maintenance of the slow sand filter is minimal but important. First, keep the filter covered to prevent contamination. Second, maintain about 5–10 cm of water above the surface of the sand to maintain the *Schmutzdecke*. Never run chemically treated water through the filter (i.e., city water or bleach) because chlorine inhibits the biological activity. When there is a significant change in the speed of filtration, it is time to clean the filter. To do this, remove the sand and gravel and rinse them with clean, nonchlorinated water. Wash the container with clean, nonchlorinated water, and rebuild the filter, again waiting 3 days before use for a new *Schmutzdecke* layer to form.

18.6.3 Filtration Through Cloth, Paper, or Nylon

Locally produced cloth and paper filters generally allow smaller viruses and bacteria to pass through the pore openings and should not be seen as a reliable means of ensuring a

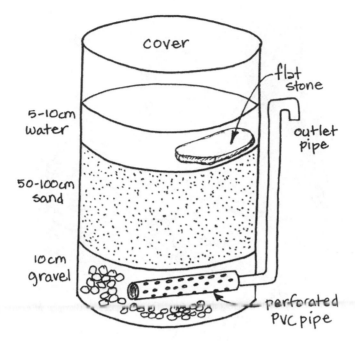

Figure 18-8. Example Design of a Household Slow Sand Filter Used by Peace Corps Cameroon.

safe water supply (Fig. 18-9). However, these filters do have an important role to play. They are reliable at reducing turbidity and eliminating larger pathogens (see Fig. 18-2), and also prevent and reduce the spread of specific diseases (e.g., Guinea worm and cholera). It is important that cloth or paper filters always be used with the same side up. Easy ways to ensure this are to have a logo on one side of the filter, a cloth with different patterns on each side, or two different cloths sewn together. Inexpensive cloth provides a mesh size of 20 µm if folded 4–8 times. This mesh size increases to 100–150 µm in older cloth that has been worn and washed. Nylon filters are available at specific mesh sizes, for example, 150 µm (Colwell et al. 2003). Their advantage is that they retain their stated mesh size.

A study done in Bangladesh showed that cholera could be reduced by half by using old sari cloth folded three times to produce an eight-layer filter (Colwell et al. 2003). A single layer of the cloth was less efficient. The performance of the cloth reduced over time as pore sizes got larger through washing and use. The cloth was decontaminated by being rinsed in the source water and then completely dried in the sun. During monsoon season, the cloths needed to be rinsed with a disinfectant solution, such as bleach.

The transmission of Guinea worm can also be prevented through the use of tightly woven polyester or monofilament nylon cloth filters in conjunction with education and behavior change. The filter should have a 0.15 mm pore size. In comparison, the parasites that cause Guinea worm disease (dracunculiasis) are present in a predatory genus *Cyclops* that is 1–2 mm in size. Personal filters, which are worn around the neck and drunk through like a straw, operate on the same principle (Fig. 18-10). Lifestraw brand personal

Figure 18-9. Filtration Through Overlaid Layers of Fabric Cloth or Nylon Mesh Filters Can Remove Larger Particles.

Figure 18-10. Personal Filters Used to Remove Guinea Worms.

filters remove particles ≥ 15 μm, remain effective for at least 700 L, and are reported to remove 99.999% of all bacteria (Lifestraw 2008). Similar straw filters have been used in the Carter Center's Guinea worm eradication campaign (Carter Center 2007).

))) 18.7 Disinfection

Disinfection refers to the process of inactivating microbiological contaminants. Several methods of disinfection exist, including free chlorine, ozone, chlorine dioxide, iodine, heat, and UV light.

18.7.1 Chlorine Disinfection

Chlorine disinfection has the advantages that it is simple and that residual disinfection capacity remains after water is treated. Here we discuss methods of disinfection with liquid and powder forms of chlorine. One advantage is that liquid laundry bleach is a readily available source of chlorine in many developing communities. The powder form of chlorine is calcium hypochlorite, and includes chlorinated lime, tropical bleach, bleaching powder, and high-test hypochlorite (HTH). Calcium hypochlorite can be found in 30–70% solutions. Liquid bleach is in the form of sodium hypochlorite and contains between 1 and 18% chlorine.

Chlorination is effective for water with pH below 8.0 and low turbidity. Turbidity should preferably be less than 1–5 NTU, but in emergencies, turbidity less than 20 NTU may be acceptable. When chlorine gas is used to disinfect a water supply, it reacts with water to form hypochlorous acid (HOCl) and hydrochloric acid (HCl), according to the following equation:

$$Cl_2 + H_2O \rightarrow HOCl + HCl \qquad (18\text{-}14)$$

Hydrochloric acid is a strong acid and dissociates completely in water to H^+ and Cl^-. Hypochlorous acid dissociates partially, depending on the pH of the water, and forms the base hypochlorite ion.

$$HOCl \Leftrightarrow H^+ + OCl^- \qquad (18\text{-}15)$$

This reaction is the same chemical reaction that occurs when sodium hypochlorite or calcium hypochlorite is added to water. In this case, the sodium (Na^+) and calcium (Ca^{2+}) ions dissociate, and the chemistry is like adding the base OCl^- to the water.

Hypochlorous acid and hypochlorite ion together are often referred to as free chlorine (free chlorine = $HOCl + OCl^-$). Both of these chemical species are active disinfecting agents. However, hypochlorous acid (HOCl) is much more effective than OCl^- for disinfection. The equilibrium constant for the reaction shown in Eq. 18-15 is $10^{-7.5}$. At a pH of 7.5, there are equal amounts of HOCl and OCl^-, and as can be seen from Fig. 18-11, HOCl is the predominant chemical species at a pH level less than 7.5. Chlorination is thus more effective in waters with pH less than 7.5. Fortunately, most natural waters have a pH in the range of 6.5 to 8.5.

Disinfection by chlorine occurs in two ways: *primary disinfection*, which involves the inactivation of bacterial pollution, and *secondary disinfection*, which results from residual chlorine that remains in the treated water. *Chlorine demand* refers to the amount of

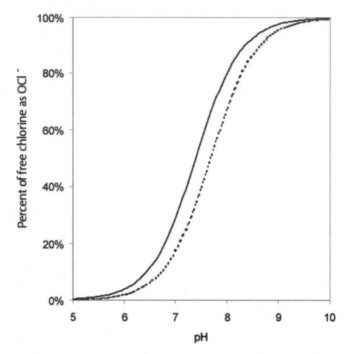

Figure 18-11. Effect of pH on the Fraction of Free Chlorine (HOCl + OCl⁻) Present as the Less-Effective Chemical Species Hypochlorite Ion OCl⁻. Dashed and Solid Lines Provide Effect for Temperatures of Approximately 10–25 °C.

chlorine required for primary disinfection. Once all the bacterial pollutants are destroyed after primary disinfection, excess chlorine remains as *residual chlorine*. This residual chlorine provides capacity to deactivate pathogens resulting from subsequent contamination of treated water, for example, contamination that occurs in pipes between the source and the tap. Disinfection requires a free residual chlorine level of more than 0.5 mg/L for more than 30 min. The WHO guideline for chlorine is 0.5 mg/L which means the WHO considers concentrations below this value (but > 0.5) to be safe. However, this level is well above the taste threshold of 0.6–1 mg/L.

Chlorine Disinfection at the Community Scale

Disinfection of community supplies can be achieved by chlorinating a well or chlorinating water in a storage tank. Hand-dug wells can be chlorinated by lowering a chlorination pot into the well or directly injecting chlorine into the well on a daily basis. Some communities inject chlorine into a well as needed to reduce contamination. However, disinfection will be short-lived with this method if the pathogens are present in the groundwater source.

Chlorination pots use bleaching powder or chlorinated lime. An example design of a chlorination pot is shown in Fig. 18-12. The chlorination pot can be made locally out of an earthen pot, ferrocement, or a plastic jug. Holes of 6- to 8-mm diameter are drilled into the bottom of the pot, and the pot is filled part way with pebbles and pea gravel

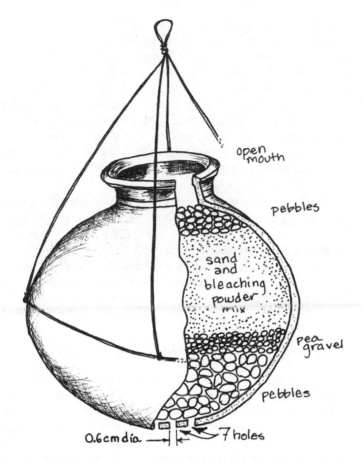

Figure 18-12. Chlorination Pot Using Bleaching Powder for Hand-Dug Wells.

Source: Redrawn with permission of Cairncross and Feachem (1983).

of 20–40-mm diameter. Above the pea gravel, a 1:2 bleaching powder:sand mixture is placed, and the pot is filled to the neck with more pebbles.

According to Huisman et al. (1981), 1.5 kg of bleaching powder in such a system should provide one week's worth of chlorination for a well from which water is drawn at a rate of 1,000–1,200 L/day. Variations of this design exist, and adjustments should be made so that the free chlorine residual concentration remains between 0.5 and 1 mg/L (or less if the users complain of taste).

Pressed calcium hypochlorite tablets have also been used successfully for chlorination of hand-dug wells (Libessart and Hammache 2000; Garandeau et al. 2006). This process involves using a press to make a 70-g press cake and placing it in the center of a durable plastic bag filled with 2 L of sand.

Chlorinators are sometimes added on top of a water storage tank to drip a bleach solution into the water, as was discussed in Chapter 14. For a batch treatment of water stored in a tank, a 1% chlorine stock solution can be added, with the dose calculated based on the volume of the tank to be disinfected. To make the 1% stock solution, mix the quantity shown in Table 18-8 for the appropriate chemical source with water to a total of 1 L.

Once the 1% stock solution is made, it is necessary to determine the dosage for disinfection of the water to be treated because the water will have biological and chemical constituents that react with the chlorine. Box 18-3 provides instructions for determining the dosage of 1% stock solution to drinking water. If measurement of residual chlorine level is infeasible, an approximate dose of 10 drops of 1% solution per quart or liter may be used, where 1/8 teaspoon is roughly equivalent to 8 drops (EPA 2006).

The chlorine residual should be measured daily, if possible. At a community scale, this task of dosing and monitoring the free residual chlorine level should be carried out by an individual trained for this task.

More advanced methods of chlorination include the drip chlorinator, solution feed devices, and proportioning devices for pumped supplies (Huisman et al. 1981; White 1999).

Chlorination at the Household Level

A 1% chlorine stock solution can also be used at the household level, with the appropriate dosage being calculated as demonstrated in Box 18-3. The small doses required for household treatment require that an appropriate measuring device be found. Plastic water bottle caps may be one possible measuring tool that people can find relatively easily.

18.7.2 Safety Considerations for Chlorine Disinfection

One disadvantage of chlorine disinfection is the dangers inherent in handling chemicals. Chlorine gas, which is given off by all concentrated compounds, may burn the eyes and skin and can start fires or explosions. A trained person should therefore handle concentrated forms of chlorine with care. Additionally, reaction of chlorine with natural organic matter in water and the presence of bromide ion (Br^-) results in the formation of disinfection by-products, such as chloroform and trihalomethanes. The health effects of disinfection by-products are not fully known, however. These by-products, as well as the residual chlorine in treated drinking water, cause concern for some water users. Additionally, the taste and odor of treated water with residual chlorine may actually discourage users from drinking the water, resulting in beneficiaries finding other, more contaminated, untreated sources of water.

Table 18-8. Amounts of Various Chlorine Sources for Preparation of a 1% Stock Solution in 1 L of Water

Chlorine Source	Percentage Available Chlorine	Quantity Required	Approximate Measures
Bleaching powder	35	30 g	2 heaping tablespoons
Stabilized or tropical	25	40 g	3 heaping tablespoons
High-test hypochlorite	70	14 mL	1 tablespoon
Liquid laundry bleach	5	200 mL	1 teacup or 6-oz milk tin
Liquid laundry bleach	7	145 mL	10 tablespoons
Javelle water	1	Already a 1% stock solution	

Source: Wisner and Adams 2002.

⟩⟩⟩ Box 18-3. Determination of Chlorine Dosage Using a 1% Stock Solution

1. Place 10 L of the water to be treated in each of four nonmetallic containers.
2. Add the following doses of 1% chlorine to each container:

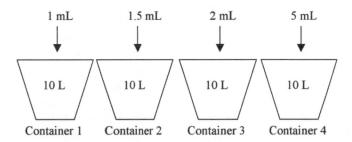

3. Wait for 30 min, and then measure the residual free chlorine concentration. This can be done using a comparator or test strip.
4. The container with a residual chlorine concentration of 0.4–0.5 mg/L has the appropriate dosage.
5. Calculate the dosage required for the quantity of water to be treated:

$$\text{Volume to be treated} \times \frac{\text{dose added to test container (mL)}}{10\,L}$$

For example, if the dosage test reveals that container 2 contains a residual chlorine concentration between 0.4 and 0.5 mg/L, and a tank with a volume of water of 1,000 L is to be treated, the dosage is calculated to be

$$\text{Dosage} = 1,000\,L \times \frac{1.5\,mL}{10\,L} = 1,500\,mL = 1.5\,L$$

Source: Wisner and Adams 2002.

18.7.3 Disinfection with Heat

Heat can also be used to disinfect water. There are two basic approaches to this: boiling and pasteurization. Both work, regardless of the turbidity of the water. Neither method offers residual protection, so take measures to prevent recontamination. Storing water in the same container (with a lid) in which it is boiled or pasteurized helps to reduce opportunities for recontamination. Also, consume water on the same day it is boiled or pasteurized.

Boiling is capable of destroying all types of water-borne pathogens. Bring water to a rolling boil to ensure that all pathogens have been destroyed. Although this is an extremely effective way to disinfect water, it has several disadvantages. The main disadvantage is that fuel is required to boil the water. This fuel can require significant time to gather and can also place additional stress on a region that is already experiencing deforestation. Additionally, using wood for fuel increases exposure to the environmental risk factor indoor smoke from burning of solid fuels. Also, water requires time to cool after it is boiled. Boiled water can have an unpleasant, "flat" taste, which can be improved by shaking the storage container to allow more air to enter the water.

Boiling is often recommended to treat water but is actually overkill because water does not need to reach 100 °C to be disinfected. However, bringing water to a rolling boil provides a built-in indicator that a sufficiently high temperature has been reached. If alternative indicators are available, water only needs to be treated to pasteurization temperatures. For example, *water pasteurization indicators* (WAPIs) are small devices filled with wax that melts at a certain temperature, which can tell users when water can be considered safe to use. Several recommendations exist as to the temperature that must be reached and the length of time that water must stay at that temperature. A conservative recommendation is that water should stay at 70 °C for 10–15 min (Laurent 2005). Figure 18-13 shows that at lower temperatures, water must be heated for longer periods of time. Water can be heated over a fire or by using a solar cooker as well (see www.solarcooking.org).

18.7.4 Solar UV Disinfection

Lower wavelengths of light disinfect water by inactivating the DNA of bacteria, viruses, and other pathogens (Gadgil and Shown 1995). In addition to the effects of direct absorption of the radiation by the bacteria, light radiation also produces reactive forms of oxygen that kill microorganisms. Ultraviolet light is most effective and is divided into three ranges: UV-A (315–400 nm), UV-B (280–315 nm), and UV-C (100–280 nm) (Gadgil and Shown 1995). The most lethal wavelength for destruction of pathogens is between 200 and 300 nm, so UV-C light is the best germicidal wavelength (Crittenden et al. 2005). A mercury lamp, similar to a fluorescent lamp, provides light of wavelengths around 254 nm, an appropriate range for destruction of germs, and this method is commonly used in drinking water treatment in developed countries.

In the case where electricity is expensive or unavailable, the next-best option may be to use solar radiation. Although the smallest wavelengths of radiation do not reach the earth, wavelengths in the UV-A range (also called the near-ultraviolet region) do reach the surface of the earth and have disinfection potential. Additionally, if the water temperature reaches 45 °C, synergy between UV radiation and temperature occurs, improving treatment. In fact, if the temperature exceeds 50 °C, the treatment process is three times faster (EAWAG 2002).

Solar disinfection (SODIS) is a simple treatment method that takes advantage of the bacterial destruction potential of sunlight. Treatment involves placing clear bottles of water to be treated in direct sunlight for a determined amount of time.

SODIS is mainly limited by the initial water quality and availability of clear and clean plastic bottles. In tropical regions, where daylight is consistent throughout the year,

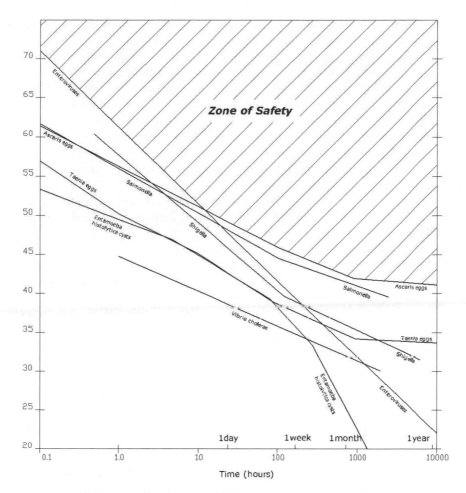

Figure 18-13. Required Temperatures for Complete Inactivation of Pathogens.

Source: Redrawn with permission of Cairncross and Feachem (1983).

light should not pose a problem. Even on 100% cloudy days, the method works with a longer exposure time. Regions between 15° N/S and 35° N/S are considered to be most favorable for solar disinfection because they generally have semiarid climates with little cloud cover. However, the region between the equator and 15° N/S is also favorable (EAWAG 2002). SODIS can be used for microbially contaminated water that has low turbidity (<30 NTU) and is free of chemical contamination (EAWAG 2002).

Because the most effective germicidal wavelengths are below the visible light range (<400 nm), clear glass or plastic bottles are the best option for solar disinfection. Clear containers transmit light in the near-ultraviolet range, as well as in the visible range. Another important factor is the material of the bottle. Glass and plastic are the only real options. The advantages to glass are that it is more resistant to scratching (which reduces light penetration), it has no photoproducts, and it is heat resistant.

Obvious disadvantages to glass are its ability to break, higher costs, and weight. Also, ordinary glass will not transmit UV-radiation if it is thicker than 2 mm. Pyrex, Corex, Vycor, and quartz glasses transmit UV-radiation significantly more than window glass. These types of glass are more costly than ordinary window glass. Plastic bottles are lightweight and less breakable, although less heat resistant (EAWAG 2002). Because they are also generally less expensive, plastic bottles seem to be the best option.

Plastic drinking bottles are either made of polyethylene terephtalate (PET) or polyvinyl chloride (PVC). Although both kinds of plastics contain additives like UV stabilizers, PET bottles contain fewer and so are chemically more stable (EAWAG 2002). In many places, plastic bottles are used for distributing commercial bottled water and are thus readily available for reuse, either sold as used bottles in markets or collected from individuals who consume bottled water.

The procedure for solar disinfection is as follows:

1. Use SODIS for water with no chemical contamination and with turbidity less than 30 NTU. Pretreat the water by filtration or sedimentation to remove turbidity.
2. Wash a 1–2-L PET bottle well the first time the bottle is used. Use four bottles per person, allowing two bottles to be used for consumption and two bottles to be used for treatment.
3. Fill the bottle 75% full of water to be treated, cover it, and shake it for 20 s to aerate the water and increase the dissolved oxygen. Higher oxygen content results in more efficient disinfection because disinfection results from oxygen free radicals and hydrogen peroxides that are produced by the sunlight in water. EAWAG recommends, however, that aeration only take place at the beginning of the SODIS process because continuous shaking throughout exposure reduces efficiency.
4. Fill the bottle fully and replace the cover.
5. Place it in the sun for at least 6 h. Table 18-9 provides the required exposure times for three meteorological conditions. To reduce the risk of breaking or contamination of the bottles, place them out of reach of children and off the ground. One good place to place the bottles is on corrugated metal roofs, in between the grooves.
6. The water is ready for consumption after the appropriate exposure time.
7. Replace old or scratched bottles.

Table 18-9. Required Exposure Time for Solar Disinfection (SODIS) Treatment of Water

Conditions	Required Exposure Time
Sunny to 50% cloudy	6 h
50% to 100% cloudy	2 full days
Continuous rainfall	SODIS is not suitable
Water temperature at or above 50 °C	1 h

Source: EAWAG 2002.

⟩⟩⟩ 18.8 Storage

Water storage is a critical issue (Box 18-4). Water that is clean at the source is often contaminated by the time it is consumed because of poor storage practices. Risk factors contributing to higher contamination include containers with wide openings; water being touched by hand, cups, or dippers that can be contaminated by fecal matter; high storage temperatures; increased storage times; high levels of airborne particles (dust storms); and inadequate hand washing (Sobsey 2002).

⟩⟩⟩ Box 18-4 How Safe Is Household Storage?

The drinking water guidelines established by the World Health Organization (WHO), state that the water source should not contain any microbiological agents that are pathogenic to humans (WHO 2006). However, these drinking water guidelines are based on water quality at the point of delivery, not through the point of actual consumption (Wright 2004).

Microbiological contamination of drinking water during collection and storage in the home has been examined by several researchers (VanDerslice and Briscoe 1995; Clasen and Bastable 2003). Agard et al. (2002) examined the microbial quality of water sources supplied to the San Fernando community in southern Trinidad and found that out of the 104 drinking water samples obtained from households, 80.8% tested positive for total coliforms, 53.8% tested positive for thermotolerant coliforms, and 67.3% tested positive for E. coli. Out of the 81 water samples collected from the Water and Sewerage Authority distribution point, 46.9% tested positive for total coliforms, 16% tested positive for thermotolerant coliforms, and 33.3% tested positive for E. coli. As the level of residual chlorine decreased, there was a statistically significant increase in the prevalence of total coliforms in water from 0.0% in treated reservoir to 80.0% in household drinking water. Agard et al. concluded that the level of household water contamination presented a public health concern to residents.

Brick and Primrose (2004) examined the effects of household storage on water quality in a southern town in India. The study showed that two-thirds of the water sources became increasingly contaminated within nine days of current household storage practices, in spite of receiving safe drinking water from municipal plants. However, the use of brass storage containers significantly decreased contamination of water. Trevett et al. (2004) evaluated the drinking water quality in three rural Honduran communities that used either a protected hand-dug well or bore hole supply. Water quality was examined in 43 households with observations made of household collection and storage practices over a two-year period. There was frequent and substantial water quality deterioration between the points of supply and consumption. Additionally, it was concluded that none of the storage factors examined made any significant difference to the stored water quality and that the contamination could have occurred at several points.

Based on what is reported in the literature, it is thus necessary to take every possible precaution to prevent contamination of water during collection, transport, and household storage.

>>> **Box 18-5 How Safe Are Plastic Containers and Tubing?**

When designing a water distribution, treatment, or storage system, engineers often are faced with the need to weigh the need for biological treatment of pathogens against potential health outcomes from chemical exposure. For example, some plastic bottles contain the chemical bisphenol A (BPA). BPA is known to be a hormone disrupting chemical in animal studies and has been associated with reproductive abnormalities, precancerous changes in the breast and prostate, and obesity and insulin resistance (NRDC 2008). In the United States, where plastics have been used for decades for storing consumable liquids, more than 93% of the population has some form of BPA in their bodies (CDC 2008).

The Natural Resources Defense Council (NRDC 2008) has recommended that plastics numbered as 1 (PETE), 2 (HDPE), 4 (LDPE), and 5 (PP) are safe and that plastics numbered 7 (other, usually polycarbonate), 3 (PVC or V), and 6 (polystyrene) should be avoided. The NRDC (2008) also recommends that infants in particular should not be given drinks from polycarbonate bottles or cups.

Water that is stored in the home should always be covered. Ideally, the storage container should also have a spout through which water can exit. Otherwise, a dipper can contaminate the water every time it is used. If a spout is not possible, the dipper should be hung from a hook in a clean place (not set on a surface) and it should have a handle, so that no hands (or surfaces that come in contact with hands) ever enter the water (Box 18-5).

Optimal storage containers (Fig. 18-14) have the following characteristics:
- 10–25-L capacity,
- one or more handles,

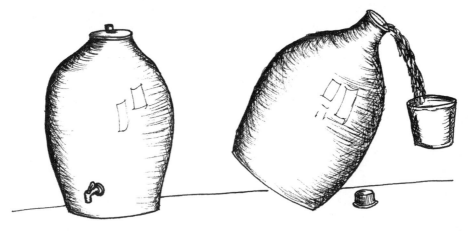

Figure 18-14. Examples of Good Water Storage.

Note: The left container has a lid, narrow neck to prevent dipping cups into it, and a spigot for drawing water. Although the right container does not have a spigot, it is easily poured, and the narrow neck prevents contamination by cups used to draw water. The containers would be easier to use if they had handles.

- flat bottoms,
- construction of light-weight, oxidation-resistant plastic (e.g., polyethylene or polypropylene),
- a 6–9-cm screw cap (big enough to allow cleaning but small enough that it discourages using hands or dippers to access the water),
- a durable and easily closed spigot or spout to dispense water, and
- pictorial and written instructions permanently attached to the container (Sobsey 2002).

Storage containers also need to be compatible with the water treatment methods being used by the household.

⟫⟫ 18.9 Measuring Turbidity with a Turbidity Tube

Turbidity is easily measured in the field by a turbidity tube, which is simple to construct (Myre and Shaw 2006). The procedure for measuring turbidity is depicted in Fig. 18-15. Pour water into the tube until the black-and-white quartered pattern located at the bottom of the tube can no longer be seen. The height of the column of water when the pattern disappears corresponds to the turbidity of the water, as shown in Table 18-10. A shadow should be cast on the tube while measurements are being taken, for example by standing between the sun and the tube.

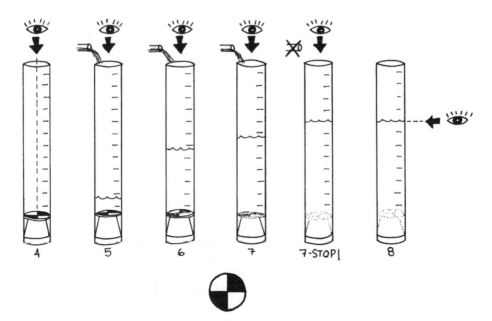

Figure 18-15. Measuring Turbidity with a Turbidity Tube.

Note: If a water sample is more turbid, the quartered disk located at the bottom will disappear with a smaller volume of water added to the tube.

Table 18-10. Conversion of Length to Turbidity When Using a Turbidity Tube

Centimeters	NTU
6.7	240
7.3	200
8.9	150
11.5	100
17.9	50
20.4	40
25.5	30
33.1	21
35.6	19
38.2	17
40.7	15
43.3	14
45.8	13
48.3	12
50.9	11
53.4	10
85.4	5

Source: UW Extension 2003.

⟫⟫ 18.10 Arsenic

18.10.1 Arsenic Occurrence and Chemistry

The occurrence of arsenic in groundwater in many parts of the world is significant enough to require removal. Exposure to high levels of arsenic rarely causes acute poisoning, but long-term exposure can result in skin diseases that can eventually lead to urinary bladder and lung cancer, damage to internal organs, gangrene in the legs, and hypertension. When developing a source of groundwater in regions where arsenic is known to exist, it is important to know whether treatment is necessary.

Arsenic poisoning may be difficult to diagnose in cases where internal organs are affected first. However, visible symptoms exist, such as darkening of the skin, hardening of the palms of the hands and the soles of the feet, or skin depigmentation. Naturally occurring arsenic in drinking water supplies has been found throughout the world (Table 18-11), but the problem is most serious in Bangladesh and West Bengal, India, where millions of wells were installed in the 1970s and 1980s to provide what was thought to be safe drinking water.

Arsenic occurs in both particulate and soluble forms. Particulate arsenic can be removed by a 0.45-μm filter (Petrusevski et al. 2007). Soluble arsenic is more difficult to remove and requires an understanding of the chemistry of soluble arsenic to plan treatment methods. Treatment of arsenic in drinking water is the subject of current research and technology development (Amy et al. 2005; SenGupta 2005; Petrusevski et al. 2007).

Table 18-11. Countries Where Arsenic Has Been Reported in Groundwater

Continent	Countries
Asia	Bangladesh, Cambodia, China (including Taiwan and Inner Mongolia), India, Iran, Japan, Myanmar, Nepal, Pakistan, Thailand, Vietnam
Americas	Argentina, Chile, Dominica, El Salvador, Honduras, Mexico, Nicaragua, Peru, United States
Europe	Austria, Croatia, Finland, France, Germany, Greece, Hungary, Italy, Romania, Russia, Serbia, United Kingdom
Africa	Ghana, South Africa, Zimbabwe
Pacific	Australia, New Zealand

Source: Petrusevski et al. 2007.

In natural waters, soluble arsenic occurs primarily in the oxidized pentavalent form [As(V)] (which occurs mostly in surface water) and the more toxic trivalent form [As(III)] (most often found in groundwater). Table 18-12 shows the many forms of soluble arsenic found in groundwater and surface water.

Table 18-13 lists treatment processes used to remove arsenic from water. Conventional treatment technologies involve processes that are heavily dependent on surface charge, and therefore speciation. As Table 18-12 shows, As(V) occurs primarily as an anion in natural surface waters, and therefore, removal of As(V) is easier than removal of As(III). Most arsenic removal technologies involve chemically oxidizing As(III) to As(V), followed by conventional adsorption and coprecipitation.

Table 18-12. The Forms of Soluble Arsenic Found in Groundwater and Surface Water

Valency	Speciation	Form	Primary Occurrence	pH at Which Ionic Forms Dominate
Arsenite [As(III)]	$H_4AsO_3^+$ H_3AsO_3 $H_2AsO_3^-$ $HAsO_3^{2-}$ AsO_3^{3-}	Reduced	Groundwater, assuming anaerobic conditions[a]	pH > 9
Arsenate [As(V)]	H_3AsO_4 $H_2AsO_4^-$ $HAsO_4^{2-}$ AsO_4^{3-}	Oxidized	Surface water	pH > 3

Note: Treatment technologies require that ionic species dominate. Because As(V) is ionic at natural pH, it is easier to remove than As(III).

[a] The generalization that As(III) is most often dominant in groundwater is less universal than the rule that As(V) dominates surface water. As(V) has been found in groundwater.

Source: Petrusevski et al. 2007.

Table 18-13. Proven Processes Used to Remove Arsenic That Can Be Feasible for Developing Communities

Process	Technologies
Precipitation	Coagulation and flocculation
	Coagulation-assisted microfiltration
	Enhanced coagulation
	Lime softening
	Enhanced lime softening
Adsorption	Activated alumina
	Activated carbon
	Iron and manganese oxide based or coated filter media
Ion exchange	Anion exchange
Membrane filtration	Nanofiltration
	Reverse osmosis
	Electrodialysis

Note: Other technologies exist but would either be too costly or are still in the development stage.

Source: Petrusevski et al. 2007.

18.10.2 Arsenic Treatment

In the absence of treatment, arsenic-contaminated wells can simply be painted with a distinguishing color. Arsenic removal in developing communities uses conventional precipitation and adsorption processes. A typical precipitation process to remove arsenic in groundwater would involve aeration, followed by coagulation and flocculation with aluminum or iron salts, followed by sedimentation and rapid sand filtration. If As(III) is known to be a significant portion of the arsenic in the water, a chemical preoxidation through chlorination or ozonation may be required. However, this process increases treatment costs and may result in oxidation by-products (Petrusevski et al. 2007). After treatment, the liquid waste should be treated as toxic. Adsorption processes have in the past used activated alumina; more recently, iron-based adsorbents are being used. A typical iron-based adsorption process involves a series of adsorptive filters and has no requirement for chemical addition (Petrusevski et al. 2007). Iron oxide coated sand can also be used as an arsenic adsorbent.

Table 18-14 lists some common systems that have been used with limited success at the household level. Figure 18-16 shows a treatment unit that uses granular activated alumina. The exhausted activated alumina is disposed of below the ground in a concrete-lined vault.

))) 18.11 Fluoride

Fluoride in drinking water can have negative or positive effects on human health, depending on concentration. Industrialized drinking water treatment often includes addition of fluoride because concentrations of approximately 1 mg/L are associated with reduced

Table 18-14. Common Processes for Point-of-Use Arsenic Removal and Examples of Systems That Use Them

Process	System	WaterAid (2001a, b) Test Result
Passive sedimentation	Passive sedimentation	Failed
Passive sedimentation and adsorption	Ardasha filter	Failed
Adsorptive filtration	Alcan enhanced activated alumina	Passed
	BUET activated alumina filter	Passed
	Apyron arsenic treatment unit	Not tested
	Read-F arsenic removal unit	Not tested
	Wellhead arsenic removal systems developed by Dr. Arup K. SenGupta and others at Lehigh University	Not tested by WaterAid (2001), but proven successful by Sarkar et al. (2005)
	UNESCO-IHE family filter (iron oxide coated sand)	Not tested by WaterAid (2001), but proven successful by Khan (2004)
Coagulation	DPHE-Danida bucket treatment unit technology system	Passed under certain conditions
	Garnet filter	Passed under certain conditions
	Stevens Institute	Passed
Ion exchange	Tetrahedron	Passed
Coagulation and adsorption	Sono 3-Kalshi filter	Passed

Note: Nine of these systems were tested by WaterAid for successful removal to below the Bangladeshi standard of 50 μg/L.

Sources: Khan (2004); Sarkar et al. (2005); and Petrusevski et al. (2007) with data from WaterAid (2001a,b).

incidence of dental cavities. Excessive levels of fluoride are associated with a number of negative health effects, including dental fluorosis (staining of teeth and erosion of enamel) and skeletal fluorosis (resulting in osteosclerosis, ligamentous and tendinous calcification, and extreme bone deformity).

Fluoride exists in natural waters primarily as the fluoride ion (F^-) or as a complex with aluminum, beryllium, or ferric iron (Crittenden et al. 2005). Because fluoride ions have the same charge and almost the same radius as hydroxide ions, they form mineral complexes with many cations (Fawell et al. 2006). Fluoride is therefore found abundantly in the Earth's crust, occurring in a variety of minerals. When calcium is present, fluorite is common (CaF_2), which has low solubility. Therefore, higher concentrations of fluoride in solution occur in calcium-poor aquifers where fluoride-bearing minerals are common.

Water is not the only exposure pathway for humans to fluoride. Fluoride can also be found in air, dental products, and foods and beverages other than water. Therefore, if dental or skeletal fluorosis exists, it may not necessarily mean that fluoride removal from

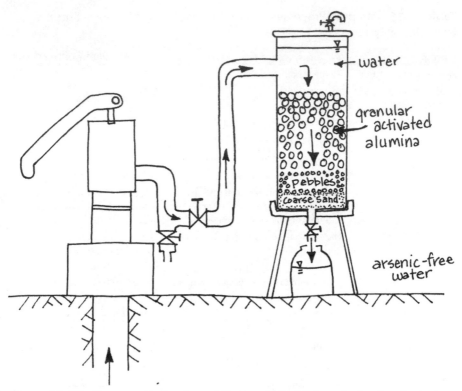

Figure 18-16. Community Treatment System to Treat Arsenic-Contaminated Water Supplies.

Note: This unit is installed at hand-pumped wells and can be regenerated with caustic soda every 4 months.

Source: Courtesy of Dr. Arup SenGupta and others at Lehigh University.

water is required. The WHO guideline for fluoride in drinking water is 1.5 mg/L. However, local conditions should be evaluated to determine whether fluorosis is a problem and if there are other sources of fluoride exposure that can be reduced or eliminated.

In the past, fluoride removal initiatives have met with limited success. Therefore, the WHO recommends that, whenever possible, an alternative source of water be used or blended with the fluoride-containing source (Fawell et al. 2006). Where this is not possible, several treatment processes can be used: sorption, coprecipitation, ion exchange, and contact precipitation (Table 18-15). Even where centralized water distribution systems exist, point-of-use treatment is often preferable because removal is only necessary for water used for drinking or cooking. Minimizing the amount of water to be treated by not treating water for other uses will also minimize the amount of toxic sludge produced by treatment.

Table 18-15. Processes Used for Removal of Fluoride in Drinking Water

Process	Technologies
Sorption	Bone charcoal Activated alumina Clay
Coprecipitation	Coagulation and flocculation (Nalgonda technique) Polyaluminum chloride Lime
Contact precipitation	Calcium and phosphate compounds
Ion exchange	Clay or other naturally occurring mineral

Source: Fawell et al. 2006.

))) References

Agard, L., Alexander, C., et al. (2002). "Microbial quality of water supply to an urban community in Trinidad." *J. of Food Protection,* 65(8), 1297–1303.

American Water Works Association (AWWA). (1990). *Water quality and treatment,* McGraw Hill, Inc., New York.

Amy, G., Chen, H. W., Drizo, A., von Gunten, U., Brandhuber, P., Hund, R., Chowhury, Z., Kommeni, S., Shahnawaz, S., Jekel, M., and Banerjee, K. (2005). *Adsorbent treatment technologies for arsenic removal,* American Water Works Association, Denver, Colo.

Brick, T., and Primrose, B. (2004). "Water contamination in urban south India: Household storage practices and their implications for water safety and enteric infections." *International J. of Hygiene and Environmental Health,* 207(5), 473–480.

Cairncross, S., and Feachem, R. G. (1983). *Environmental Health Engineering in the Tropics,* John Wiley, New York.

Carter Center. (2007). *Guinea worm eradication program,* <http://www.cartercenter.org/health/guinea_worm/index.html> (Jan. 30, 2008).

Centers for Disease Control and Prevention (CDC). (2008). "National report on human exposure to environmental chemicals: Spotlight on bisphenol A." Centers for Disease Control and Prevention, Atlanta, Ga. <http://www.cdc.gov/exposurereport/pdf/factsheet_bisphenol.pdf> (Jan. 20, 2009).

Clasen, T. F., and Bastable, A. (2003). "Faecal contamination of drinking water during collection and household storage: The need to extend protection to the point of use." *J. of Water and Health,* 1(3), 109–115.

Clasen, T., Roberts, I., Rabie, T., Schmidt, W., and Cairncross, S. (2006). "Interventions to improve water quality for preventing diarrhea (A Cochrane review)." *The Cochrane Database of Systematic Reviews,* Issue 3. Art. No. CD004794. DOI: 10.1002/14651858.CD004794.pub2.

Colwell, R. R., Huq, A., Sirajul Islam, M., Aziz, K. M. A., Yunus, M., Huda Khan, N., Mahmud, A., Bradley Sack, R., Nair, G. B., Chakraborty, J., Sack, D. A., and Russek-Coheni, E. (2003). "Reduction of cholera in Bangladeshi villages by simple filtration." *Proceedings of the National Academy of Sciences.* 100(3), 1051–1055.

Crittenden, J. C., Trussell, R. R., Hand, D. W., Howe, K. J., and Tchobanoglous, G. (2005). *Water treatment: Principles and design,* Second ed. John Wiley & Sons, Hoboken, N.J.

Eigenössische Anstalt für Wasserversorgung, Abwasserreinigung und Gewässerschutz (EAWAG). (Swiss Federal Institute for Environmental Science and Technology). (2002). *Solar water disinfection: A guide for application of SODIS*. SANDEC Report No. 6/02, <http://www.sodis. ch/files/SODIS_Manual_english.pdf> (Jan. 20, 2009).

Environmental Protection Agency (EPA). (2006). *Emergency disinfection of drinking water*. <www .epa.gov/safewater/faq/emerg.html> (Feb. 1, 2008).

Fawell, J., Bailey, K., Chilton, J., Dahi, E., Fewtrell, L., and Magara, Y. (2006). *Fluoride in drinking-water,* World Health Organization, IWA Publishing, Seattle. <http://whqlibdoc.who.int/ publications/2006/9241563192_eng.pdf> (Jan. 20, 2009).

Fuglie, L. J. (2001). *The miracle tree,* Moringa oleifera: *Natural nutrition for the tropics*. Training manual of Church World Service, New York.

Gadgil, A. J., and Shown, L. J. (1995). *To drink without risk: The use of ultraviolet light to disinfect drinking water in developing countries,* Center for Building Science, Lawrence Berkeley Laboratory, Berkeley, Calif.

Garandeau, R., Trevett, A., and Bastable, A. (2006). "Chlorination of hand-dug wells in Monrovia." *WaterLines,* 24(3), 19–21.

Huisman, L., De Azevedo Netto, J. M., Sundaresan, B. B., and Lanoix, J. N. (1981). *Small community water supplies: Technology of small water supply systems in developing countries,* E. H. Hofkes, ed., International Reference Centre for Community Water Supply and Sanitation, The Hague, Netherlands.

Khan, M. S. A. (2004). *Field-testing of improved IHE family filter in Bangladesh.* UNESCO-IHE, Delft, Netherlands. (MSc. Thesis; no. SE 04.16).

Laurent, P. (2005). *Household drinking water systems and their impact on people with weakened immunity.* Medecins sans Frontieres-Holland, Amsterdam.

Libessart, Y., and Hammache, Y. (2000). "Integrated chlorination campaign in Mogadishu." *Proceedings of the 26th WEDC Conference in Dhaka, Bangladesh,* WEDC: Loughborough, U.K., pp. 237–239.

Lifestraw. (2008). *Lifestraw personal claims.* <http://www.lifestraw.com> (Jan. 30, 2008).

Metcalf and Eddy, Revised by Tchobanoglous, G., Franklin, B. L., and Stensel, H. D. (2003). *Waste-water engineering: Treatment and reuse,* 4th ed. McGraw Hill, Boston.

Mihelcic, J. R., and Zimmerman, J. B. (2009). *Environmental engineering: Fundamentals, sustainability, design,* John Wiley & Sons, New York.

Myre, E. A., and Shaw, R. P. (2006). *The turbidity tube: Simple and accurate measurement of turbidity in the field.* <http://cee.eng.usf.edu/peacecorps/Resources.htm> (Jan. 20, 2009).

Natural Resources Defense Council (NRDC). (2008). "Chemicals in plastic bottles: How to know what's safe for your family." Natural Resources Defense Council, New York <http://www. nrdc.org/health/bpa.pdf> (Jan. 20, 2009).

Ovanedel-Craver, V. A., and Smith, J. A. (2008). "Sustainable colloidal-silver-impregnated ceramic filter for point-of-use water treatment." *Environ. Sci. and Technol.,* 42(3), 927–933.

Petrusevski, B., Sharma, S., Schippers, J. C., and Shodt, K. (2007). *Arsenic in drinking water. Thematic overview paper 17.* IRC WHO Collaborating Centre, Delft, Netherlands.

Pickford, J. (ed). (1991). *The worth of water: Technical briefs on health, water and sanitation.* Intermediate Technology Publications, London.

Potters for Peace website. (n.d.). <www.pottersforpeace.org> (Jan. 20, 2009).

Sarkar, S., Gupta, A., Biswas, R. K., Deb, A. K., Greenleaf, J. E., and SenGupta, A. K. (2005). "Well-head arsenic removal units in remote villages of Indian subcontinent: Field results and performance evaluation." *Water Res.,* 39(10), 2196–2206.

SenGupta, A. (2005). "Arsenic crisis in Indian subcontinent: An indigenous solution." <http://www .lehigh.edu/~aks0/arsenic.html> (Jan. 20, 2009).

Shaw, R. (ed). (1999). *Running water: More technical briefs on health, water and sanitation.* Intermediate Technology Publications, London.

Sobsey, M. (2002). *Managing water in the home: Accelerated health gains from improved water supply.* World Health Organization, Geneva, Switzerland. <http://www.who.int/water_sanitation_health/dwq/wsh0207/en/> (Jan. 9, 2009).

Trevett, A. F., Carter, R., and Tyrrel, S. (2004). "Water quality deterioration: A study of household drinking water quality in rural Honduras." *International J. of Environmental Health Research,* 14(4), 273–283.

UW Extension. (2003). "Turbidity: A water clarity measure." *Water action volunteers: Volunteer monitoring factsheet series.* University of Wisconsin, Madison, Wis. <http://watermonitoring.uwex.edu/pdf/level1/FactSeries-Turbidity.pdf> (Feb. 1, 2008).

VanDerslice, J., and Briscoe, J. (1995). "Environmental interventions in developing countries: Interactions and their implications." *American J. of Epidemiology,* 141(2), 135–144.

Visscher, J. T., Paramasiram, R., Raman, A., and Heijnen, H. A. (1987). *Slow sand filtration for community water supply: Planning, design, construction, operation, and maintenance.* Technical Paper Series No. 24. IRC International Water and Sanitation Centre, The Hague, Netherlands.

WaterAid. (2001a). *Rapid assessment of household level arsenic removal technologies. Phase II–Report, January 2001.* WS/Atkins, BAMWSP/DFID/WaterAID, Dhaka, Bangladesh.

WaterAid. (2001b). *Rapid assessment of household level arsenic removal technologies. Phase II–Report, March 2001.* WS/Atkins, BAMWSP/DFID/WaterAid, Dhaka, Bangladesh.

White, G. C. (1999). *Handbook of chlorination and alternative disinfectants,* 4th ed., Wiley-Interscience, New York.

Wisner, B., and Adams, J., eds. (2002). *Environmental health in emergencies and disasters: A practical guide.* World Health Organization, Geneva, Switzerland.

World Health Organization (WHO). (2006). *Guidelines for safe drinking water quality: first addendum to third addition, Volume 1: Recommendations.* WHO Press, Geneva, Switzerland. <http://www.who.int/water_sanitation_health/dwq/gdwq0506.pdf> (Jan. 20, 2009).

Wright, J., Gundry, S., and Conroy, R. (2004). "Household drinking water in developing countries: A systematic review of microbiological contamination between source and point-of-use." *Tropical Medicine and International Health,* 9(1), 106–117.

))) Further Reading

Davis, Jan, and Lambert, Robert. (2002). *Engineering in emergencies: A practical guide for relief workers,* 2nd ed. RedR/ITDG Publishing: Warwickshire, U.K.

Wegelin, M. (1996). *Surface water treatment by roughing filters: A design, construction, and operation manual,* SANDEC Report No. 2/96, Swiss Federal Institute for Environmental Science and Technology (EAWAG), Department of Water and Sanitation in Developing Countries (SANDEC), Duebendorf, Switzerland. <http://www.eawag.ch/organisation/abteilungen/sandec/schwerpunkte/ws/documents/surface_water_treatment> (Jan. 20, 2009).

Wastewater Treatment

19

Wastewater Composition and Generation

))) 19.1 Access to Adequate Sanitation

The fact that water supply is essential to human survival often leads to the assumption that it should be given a higher priority than sanitation. However, improvements in sanitation also result in healthier children and fewer incidences of diarrhea (see Chapter 2). Unfortunately, more than three times more people now lack access to improved sanitation compared to access to an improved water supply (Box 19-1). Several reasons for this are the following:

- The advantages of an improved water supply are often more apparent than those of sanitation. This situation means that people are more willing to pay or donate their time to increase their access to cleaner water.
- Water projects are often highly political at all levels of governance. Donors and political leaders, whether they represent a rural community or a larger region, usually prefer to have their names associated with water projects instead of sanitation projects (e.g., latrines).
- Typically, the costs of water supply projects are distributed among the whole community, whereas individual households must pay the entire cost of their own sanitation facility. If not all households buy into a sanitation plan, the overall health of a community may not improve as much as expected because an effective sanitation plan requires almost 100% coverage in a community.

The link between wastewater, human health, and environmental degradation is widely recognized. The United Nations has accordingly included sanitation targets in the Millennium Development Goals (see http://www.un.org/millenniumgoals/). However, the proper disposal and treatment of wastewater remains a significant challenge around the world.

Wastewater is the collective term applied to the liquid and fecal waste generated by a community. There are two main sources of domestic wastewater:

1. *blackwater:* water used to convey human excreta (both urine and feces), and
2. *greywater:* water used for cleaning, bathing, and cooking.

Jennifer R. McConville contributed to this chapter.

))) Box 19-1. Sanitation Technologies Considered to Be Improved and Unimproved

Improved	Unimproved
Connection to a public sewer	Service or bucket latrines (excreta are moved manually)
Connection to a septic system	Public latrines
Pour-flush latrines	Open latrines
Ventilated improved pit latrines	
Composting latrines	
Simple pit latrines	

Source: Definitions from WHO/UNICEF 2006.

Wastewater is contaminated with a variety of microbial pathogens (Table 19-1) and other organic and inorganic pollutants. These can contribute to environmental and public health problems if they are not controlled through wastewater treatment. Subsequent chapters present a variety of technologies for managing either separate or combined flows of greywater and blackwater.

A lack of wastewater treatment services compounds problems of environmental degradation. Aside from microbial pathogens, wastewater contains suspended solids (measured as turbidity and/or total suspended solids (TSS)), oxygen-depleting organic matter (measured as biochemical oxygen demand (BOD) or chemical oxygen demand (COD)), and nutrients (i.e., nitrogen and phosphorus), all of which adversely affect receiving water bodies.

Because domestic water use is considered nonconsumptive, the process returns most of the extracted water back to the environment, where it lowers water quality. In fact, it has been estimated that every cubic meter of untreated wastewater discharged to surface water makes 8–10 m^3 of freshwater unusable (Shiklomanov 2000). Surface water pollution resulting from low sanitation coverage is especially critical in locations where people living in poverty depend on the environment for a significant fraction of their economic welfare, notably from food provided by fields, forests, and fisheries.

))) 19.2 Wastewater Composition

Designing and implementing appropriate wastewater treatment systems requires a basic understanding of wastewater composition and the key parameters to be controlled. Wastewater is characterized by its biological, chemical, and physical properties (discussed in detail elsewhere, e.g., Mihelcic 1999; Tchobanoglous et al. 2003; Mihelcic and Zimmerman 2009). Health concerns over domestic wastewater stem mainly from the microbiological constituents of fecal matter. Pathogens are the microorganisms that cause sickness and disease. They include many classes of microorganisms: viruses, bacteria, protozoa, and helminths. Their sizes range from less than a micrometer to several millimeters.

Table 19-1. Pathogenic Organisms Found in Untreated Wastewater and Associated Diseases

Organism	Disease
Bacteria	
Campylobacter jejuni	Gastroenteritis
Escherichia coli (enteropathogenic)	Gastroenteritis
Legionella pneumophilia	Legionnaires' disease
Leptospira	Leptospirosis
Salmonella ($\approx$ 2,100 serotypes)	Salmonella
Salmonella typhi	Typhoid fever
Shigella	Shigellosis
Vibrio cholerae	Cholera
Yersinia enterocolitica	Yersiniosis
Protozoa	
Balantidium coli	Balantidiasis
Cryptosporidium parvum	Cryptosporidiosis
Cyclospora cayetanensis	Cyclosporasis
Entamoeba histolytica	Amebiasis (amoebic dysentery)
Giardia lamblia	Giardiasis
Helminths	
Ascaris lumbricoides	Ascariasis
Enterobius vermicularis	Enterobiasis
Fasciola hepatica	Fascioliasis
Hymenolepis nana	Hymenolepiasis
Taenia saginata	Taeniasis
Taenia solium	Taeniasis
Trichuris trichiura	Trichuriasis
Viruses	
Andenovirus (31 types)	Respiratory disease
Enteroviruses (more than 100 types)	Gastroenteritis, heart anomalies, meningitis
Hepatitis A virus	Infectious hepatitis
Norwalk agent	Gastroenteritis
Parvovirus (2 Types)	Gastroenteritis
Rotavirus	Gastroenteritis

Source: From Tchobanoglous et al. 2003.

Human feces contain an abundance of pathogens. Table 19-2 provides typical concentrations of pathogens in untreated wastewater, as well as the infectious dose for these pathogens. In general terms, one gram of feces can contain 10 million viruses, 1,000,000 bacteria, 1,000 parasite cysts, and 100 parasite eggs. The major risk from urine is not from the urine itself, but from fecal cross-contamination. Urine is normally sterile in the bladder but may pick up microorganisms that are present in the lower parts of the urinary

Table 19-2. Concentrations of Microorganisms Found in Untreated Wastewater and the Corresponding Infectious Doses

Organism	Concentration (MPN/100 mL)	Infectious Dose (Number of Organisms)
Bacteria		
Bacterioides	10^7–10^{10}	
Coliform, total	10^7–10^9	
Coliform, fecal	10^6–10^8	10^6–10^{10}
Clostridium perfringens	10^3–10^5	1–10^{10}
Enterococci	10^4–10^5	
Fecal streptococci	10^4–10^7	
Pseudomonas aeruginosa	10^3–10^6	
Shigella	10^0–10^3	10–20
Salmonella	10^2–10^4	10^1–10^8
Protozoa		
Cryptosporidium parvum oocysts	10^1–10^3	1–10
Entamoeba histolytica cysts	10^{-1}–10^1	10–20
Giardia lamblia cysts	10^3–10^4	<20
Helminth		
Ova	10^1–10^3	
Ascaris lumbricoides	10^{-2}–10^0	1–10
Viruses		
Enteric virus	10^3–10^4	1–10
Coliphage	10^3–10^4	

Note: MPN stands for most probable number.

Source: Tchobanoglous et al. 2003.

tract. The following chapter contains a table (Table 20-5) that provides a list of microorganisms found in urine, their potential transmission route, and their importance.

Coliform bacteria are the most abundant fecal microorganism, comprising up to one-third the total weight of feces. They are therefore commonly used as indicators for the presence of fecal contamination. Water is typically tested for *total coliforms* (a group that includes some species widely distributed in the environment) and *fecal coliforms* (a group that is more specific to fecal contamination). *Escherichia coli (E. coli)* is one member of the coliform group. It is a normal inhabitant of the digestive tract of humans and other warm-blooded animals and is more resistant to disinfection and environmental exposure than several other pathogens.

The environmental effects of wastewater are closely linked to its chemical composition, specifically the concentrations of nitrogen (N), phosphorus (P), and biochemical oxygen demand (BOD) (also measured as chemical oxygen demand (COD)). N and P are key nutrients for plants and are therefore the main causes of *eutrophication* (the state of a lake, for example, having an abundant accumulation of nutrients that support a dense growth of plant and animal life, the decay of which depletes the shallow waters of oxygen in summer). BOD is a measure of the amount of dissolved oxygen that will be consumed during microbial decomposition of the organic matter in the wastewater. Discharging wastewater with high levels of N, P, and BOD into surface waters can result in algal blooms, fish kills, loss of aquatic biodiversity, and general water quality degradation (for more details, see Mihelcic and Zimmerman 2009).

One important physical characteristic for designing wastewater collection and treatment systems is the total solids concentration of the wastewater. Solids entering the wastewater system from a home can include feces, anal cleansing materials (e.g., toilet paper), and food waste. In a pit latrine, additional solids used for anal cleansing can include newspaper, corncobs, leaves, and stones. A compost latrine receives additional solids in the form of wood ash, grass clippings, and sawdust. The concentration of these solid materials changes, depending on whether the wastewater is conveyed by large volumes of greywater, versus a pit latrine, where there is little input of water. The fraction of solids to liquids determines possibilities for transport of the wastewater. Solids accumulation also influences the sizing of a latrine pit.

))) 19.3 Wastewater Production

The production of wastewater within a community is closely linked to water usage. In the developed world, the volume of wastewater generated is typically 60–80% of the water usage. In the developing world where sewers are used to collect blackwater and greywater, the volume of wastewater generated is generally assumed to be close to the amount of water used. However, the mass and type of pathogens, nutrients, and solids varies in these two situations, depending on the nutrition habits, sanitation practices, age of population, and the socioeconomic situation of the population.

As we will see for latrine design in Chapter 20, solids accumulation can range from 40 L/capita-yr (0.04 m^3/capita-yr) to 90 L/capita-yr (0.06 m^3/capita-yr), depending on the type of materials used for anal cleaning as well as the pit design. Greywater generally has lower pathogen and nutrient concentrations, and solids derived from food wastes are negligible. However, the use of phosphorus-containing detergents can contribute 10–30% of the total P loading to the wastewater system (WHO 2006).

Whereas feces, urine, and greywater are often seen as wastes that require collection and treatment, they should instead be viewed as a resource. After proper collection and possible treatment, all can be used to enhance agricultural productivity. Greywater can also be reused in pour-flush latrines.

The majority of biological contaminants are found in fecal matter, and a significant portion of the nutrient loading is found in urine. On average, an adult produces between 25 and 50 kg of feces/yr, containing 0.55 kg N and 0.18 kg P. In contrast, the average adult produces 400 L of urine/yr, containing 4.0 kg N and 0.4 kg P (Jönsson 1997). In the developed world, many of the nutrients found in urine are diluted by excessive water usage. Safe separation and storage of urine should be viewed as a method of returning nutrients back into agricultural use. Composting toilets can be designed to separate urine from feces. In this case, both the urine and composted feces can be used to amend agricultural soils.

⫸ References

Jönsson, H. (1997). "Assessment of sanitation systems and reuse of urine." Ecological alternatives in sanitation. Publications on Water Resources No. 9, Sida, Stockholm, Sweden.

Mihelcic, J. R. (1999). *Fundamentals of environmental engineering,* John Wiley & Sons, New York.

Mihelcic, J. R., and Zimmerman, J. B. (2009). *Environmental engineering: Fundamentals, sustainability, design,* John Wiley & Sons, New York.

Shiklomanov, I. A. (2000). "Appraisal and assessment of world water resources." *Water International,* 25(1), 11–32.

Tchobanoglous, G., Burton, F. L., and Stensel, H. D. (2003). *Wastewater engineering,* McGraw Hill, Boston.

World Health Organization (WHO). (2006). "Volume 4: Excreta and greywater use in agriculture." *Guidelines for the safe use of wastewater, excreta, and greywater,* WHO Press, Geneva, Switzerland.

WHO/UNICEF. (2006). *Meeting the MDG drinking water and sanitation target: The urban and rural challenge of the decade.* WHO Press, Geneva, Switzerland, <http://www.who.int/water_sanitation_health/monitoring/jmpfinal.pdf> (Jan. 18, 2009).

⫸ Further Reading

Drangert, J. (1998). "Fighting the urine blindness to provide more sanitation options." *Water SA,* 24(2), 157–164.

Shaw, R. (2009). *Separating urine to use as fertilizer,* <http://cee.eng.usf.edu/peacecorps/Resources.htm> (Jan. 24, 2009).

Snauffer, A. (2008). *Development of a decision support tool for on-site wastewater treatment system,* <http://cee.eng.usf.edu/peacecorps/Resources.htm> (Jan. 17, 2009).

20

Latrines

》》》 ## 20.1 Sanitation and Latrines

Latrines are a proven and acceptable sanitation technology. To place them in context with other sanitation technologies, Table 20-1 provides a list of excreta disposal methods considered to be improved or unimproved. From a public health standpoint, the main goal of a latrine is to engineer a barrier between people and the pathogens found in human feces. The F-diagram (Fig. 2-2 from Chapter 2) showed routes of contamination that remain open even if latrines are implemented as a sanitation technology. Hygiene education is therefore also essential, as it is with all sanitation technologies applied to human excrement.

The perception of sanitation varies significantly from culture to culture. Improvements in health are not the only reasons that communities accept sanitation projects. A survey of rural households in the Philippines identified the reasons that people were satisfied with their newly built latrines (Cairncross 1992):

- lack of smell and flies,
- cleaner surroundings,
- privacy,
- less embarrassment when friends visit, and
- fewer incidences of gastrointestinal disease.

This list shows the importance of understanding a community's motivation for accepting an improved sanitation technology.

Children should be given particular consideration when designing a latrine. Children often avoid latrines if they are too far away (especially at night), the wait is long, the hole is too large, the building is dark, or there are insects and bad odors present. Women need to be consulted on use of the latrine, especially regarding the privacy shelter or any privacy partitions. They may be used to meeting as a group in fields outside the community in the morning and could be resistant to losing some additional social time with other female community members. Men may be reluctant to embrace a latrine because their only experience of latrines may have been in a cramped and dirty inner-city latrine.

20.1.1 Traditional Pit Latrines

A traditional pit latrine (Fig. 20-1) is the most common type of latrine in the developing world because it is the simplest and the least expensive to construct. It requires little maintenance and is easy to operate. A *traditional pit latrine* consists of:

Josephine Kaiser, Daniel M. Hurtado, Jessica A. Mehl, Jason N. Huart, Eric Tawney, and James Dumpert contributed to this chapter.

Table 20-1. Sanitation Technologies Considered to Be Improved and Unimproved

Improved	Unimproved
Connection to a public sewer	Service or bucket latrines (excreta are moved manually)
Connection to a septic system	Public latrines
Pour-flush latrines	Open latrines
Ventilated improved pit latrines	
Composting latrines	
Simple pit latrines	

Source: WHO/UNICEF 2006.

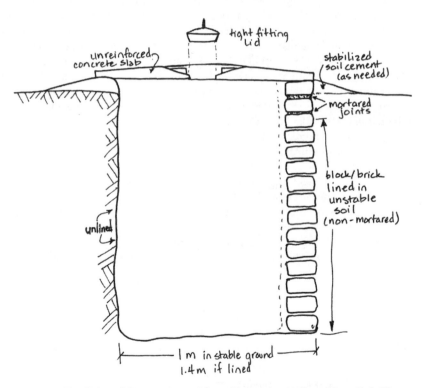

Figure 20-1. Traditional Pit Latrines Often Cover the Hole with a Slab That Consists of Logs, Soil-Covered Wood Planks, or Concrete Slab That Can Be Constructed with Reinforcing or Without.

Note: A superstructure of local materials is typically added to provide privacy. Pits should be lined for the first 0.5 m. Below that, they are unlined if the soil is deemed unlikely to collapse (the figure provides both unlined (left) and lined (right) options).

Source: Redrawn with permission from Davis and Lambert (2002).

- a hole dug in the ground;
- a slab covering the hole, constructed of logs or planks covered by compacted soil or of concrete that uses reinforcement or is constructed without reinforcement (e.g., domed designs that do not require reinforcement);
- an opening in the slab, which might have a seat on top of it (depending on cultural preference); and
- an enclosing structure constructed of local materials that provides some level of privacy.

The traditional pit latrine has several problems, including odors, the breeding of flies, risk of latrine collapse because of dampness and rotting of wood planks, termites and their undermining of the soil, hookworm and other pathogens breeding on wet and dirty slabs, and the danger of children or animals falling into a hole that is too large. Flies can make using a latrine an unpleasant and even frightening experience, particularly when they are coming out of the latrine pit toward the user. They also present a significant health risk because the small amounts of feces that stick to a fly's legs can contaminate food, water, eyes, and other surfaces.

20.1.2 Upgrading an Existing Pit Latrine

Upgrading an existing latrine should be given serious consideration before a new latrine is built. A rule of thumb is that if a pit has at least three years of useful life remaining, it may be worth upgrading the existing latrine (Wegelin-Schuringa 2000). Space constraints may also make building a new facility impractical. Traditional methods of upgrading latrines should be thoroughly investigated. Some of these methods might only be known to individual families because of taboos on discussing sanitation.

It is wise to build a new latrine in a different location if the current latrine is near a drinking water well, particularly in an area with a high water table. Also, if constructing the new latrine would only be slightly more expensive than upgrading an existing one, then it is often better to build a new one. The structural stability of the slab is an additional factor to consider.

Figure 20-2 shows several of the options to make a floor slab more hygienic. Footrests can be added to help ensure proper positioning. If the slab is already made of concrete, a smooth layer of cement mortar (1 : 3 cement to sand by volume) can make it easier to clean. Other options include covering earthen floors with thin, wet cement or using local recipes for creating a hard floor (e.g., mixing clay and cow dung or cassava and soil) (Wegelin-Schuringa 2000). If the slab is structurally sound but not hygienic, a prefabricated slab can also be added on top of the existing floor.

Odors and flies can be reduced in a number of ways. A tight-fitting lid can control both. Ash or sawdust can be used to cover excreta each time the latrine is used, reducing insect breeding and smells. Hot ashes (i.e., straight from a cooking fire) have the additional advantage of killing maggots, but take care to ensure that the ashes do not escape and burn surrounding wooden structures. An existing latrine can also be improved by preventing rainwater from entering the pit, either by improving the roof of the superstructure or by mounding soil around the latrine to create a berm between any surface runoff and the pit.

Figure 20-2. Several Methods Can Make an Existing Latrine Floor (or Slab) More Hygienic.

Note: Footrests can be added to help ensure proper positioning. A concrete slab can be added, and a tight cover can be placed over the hole. If the slab is already concrete, a smooth layer of cement mortar (parge) can make the slab easier to clean.

20.1.3 Siting a Latrine

Proper siting of a latrine is important to its functionality, ease of use, and ability to positively affect the health of a community. The top of Table 20-2 lists issues that need to be taken into account when choosing the location of a latrine. In many cases, the requirements of a site (and cultural considerations) narrow the list of appropriate latrine types.

The distance from the bottom of a latrine pit to the water table is important in areas where groundwater is used as a source of drinking water. In densely populated urban areas where the groundwater is already acknowledged to be contaminated and a dependable and affordable piped water supply exists, regional decisions can be made that allow further contamination. Latrines can be built on earth fill to increase the distance from the bottom of the pit to the water table. Alternatively, a shallower pit can be dug, with the

Table 20-2. Selection of the Appropriate Latrine Technology Is Based on Location, User Comfort, Cost, Maintenance, Materials and Water Use, and Expected Life

Factors Involved in Selecting a Latrine	Traditional Pit Latrine	VIP Latrine	Pour-Flush Latrine	Composting Latrine
Location				
Minimum distance to house	10–20 m	10–20 m	Not applicable	Not applicable
Minimum distance to water source	Downhill or 30 m	Downhill or 30 m	Downhill or 30 m	Not applicable
Minimum distance from pit bottom to groundwater table	1.5 m	1.5 m	3 m[a]	Not applicable
Ability to be built in areas with a high water table	No	No	No	Yes
Ability to be built in floodplains	No	No	No	Yes
Required soil conditions	Uniform and free from cracks	Uniform and free from cracks	Low clay content, stable and firm	Not applicable
Additional notes		Location away from anything that can reduce wind flow, like a building or tree	Can be located in a house	Greatest flexibility in location
User Comfort				
Produces odors?	Significant amount	Some	No	Minimal
Breeding ground for insects?	Significant amount	Some	No	Minimal
Possibility of falling in the latrine?	No	No	No	No
Cost				
Construction materials	Lowest	Lower	Higher	Higher
Requires outside materials?	No	Yes	Yes	Yes
Maintenance				
Time requirement	Minimal	Minimal	Some	More
Knowledge requirement	Minimal	Minimal	Some	More
Significance that maintenance has on latrine operation	Minimal	Minimal	Some	More

Table 20-2. Selection of the Appropriate Latrine Technology Is Based on Location, User Comfort, Cost, Maintenance, Materials and Water Use, and Expected Life (*Continued*)

Factors Involved in Selecting a Latrine	Traditional Pit Latrine	VIP Latrine	Pour-Flush Latrine	Composting Latrine
Materials				
Materials required during use	Ash, sawdust, soil (optional)	Ash, sawdust, soil (optional)	1–5 L water per use	1–2 cups (handfuls) of dry organic material per use
Anal cleansing materials may be used	Any	Any	Water or small pieces of soft paper	Paper and organic material, or water in a separate hole
Expected Life	10 years	10 years	10 years	20 years (life of concrete)
Water Requirements[b]	None	None	1–5 L per use	None

[a]Pour-flush latrines need to be sited in soils that allow sufficient liquid filtration as determined by a filtration test.

[b]Water requirements do not include water for anal cleansing.

superstructure and slab built in such a way that they can be moved when the pit is full. One advantage of composting latrines is that they can be sited in areas where groundwater tables are close to the ground surface (e.g., coastal areas).

Locate latrines close to the household so that children are not discouraged from using them because of the time required to walk to them. In the case of latrines that serve more than one family, balance the community's desire for privacy with the safety of women and children. If community members feel embarrassment from being seen returning from using a latrine, a common household errand can be incorporated into the process of using the latrine. For example, wood can be placed near the latrine or on the return path from the latrine so that the user appears to be collecting daily firewood.

20.1.4 The Pit

Traditional pit latrines and ventilated improved pit (VIP) latrines are constructed over a pit. The pour-flush latrine can be built either over or offset from a pit. The latter option eliminates the risk of the slab collapsing or small children falling through the hole. Composting latrines have aboveground chambers that are built to collect and recycle human waste. The purpose of the pit is not to treat the waste but instead to safely contain the waste material and serve as a barrier to exposure between humans and pathogens.

The first step in calculating the required pit depth is to determine the solids accumulation rate. Solids accumulation rates range from 0.02 to 0.09 m³/capita-year. The specific rate depends on both the amount of excreta produced by individuals (affected by diet, activity level, and water consumption), the material that is used for anal cleansing (e.g., water, leaves, stones, corncobs, paper), and environmental factors in the pit (e.g., temperature, whether the majority of the waste is above or below the water table) (Franceys et al. 1992). Accurate estimation of the solids accumulation rate must be based on local, direct measurement. If these data are unavailable, the following long-term values can be used (Franceys et al. 1992):

- For pits dug above the water table, with biodegradable materials used for anal cleansing (e.g., water, toilet paper), assume 0.06 m³/person/year.
- For pits dug above the water table, with bulky or nonbiodegradable materials used for anal cleansing (e.g., corncobs, stones, newspaper), assume 0.09 m³/person/year.
- For pits dug below the water table, with biodegradable materials used for anal cleansing, assume 0.04 m³/person/year.
- For pits dug below the water table, with bulky or nonbiodegradable materials used for anal cleansing, assume 0.06 m³/person/year.

The plan (cross-sectional) area of a pit is based on local preference. Pits are often 1–1.5 m wide to allow for easy excavation. Rectangular pits are easier to dig but are structurally weaker than circular pits. In unstable soil, use a circular pit. Soil is considered stable if it contains more than 30% clay (see Chapter 8). Dig the pit walls as plumb as possible to maintain structural integrity. If the pit will be lined, only the volume inside the lining material should be included in calculations.

The depth for solids storage is determined as

$$\text{Solids storage depth} = \frac{\text{Number of users} \times \text{Accumulation rate} \times \text{Design life}}{\text{Cross-sectional area}} \quad (20\text{-}1)$$

The number of users is related to the number of people living in a household, and the design life is typically 10–20 years. The cross-sectional area for a 1-m-diam. circular pit is approximately 0.80 m², and for a 1.5-m-diam. circular pit is approximately 1.8 m². Equation 20-1 can also be used to determine the life of a pit for a given area and depth.

After calculating the required depth, add an additional 0.5 m to determine the total depth of the pit. When the excreted solids in the pit reach 0.5 m below the ground surface, take the latrine out of commission and fill in the remaining space with soil to provide a sanitary cap for the pit.

Line latrine pits with impervious material for the top 0.5 m. This procedure helps to prevent rainwater from infiltrating into the pit and rodents from excavating under the slab to enter the pit. The remaining depth of the pit should be lined, unless there is evidence that it is not needed, such as unlined shallow wells in the area that have not collapsed (Franceys et al. 1992). Figure 20-1 shows both an unlined and a lined pit.

Use local materials for the lining, including concrete block, bricks, masonry, stones, or hardwood. If using blocks, bricks, masonry, or stones, fill the joints between them completely with mortar in the top 0.5 m of the pit. For the remainder of the pit, leave the vertical joints without mortar to provide the liquid in the pit the ability to soak into the soil.

If a pit will later be excavated to remove contents of a latrine pit for soil amendment, line the pit to prevent collapse during reexcavation. After two years of being sealed in the pit, the contents are a safe soil amendment (WHO 2006). This time allows destruction of pathogens and biochemical decomposition of some nutrients (e.g., phosphorus, carbon), although much of the nitrogen, located in urine, is lost through infiltration into surrounding soils (WHO 2006).

20.1.5 The Slab

Latrine slabs can be rectangular or circular and are reusable after the pit has filled with solids, if well-constructed. Slabs are typically constructed from steel-reinforced concrete (Fig. 20-3) and should

- be structurally sound;
- be smooth and impermeable to facilitate cleaning and allow for a tight-fitting lid, if a lid is specified;
- have a sloped area immediately around the hole to improve drainage into the pit;
- have footrests to ensure proper positioning (especially at night); and
- have a hole that is keyhole-shaped (pear-shaped) to prevent fouling, and that is narrow enough to ensure that children will not fall through the opening.

Slabs can be built at the site or prefabricated off site. Typical thickness of a reinforced slab is 65–70 mm using a concrete mix of 1:2:4 (cement:sand:gravel) (by volume) and 6- or 8-mm rebar. A Mozambique slab (diameter of 1.2 to 1.5 m) can be constructed from non-reinforced concrete. The mix is 1:2:3 (by volume), and the slab thickness is 50 mm. This slab is circular and dome shaped to provide structural strength. The slab is cast in a circu-

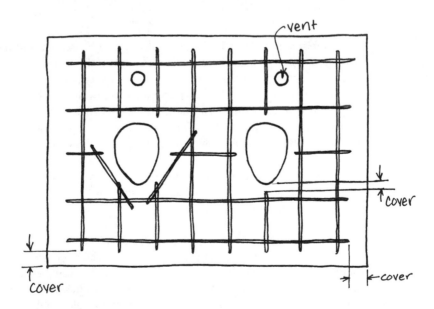

Figure 20-3. Layout of Steel-Reinforced Latrine Slab Showing Rebar Spacing and Need for Concrete Cover of the Reinforcement at the Edges.

Note: This particular slab is for a double-vault latrine that will have air ventilation. Similar construction can occur for round slabs.

lar form of earth, wood, or steel. Brandberg (1997) provides information on construction of the popular SanPlat system. The various SanPlat slab models have easy-to-clean surfaces, with elevated footrests, and a right-sized drop hole (see http://www.sanplat.com/).

20.1.6 Superstructures

With the exception of vent pipes and roofs that don't leak, the superstructure of a latrine is largely irrelevant to its functioning but influential in whether people are interested in owning and using the latrine. Therefore, leave the design of the superstructure as much as possible to local preference. Local materials such as wood, woven thatch, and bricks are recommended over materials imported into the community. Other materials that are used include cloth, concrete block, ferrocement, and corrugated iron. The design, construction, and economic cost of the superstructure can often be left to the owners. In the case of VIP and composting latrines, ensure that vent pipes are included, and cost subsidized if necessary.

))) 20.2 Types of Latrines

The selection of the appropriate latrine technology is based on several items, including siting conditions, user comfort, cost, maintenance, knowledge requirements, and expected life (Table 20-2). Consider the amount by which a sanitation technology can increase water usage: no effect for a dry technology such as a composting latrine, several liters per flush for a pour-flush latrine, and up to 70 L/capita-day for a sanitary sewer (Fry et al. 2008). The remainder of this section discusses ventilated improved pit, pour-flush, and composting latrines.

20.2.1 Ventilated Improved Pit (VIP) Latrines

The ventilated improved pit (VIP) latrine has features that address some of the problems associated with traditional pit latrines. As shown in Fig. 20-4, a vent pipe allows air to circulate out of the pit and away from the user, reducing odors and insects. A screen prevents flies from leaving the vent pipe, where they die and fall back into the pit. Upward air movement in the vent pipe is caused by wind blowing across the top of the pipe and/or by the heating of the pipe by the sun.

The VIP latrine requires a minimum amount of imported materials for its construction, resulting in its low cost. It requires little maintenance and no source of water. However, odors can be more of a nuisance than in pour-flush or composting latrines. It can also be difficult to find a material for the screen that holds up to the corrosive gases that emerge from the pit.

Construct the vent pipe from 100-mm-diam. PVC pipe if the average wind speed at the top of the pipe is >3 m/s, and ≥150-mm-diam. pipe if the wind speed is <3 m/s. Paint the pipe a dark color to facilitate air movement up the pipe. In lieu of PVC, use local materials, such as large-diameter bamboo, bricks, or blocks. Vent pipes of local material must have a larger diameter because the rougher interior reduces the wind speed through the pipe (Wegelin-Schuringa 2000). Install the vent pipe 0.5 m above the roof line so that the roof does not prevent wind from blowing across the top of the pipe.

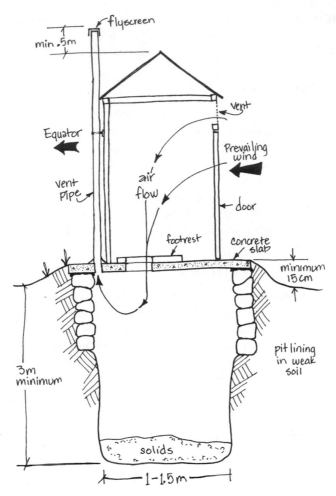

Figure 20-4. Components of the Ventilated Improved Pit (VIP) Latrine.

Source: Redrawn with permission from Pickford (1991).

Consider the vent pipe hole size and location in the slab construction (Fig. 20-3). Insert the vent pipe through the slab and seal with mortar around the pipe or cast a pipe stub, with the female end up, in the location of the vent pipe hole while pouring the slab. Roughen the pipe exterior to allow a better grip with the slab concrete. For stability, attach the vent pipe to the latrine superstructure. Bend a strip of scrap metal to fit around the pipe and attach the metal to the superstructure. The location of the vent pipe (inside or outside the superstructure) (and the privacy shelter door) depends on the type and size of superstructure (Fig. 20-5) and the direction of the prevailing wind. A vent pipe located outside the structure and positioned facing the sun can take better advantage of solar heating and induce an upward draft. Some measurements suggest that the pit contents heat up from biological activity, which may also induce air flow out of the pit.

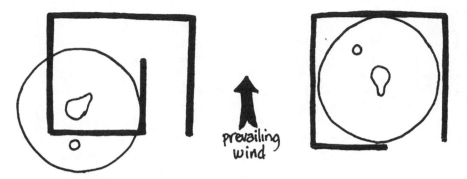

Figure 20-5. Ways That the Vent Pipe, Privacy Shelter (Spiral or Rectangular), and Slab Can Be Oriented in a VIP Latrine.

The screen at the top of the vent pipe should have a maximum mesh size of 1.2 mm by 1.5 mm to prevent flies from exiting the pipe. Because of the composition of the gas exiting the vent pipe and the screen's exposure to the weather, it should be made of corrosion-resistant material (e.g., stainless steel, plastic-coated glass-fiber mesh). Educate latrine owners to repair or replace the screen when holes form. A way to attach the screen is by placing a PVC collar on the top of the pipe to hold the screen in place (Mara 1984).

In a VIP latrine, the superstructure has an additional purpose besides privacy. It should be relatively dark inside the shelter so that the only light source visible to flies in the pit is up the vent pipe. Flies will be attracted to the light at the top of the vent pipe and fly upward, where their escape will be blocked by the screen. They will eventually die and fall back into the pit. The latrine door should face the prevailing wind direction to encourage air flow, and a small area above the door needs to be left open to allow air to enter. It has been recommended that the system be designed to allow for six air exchanges per hour, thus the volume of the privacy shelter is an important consideration. The air exchange rate is commonly used as a measurement of indoor air quality and is measured in air changes per hour, as follows:

$$\text{Number air exchanges per hr} = \frac{\text{Ventilation rate (m}^3\text{/h)}}{\text{Volume superstructure (m}^3\text{)}} \qquad (20\text{-}2)$$

The structure can be built with the traditional rectangular design, but spiral designs that require no door have been shown to better control odors and flies.

In addition to regular cleaning of the floor slab, occasionally pour a small amount of water down the vent pipe to clear the screen and the pipe of spider webs and other accumulations. Also, keep the latrine seat or squat hole in a VIP latrine uncovered when the latrine is not in use to create the necessary air circulation (Fig. 20-4), preventing odors and insect breeding. This procedure is contrary to the operation of a traditional pit latrine, so the engineer must understand that a family may be used to covering the hole if they are upgrading to a VIP latrine. Users may also be resistant to leaving the hole uncovered because of concern about animals and children falling into the pit, especially at night and if latrines are constructed with a door.

20.2.2 Pour-Flush Latrines

The pour-flush latrine uses water to flush solids from a bowl, similar in design to what is commonly used in sewered communities. A pour-flush latrine can be located either directly over the pit or offset from it (Fig. 20-6). Because of the airtight seal, the latrine can be located in the house with the offset pit located outside.

If groundwater is used for drinking, pour-flush latrines should not be used in an area with a high water table. Low-permeability soils are also not suitable for pour-flush latrines. Box 20-1 describes a simple soil filtration test that can be performed to determine the feasibility of a site for locating a pour-flush latrine. In addition, if the community uses leaves, stones, or corncobs for anal cleansing, these items cannot be discarded in the bowl because they can break the siphon (WHO 2006).

The main distinction between materials used in the pour-flush system and the other types of latrines is the presence of a bowl that has a water seal trap. The bowl is traditionally a porcelain bowl. Pour water into the bowl after each use, cleaning the bowl and passing the waste into the pit. The bowl may have a seat or may be used without a seat if the latrine is used by squatting (Fig. 20-7). The trap maintains a quantity of water in the bowl at all times, similar to flush systems commonly used throughout the world. A properly designed and operated pour-flush latrine creates a water seal between the user and the contents of the pit, reducing odors and preventing flies from entering the pit.

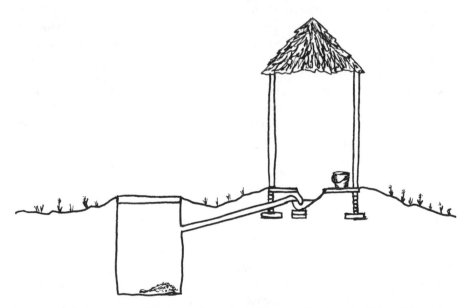

Figure 20-6. For a Pour-Flush Latrine, the Collection Pit Can Be Either Offset (as Shown Here) or Placed Directly Under the Bowl Where Defecation Occurs.

Note: The top of the pit cover should be slightly higher than the surrounding ground to ensure that water does not drain into or pool near the pit.

⟫⟫ Box 20.1 A Soil Filtration Test to Determine the Feasibility of Siting a Pour-Flush Latrine

A simple soil filtration test can determine the rate at which water can leave the pit and filter through the pores in the soil. If the filtration rate is too slow, the pit may fill too quickly and shorten the life of the latrine. On the other hand, a high filtration rate may be a sign of a fissure in the soil, which could allow untreated waste to seep directly into the groundwater via a natural conduit, or poor soil structure, which could cause the pit to collapse.

Because soil type can vary over small distances, perform this test in the exact location of each planned latrine. Test results can also vary with depth because of vertical variations in soil conditions. The most accurate tests are performed at the finished pit depth, although a minimum test depth of 0.5 m is acceptable.

To conduct the test, dig a 350-cm² hole that is at least 350 cm deep in the bottom of the excavated pit. While digging the test hole, some smudging and compaction of the dirt on the walls of the test hole may occur. To reverse this effect, scratch or score the sidewalls to provide a natural soil–water interface. Also, remove foreign or loose material from the walls of the test hole.

Then fill the test hole with water, which is allowed to soak in for 24 h. Additional water may be required to ensure that the test hole remains wetted for a complete 24 h. This time simulates how the soil will behave during the wettest part of the year while the latrine is in use.

After 24 h, insert a ruler or a marking stick along the wall of the test hole. Fill the test hole halfway with water and record the water level every hour. At the optimal infiltration rate, the water level in the hole should drop approximately 2.5 cm every hour. Combine the results of this test with other observations and engineering judgment to determine if a pour-flush latrine is suitable for a given site. If the test indicates that the area is not suitable for a pour-flush latrine, the pit can still be used for a VIP or traditional latrine, provided that other site requirements are met.

During construction, install the porcelain bowl to ensure that the bowl is level and that water in the bowl makes a complete seal. To test if the bowl is level, pour water in the bowl slowly until the trap and the bowl are full and water begins to leave the bowl to simulate in-use conditions. At this point, visually check that the water surface in the bowl is level and if it is not, adjust the bowl. Be careful not to pour too much water or spill because the water can damage the uncured slab. The drain pipe that connects the bowl to an offset pit should have a slope between $1:5$ and $1:15$ ($H:L$). Ensure that there is at least a 20-mm water seal between the pipe-to-bowl connection and the pan (Fig. 20-7).

The pour-flush latrine is often seen as a status symbol or sign of development because it resembles the indoor system found in more developed areas, however, there is a minimal amount of maintenance required.

To install a pour-flush latrine, there must be a reliable, year-round supply of water for flushing the latrine. Pour-flush latrines require 1–5 L of water per use (with an average of 2–3 L). Assuming that each person uses the latrine five times per day, an additional

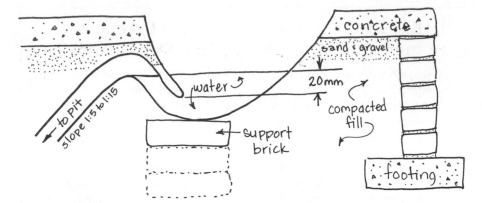

Figure 20-7. Close-Up of the Water Seal in a Pour-Flush Latrine Designed to Be Used by Squatting.

5–25 L of water is required per person per day if the latrine is flushed with every use. Do not underestimate the effect of this amount of water. If the increased water demand cannot be met on an annual or seasonal basis, pour-flush latrines will cease to function and residents may return to defecation in open areas, thereby decreasing the overall health of the households and the community. Integrating other sources of water (e.g., rainwater, greywater) into the overall household water budget may provide a solution.

Clean the surrounding floor and bowl of the latrine on a regular basis. In addition to an adequate supply of water, the key to ensuring that the latrine continues to function properly is having a large enough container of water available at all times. (The minimum required container size differs slightly, depending on the bowl and pipe setup.) Three 1-L flushes do not equal one 3-L flush in terms of moving solids out of the system and into the pit.

20.2.3 Composting Latrines

The *composting latrine* (also called *ecological sanitation* or *eco-san*) system converts human excrement into a soil amendment that can improve both the physical structure and the nutrient content of soils. The latrine consists of two chambers that are used alternately (Fig. 20-8). While one chamber is in use, waste is being composted and pathogens are dying off in the other.

Urine must be separated from feces for composting to function properly. Urine has two adverse effects on the composting process. First, it keeps the excreta saturated with moisture, which minimizes air circulation and thus the transfer of oxygen in the compost pile. Second, because of urine's high nitrogen content (its $C:N$ ratio is $0.8:1$), it lowers the $C:N$ ratio of the compost pile below the optimal ratio of $30:1$. This difference is especially important because the $C:N$ ratio of feces alone is already low ($6:1$ to $10:1$).

Because of its aboveground design, the composting latrine can be built in locations where other latrines cannot, such as in floodplains, in areas with a high groundwater table, and areas close to surface water. It does not need to be moved or rebuilt when full, thus

false floor   urine diversion

Figure 20-8. Composting Latrines Showing Two Vaults.

Note: Only one vault is in use at any given time, while waste solids are composted in the other. The left two drawings show a false-floor latrine, and the right two drawings show a urine diversion latrine. Both the soak pit and the urine receptacle options are shown. Urine can also be routed to a nearby garden. An optional vent pipe was added to the left drawings.

extending the life of the latrine. Another advantage is that it does not produce bad odors or provide a breeding ground for insects when properly maintained. The latrine also does not require a source of water, so it is technology appropriate for water-scarce areas.

Composting latrines require more time and attention in their operation and maintenance than other types of latrines. As opposed to the pit, VIP, and pour-flush latrines, a reservoir of dry organic material must be available for each use. Also, the owner of the latrine must be trained to monitor the contents of the latrine to ensure that the solids are composted properly. Understanding the basics of how a composting latrine functions is required not just by those who are directly responsible for its cleaning and maintenance, but by those who use it. In addition, the latrine owner must remove the composted materials from the chambers when they are ready to be harvested. People may initially have an aversion to handling the compost.

Design

The basic design for a dry composting latrine seat includes a receptacle to collect urine and a large chamber that collects and stores feces. Some composting latrines consist of two steel-reinforced concrete slabs: a slab-on-grade and an elevated (latrine floor) slab supported by masonry walls. Wall openings provide access to each chamber.

Size the chambers so that it will take at least a year for each chamber to be filled (see the section called Composting Operation below). If this size is not feasible because of cost or space concerns, it is possible to remove the compost product from the latrine and further compost it outside the latrine for an adequate time, or cover it with several centimeters of soil.

The slab-on-grade (called the *base platform*) is constructed first. After the bottom slab has cured, construct the chamber walls. Parge the inner walls to contain liquids and prevent outside water from infiltrating. After the walls are constructed, construct the elevated slab. If sufficient materials are not available for the formwork required to cast the floor in place, an alternative is to cast two separate slabs on-grade and move them on top of the block walls after the slabs have cured (see Chapter 7).

Install an access door in the back wall of each latrine compartment. Access doors that can be easily opened (compared to a mortared block door) are important because the solids compost better and at a faster rate if they are manually turned on a regular basis to promote aeration. Construct the access doors of local materials that are durable and will not rot. Use wood planks, metal panels, or concrete or masonry. A reinforced concrete or concrete block door can be attached using a weak mortar mix. The mortar can be broken at the time when the compost chamber needs to be accessed. Corrugated metal panels can be cut and the edges folded, forming a 1-cm lip to fit snugly into the opening.

Urine Separation

There are two methods to separate urine: a urine diversion system or a false floor. In a *urine diversion system*, the urine can either be collected in a pot and used as fertilizer or directed into a soak pit for subsurface disposal (Fig. 20-8). The urine from a composting latrine may also be routed to a nearby garden through an irrigation distribution system. Because urine contains 50–70% of the nitrogen in municipal wastewater and 50% of the phosphorus, it should be viewed as a valuable source of nutrients for agriculture. Composting latrines designed with a false floor allow the urine to pass through slotted openings.

Figure 20-9 shows how urine diversion can be incorporated into the seat design. Construct seats from wood or concrete. Fiberglass molds can be purchased and used to cast urine diversion toilet seats from concrete. Although not discussed here, some seats are constructed with a separate receptacle toward the back of the seat that collects anal ablution water and directs it into a soak pit. If premade composting latrine seats are unavailable, fit a regular pit latrine seat with a receptacle to collect urine. Cut a large plastic bottle (e.g., a bleach bottle) and invert it to serve as a funnel (Fig. 20-10). Keep composting latrine seats covered.

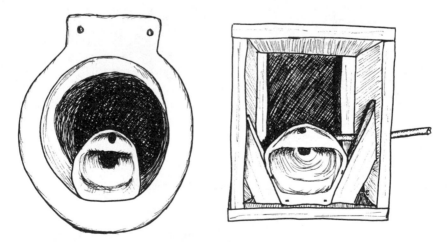

Figure 20-9. Porcelain Bowls (Left) or Wooden Seats (Right) Can Be Adapted for Urine Diversion Using a Bleach Bottle Top.

Note: Covers are not shown. (The smaller of the two holes shown in the bottle is part of the structure (handle) of the bottle itself (Figure 20-10) and is not required for the latrine.)

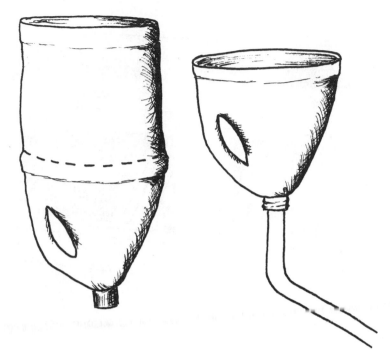

Figure 20-10. The Plastic Bleach Bottle on the Left Has Been Cut to Produce a Urine Separation Device.

Note: It is connected to a pipe that flows either into a receptacle that collects urine (to be used as fertilizer) or into a soak pit.

Three main problems that arise with a seat are the urine receptacle is located too high and touches the user, the space for the excrement is too small, making it difficult to aim properly, and the location of the receptacle makes it hard for the users, especially women, to keep urine and excrement separate. Therefore, when constructing a seat, it is important to have both women and men consulted in the design process.

As an alternative to a special seat and urine diversion piping, install a slotted false floor in the compost compartment to separate urine from feces (Fig. 20-11). The false floor retains the solids while allowing the urine to drain through it and into a collection system or soak pit (Fig. 20-12). Do not use the urine from a false floor latrine as fertilizer because it has been contaminated by untreated feces. A wooden false floor frame fits inside each chamber. On the frame, nail boards that are at least 5 cm wide and 2.5 cm thick. Space boards 1.5 cm apart (Tawney 2006).

The false floor rests on a ledge located on the wall opposite the access door inside of each chamber, allowing the false floor to slope toward the access door (Fig. 20-12). Use concrete blocks (or bricks turned on their side) or construct a concrete ledge while casting the slab-on-grade. The ledge should be approximately 10 cm high and 20 cm wide to accommodate the false floor.

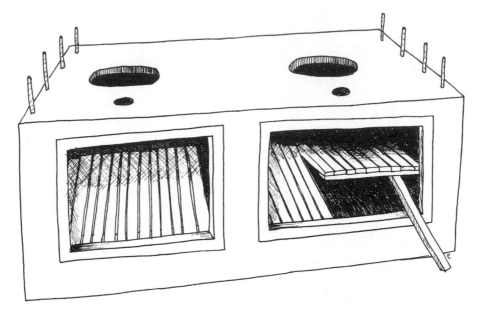

Figure 20-11. Sloped False Wooden Floors Separate Urine and Other Liquids from Solids.

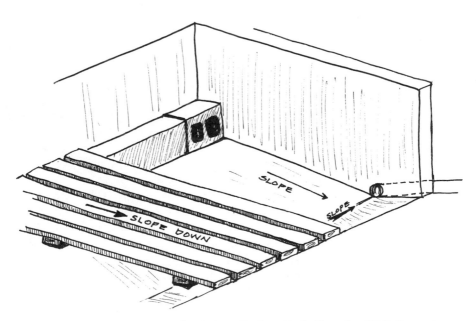

Figure 20-12. A Concrete Ledge Is Installed on Each Chamber Wall Opposite the Access Door to Prop Up One Side of the False Floor.

Note: The floor is then sloped to drain the liquid to piping that carries it into a soak pit.

Composting Operation

Before the first use of the latrine, prime the chamber by placing a thick layer (about 10 cm deep) of dry organic material onto the bottom slab or the false floor. Then put the latrine into use. In the case of a urine diversion toilet, make sure the urine falls into the urine receptacle as intended.

After each use of the latrine, add 1–2 cups (or handfuls) of a dry organic material to cover any fresh excrement. Examples of this dry material are sawdust, dry leaves, dry grass, or wood ash. Also, cover the latrine opening after finishing.

It is helpful for owners of composting latrines to have a flashlight or a lantern. Regular inspection of the chamber from above or from the access door can reveal maintenance problems before they get out of hand. Any spots on the pile that appear wet should be covered with dry material. These are the areas that cause odor and fly problems. This type of spot-checking can increase overall user satisfaction.

Pathogens are destroyed in a composting latrine by several mechanisms (WHO 2006):
- *Storage time*—Compost should be stored in the unused chamber for as long as possible; minimum times are 6 months to 1 year.
- *Temperature*—Higher temperatures lead to faster pathogen destruction. Figure 20-13 shows how temperature and contact time affect complete inactivation of specific pathogens.
- *pH*—Raise the pH of the compost over 9.
- *Moisture content*—Keep the moisture content of the compost as low as possible. The fastest pathogen destruction occurs at moisture levels <25%.

When excreta and agricultural or yard waste are composted at the community level in a controlled environment, the primary method of pathogen destruction is through elevated temperatures. However, this situation is difficult to achieve in a composting latrine, where temperatures often increase only marginally above ambient temperatures. Our experience is that most household compost latrines do not achieve a high enough temperature to effectively destroy all pathogens (Hurtado 2005; Kaiser 2006; Mehl 2008), especially *Escaris* eggs.

Composting latrine education must promote pathogen destruction through extended storage, desiccation, and some increase in pH. For ambient temperatures from 2 to 20 °C, store dry excreta for 1.5 to 2 years. For temperatures from 20 to 25 °C, store it more than a year (WHO 2006). Desiccation is also a proven method to destroy pathogens and depends on the proper application of desiccants after each use (Kaiser 2006). However, local climate and seasonal economic activities may present challenges in providing sufficient and dry desiccants over the course of a year (Example 20-1).

Because an elevated pH level will assist in pathogen destruction, locally produced materials with high pH, such as wood ash, are an excellent choice as desiccants. A pH > 9 can be achieved by adding 200–500 mL (or 1–2 cups) of wood ash after each use. This volume is enough to cover fresh excreta. Table 20-3 lists the pH of several commonly used desiccants. Coal ash will not raise pH sufficiently (WHO 2006).

Proper ventilation speeds up the desiccation process and also helps with odor control. Add vent pipes to composting latrines to aid in the evaporation of moisture. Incorporating a urine collection or diversion system into the design is one method to reduce the moisture of the compost pile.

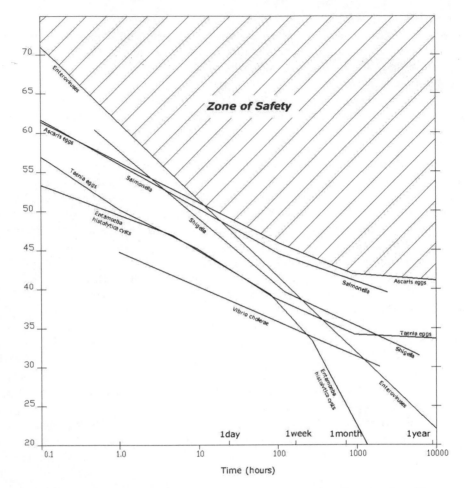

Figure 20-13. Temperature and Contact Time Required to Destroy Specific Pathogens Found in Human Feces.

The zone of safety shows that a temperature of 42 °C and a contact time of 6 months to 1 year results in destruction of all pathogens.

Source: Redrawn with permission from Cairncross and Feachem (1993).

To maintain proper air flow throughout the pile, regularly add bulky materials, such as sawdust or grass. These materials also have a higher carbon-to-nitrogen (C:N) ratio. The C:N ratios for sawdust, grass, and straw are 200–500, 12–15, and 128–150, respectively. For ideal compost, the C:N ratio should be approximately 30:1. Because feces has a low C:N ratio, it should be composted with materials that have high C:N ratios (Table 20-4). Thus, the challenge is to develop a proper mix of added organic material that provides desiccation, increase in pH, bulk to circulate oxygen into the pile, and also an appropriate C:N ratio. Other books that discuss solid waste management provide

Table 20-3. pH of Selected Desiccants Commonly Used in Composting Latrines

Material	pH
Wood ash	9.4–11.3
Rice husks	10.6
Sawdust	4.5–7.8
Lime	10.3

Source: Kaiser 2006.

examples on how to determine the appropriate mix of materials to ensure the proper $C:N$ ratio (e.g., Mihelcic and Zimmerman 2009).

))) Example 20-1. Determining the Proper Ingredients for Successful Composting

Assume that feces are 6.3% N (on a dry mass basis) and have a moisture content of 80% and $C:N$ ratio of 10. The feces are to be composted with the desiccant, sawdust, which is assumed to be 0.1% N (on a dry mass basis) and has a moisture content of 10% and $C:N$ ratio of 400. Determine the kilograms of sawdust required per kilogram of feces to attain a desired $C:N$ ratio of 30.

Solution

Assume 1 kg of feces dry mass. Let Y equal the kilograms of sawdust on a dry mass basis. The mass of carbon and nitrogen obtained from each material in the mixture is

Dry mass nitrogen from feces $= 1 \text{ kg} \times (1 - 0.8) \times 0.063 = 0.0126 \text{ kg}$

Dry mass carbon from feces $= 1 \times (1 - 0.8) \times 0.063 \times 10 = 0.126 \text{ kg}$

Dry mass nitrogen from sawdust $= Y \times (1 - 0.1) \times 0.001 \text{ N} = 0.0009 \times Y \text{ kg}$

Dry mass carbon from sawdust $= Y \times (1 - 0.1) \times 0.001 \text{ N} \times 400 \text{ kg C/kg N} = 0.36 \times Y \text{ kg}$

Table 20-4. Nitrogen and $C:N$ ratios of Various Materials That Might be Added to a Composting Latrine

Material	Nitrogen (% Dry Mass)	C:N Ratio (Dry Mass Basis)
Urine	15–18	0.8
Human feces	5.5–6.5	6–10
Wheat straw	0.3–0.5	130–150
Oat straw	1.1	48
Grass clippings	2.4–6	12–15
Sawdust	0.1	200–500

Source: Selected values from Mihelcic and Zimmerman 2009.

The desired C:N ratio is

$$30 = \frac{(\text{mass carbon from feces} + \text{mass carbon from sawdust})}{(\text{mass nitrogen from feces} + \text{mass nitrogen from sawdust})}$$

Plugging in the values we determined into this equation results in

$$30 = \frac{(0.126 + 0.36 \times Y)}{(0.0126 + 0.0009 \times Y)}$$

Solve for Y, which equals 0.76 kg. Thus, for every 1 kg of feces, 0.76 kg of sawdust must be added to obtain the desired C:N ratio of 30. The feces are a better source of nitrogen, and the sawdust is a better source of carbon. This example points out several other issues associated with proper management of composting latrines. They may be designed and operated to achieve pathogen destruction, but they are much more difficult to optimize for aerobic thermophilic biological decomposition as is done on a larger scale by municipalities. The amount of desiccant a user adds is never adjusted with the C:N ratio in mind because most composting latrines are operated with the primary purpose to destroy pathogens, not to produce agriculturally ready compost.

Sawdust is a good desiccant, but it will not raise the pH of the compost pile. It is also a poorer bulking agent compared to dry grasses, which provide carbon as well as assist aeration of the pile. The amount of sawdust required in this example may also not be realistic for a household on a short- or long-term basis. Some questions that need to be asked are the following:

- What is the primary mechanism of pathogen destruction, and how will the compost be used?
- How will introduction of liquid cooking fuels into a community affect the availability of desiccants such as wood ash?
- Is logging performed in the area, and is it performed close to the home or far into the forest?
- Is logging a seasonal activity that could limit the amount of available sawdust?
- And what about the amount of sawdust required here? The value seems quite large, and we have not even thought out the storage requirements for maintaining a sufficient supply of desiccant that also provides a source of carbon.
- How does increased development affect the availability of desiccants associated with cooking or seasonal employment?

All these questions are raised, not to discourage communities from adopting a composting latrine as a form of sanitation technology, but instead, to demonstrate the many issues that need to be addressed so the sanitation technology selected is sustainable in the short and long term. ⟩⟩⟩

Keep the inside of the latrine clean, including the pedestal. For latrines with a urine diversion device built into the bowl, wash the urine receptacle on a regular basis, making sure not to drip excessive water into the chamber. If the latrine develops a bad odor or lots of insects, the users should double the amount of dry material that is being added and check for possible leaks or blockages in the urine diversion pipe. The user may also want to turn the pile on a regular basis (e.g., twice a week), which will help ensure aerobic and well-mixed conditions. This result, of course, depends on how the access door is sealed. Hot wood ash is also helpful in reducing the number of insects in the latrine, but take care when using them near a wooden structure. In the case of a blockage in the urine tubing (in a urine diversion system), use a cloth attached to the end of a long flexible wire to clean the inside of the tube.

It is time to change chambers when one chamber is full or when a seasonal calendar event occurs, about every 6–12 months, that is associated with changing chambers. Before closing the chamber that has just been taken out of commission, fill it up with dry organic material and attempt to mix the contents as best as possible. If the latrine was designed to only have one toilet seat for the chamber that is in use, move it at this time and cover the other chamber opening.

Open the other, sealed chamber and remove the composted material with a shovel. If the chamber was sealed for the recommended time and the material has an earthy smell, appears fully composted, and is generally inoffensive, the compost can be used according to the guidelines provided in Box 20-2 and in the following section. If the material was not able to be stored for the recommended amount of time, or if it appears to not be fully composted, it should be solar dried or stored in a bin that is protected from rainfall for an additional period and then used according to the same guidelines, or at a minimum, buried below several centimeters of soil.

After removing the compost, check the chamber, making sure it is clean and the tubes and urine drainage beds are in working condition. Then reprime the chamber with dry organic material, close the access door, and begin to use it.

Compost as a Soil Amendment

If proper steps are taken to reduce risk, a composting latrine may provide a safe soil amendment that serves as a fertilizer and improves the physical structure of soil. Besides the pathogen destruction that occurs in the latrine, other activities provide additional barriers to disease transmission. If there is an area protected from access by children and animals, solar drying can provide an additional layer of protection. Once the compost is harvested, spread it out in the sun for 1 week. This process promotes pathogen destruction through elevated temperatures and desiccation.

The choice of crops on which to use the compost can also help to reduce risk. If there is a concern that the compost is not fully treated, it should not be applied to vegetables, fruits (except for fruit trees well before the tree begins to drop its fruit), or root crops that are consumed raw. Hairy, sticky, or rough crops can protect pathogens on their surface and are harder to wash. Crops that retain water (e.g., lettuce) are also associated with higher risk because rain can splash up contaminated soil (WHO 2006).

The best way to use the compost from a latrine is to mix it into the soil before planting or around the base of a fruit tree because the nutrients are helpful to small plants and roots (WHO 2006). There will also be fewer opportunities for people or edible parts of plants to come into contact with it if it is worked into the soil. Whoever is doing this work should carefully wash their hands with soap and water afterward.

>>> **Box 20.2 How Does a User Manage Operational Conflicts Between Pathogen Destruction and Aerobic Decomposition in a Composting Latrine?**

Operating a compost latrine properly may at times appear to have two conflicting goals: pathogen destruction and producing compost suitable for small-scale agricultural use. The conflict arises because the environmental conditions that optimize each goal are at times similar and at times conflicting. Aerobic decomposition that produces suitable compost requires adding sufficient carbon to raise the carbon-to-nitrogen ratio of human feces to (20 or 30 to 1), maintaining pH near neutral (6–8), providing sufficient moisture, and mixing the pile on a regular basis to ensure oxygenation of the compost pile. Pathogen destruction, though, is dependent on a variety of several factors, including the high temperature that results from aerobic decomposition, and if that is not achieved, two conflicting environmental parameters: high pH and desiccation.

To assess the relationship among temperature, high pH, desiccation, decomposition, and pathogen destruction, a six-year study was performed on 63 composting latrines in five indigenous communities in the province of Bocas del Toro in Panama that observed the use of desiccant and measured temperature, pH, percent moisture, carbon-to-nitrogen ratio (C:N), and presence of pathogens.

The temperature results support previous findings that compost latrines do not get hot enough to kill all pathogens; rather, the latrines remained close to ambient temperatures. The pH results show that many latrines were operating within the range for ideal biochemical decomposition, pH of 7.5–8.5, but only 17% of latrines measured pH 9 or above, the recommended minimum pH for pathogen destruction (Fig. 20-A). The vast majority of composting latrine users added desiccant materials (primarily sawdust and wood ash), to lower the moisture level and provide carbon for decomposition. However, it appears that insufficient desiccant materials were added because moisture levels remained above the suggested maximum of 25% for pathogen destruction, and C:N ratios were in the range of the ratio of raw human feces, even after 6 months of use. More importantly, the results of the microbiological analysis show that various pathogens, primarily hard-to-destroy helminthes (like *Ascaris*), are still present in the compost samples that had been stored for the recommended 6-month storage time.

From these results, it follows that 100% of pathogens are not being removed in composting latrines, nor is thermophilic aerobic decomposition taking place. As an improved sanitation technology, composting latrines must be operated to destroy pathogens. Storage time should be increased to a minimum of 1 year, and users should be instructed to add more desiccant materials of both the high pH type (e.g., wood ash) and the bulky type (e.g., sawdust, dry grass, or leaves).

We remain large supporters of composting latrines; however, the composting latrine design may have to be adjusted for chambers of larger volume to increase storage time. Education for latrine users should also encourage the use of different

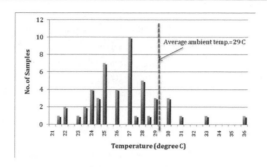

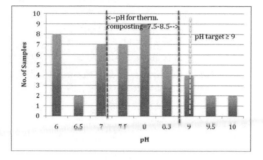

Figure 20-A. Measured Temperature and pH of Active Compost Latrines in the Province of Bocas del Toro in Panama. Temperatures of 42 °C and/or pH >9 Are Recommended, Along with Contact Times of 6 Months to 1 Year to Achieve Significant Pathogen Destruction.

desiccants—both high pH types (ash) and bulky types that also provide large inputs of carbon (sawdust and grasses). As far as harvesting and using the final compost, the following recommendations may be suitable until more field research is performed:

1. Harvest the aged compost during the dry season, or if there is no marked dry season, during the less rainy time of the year on a clear day, to prevent runoff into surface waters.
2. Allow the aged compost to solar-dry for one week. This may best be done by spreading the compost on zinc sheets in a thin layer (≤4 in.) under full sunlight.
3. Mix the solar-dried compost with soil or bury the compost under a few centimeters of soil when using in agricultural fields or near the base of local fruit trees.
4. Only use the compost on ornamentals and plants whose fruits are not low-lying or collected as roots. For example, use the compost around the base of banana trees, working it into a few centimeters of soil, but not around lettuce or potato plants.

Source: Mehl 2008.

Table 20-5. Pathogens in Urine

Pathogen	Urine as a Transmission Route	Importance
Leptospira interrogans	Usually through animal urine	Probably low
Salmonella typhi and *Salmonella paratyphi*	Probably unusual, excreted in urine in systemic infection	Low compared to other transmission routes
Schistosoma haematobium (eggs excreted)	Not directly but indirectly, larvae infect humans via freshwater	Need to be considered in endemic areas where freshwater is available
Mycobacteria	Unusual, usually airborne	Low
Viruses: CMV, JCV, BKV, adeno, hepatitis, and others	Not normally recognized other than single cases of hepatitis A and suggested for hepatitis B. More information needed	Probably low
Microsporidia	Suggested, but not recognized	Low
Venereal disease causing	No, do not survive for significant periods outside the body	Not applicable
Urinary tract infections	No, no direct environmental transmission	Low

Source: Reprinted from Schönning and Stenström (2004) with the permission of Stockholm Environment Institute.

Because it is difficult to ensure that all pathogens have been eliminated during composting, treatment in the soil should also be seen as part of the treatment process. Pathogens are usually inactivated more quickly in the soil than they are in a latrine. Hot and sunny weather is better for pathogen destruction than cool, cloudy, and rainy weather. Adding compost to soil during a dry season will also ensure that runoff does not occur into surface water. There should be at least 1 month between the last application of the compost and the harvesting of the crop. This is called a *withholding period*. The withholding period for treated feces is usually automatic because the compost is used during planting or when plants are young (WHO 2006).

Urine as a Fertilizer

The urine that is separated in a urine-diverting composting latrine should be seen as a resource and should only go into a soak pit if there is no opportunity to use it as a fertilizer. Urine can be directly routed from the latrine to a small garden plot via subsurface tubing commonly used for irrigating small gardens. The majority of excreted nitrogen and phosphorus (two nutrients required for plant growth) are found in urine. Urine can be used to fertilize most non-nitrogen-fixing crops. Crops such as spinach, cauliflower, and maize have high nitrogen requirements (WHO 2006). Urine fertilization is not as helpful for legumes because they fix nitrogen on their own.

The major risks of urine usage come from fecal cross-contamination, so take measures to reduce the chances of fecal matter contaminating the urine. Fewer pathogens actually come from the urine itself (Table 20-5), but those that do include *Ascaris* ova and *Schistosoma* eggs (Drangert 1998). If urine is stored before use, it should be in a sealed container. This step reduces odor as well as the loss of nitrogen through transfer of ammonia to the gaseous phase. The breakdown of organic nitrogen found in fresh urine to odorous inorganic ammonia nitrogen occurs because of a hydrolysis reaction with water. Urine should not be diluted with water during storage because concentrated urine is a harsher environment for pathogen destruction (WHO 2006).

No storage period is required when urine is used as a fertilizer at the household level, unless there is fecal cross-contamination. (If urine were collected from a community, additional precautions would be necessary.) For vegetables, fruits (except fruit trees), and root crops that are eaten raw, there should be a one-month period between the last application of urine and the time when the crop is harvested (WHO 2006).

Dilute urine before applying it to fields. A typical dilution is 1 part urine to 4 parts water. If a community is not familiar with using urine as a fertilizer, an experiment can be done to demonstrate its effectiveness. A small portion of a field planted with a particular crop should be selected to receive the urine fertilizer once a week, while the rest of the field receives only water. A difference should quickly become apparent between the fertilized and the unfertilized sections. After community members have seen the results, the dilution ratio can be adjusted based on further experiments to allow more efficient use of the urine, because not all plants require such a strong solution.

⟫⟫ References

Brandberg, B. (1997). *Latrine building: A handbook for implementation of the SanPlat system.* ITDG Publishing, London.

Cairncross, S. (1992). *Sanitation and water supply: Practical lessons from the decade* (Water and sanitation discussion paper series), International Bank for Reconstruction and Development, World Bank, Washington, D.C.

Cairncross, S., and Feachem, R. G. (1993). *Environmental health engineering in the tropics: An introductory text.* Wiley, New York.

Davis, J., and Lambert, R. (2002). *Engineering in emergencies,* ITDG Publishing, Rugby, Warwickshire, U.K.

Drangert, J. (1998). "Fighting the urine blindness to provide more sanitation options," *Water SA,* 24(2), 157–164.

Franceys, R., Pickford, J., and Reed, R. A. (1992). *A guide to the development of on-site sanitation,* WHO Press, Geneva, Switzerland, <http://www.who.int/water_sanitation_health/hygiene/envsan/onsitesan/en/> (Oct. 5, 2007).

Fry, L. M., Mihelcic, J. R., and Watkins, D. W. (2008). "Water and non-water-related challenges of achieving global sanitation coverage," *Environ. Sci. and Technol.,* 42(4), 4298–4304.

Hurtado, D. (2005). *Composting latrines in rural Panama: Design, construction and evaluation of pathogen removal,* <http://cee.eng.usf.edu/peacecorps/Resources.htm> (Jan. 17, 2009).

Kaiser, J. (2006). *An analysis of the use of desiccant as a method of pathogen removal in compost latrines in rural Panama,* <http://cee.eng.usf.edu/peacecorps/Resources.htm> (Jan. 17, 2009).

Mara, D. D. (1984). *The design of ventilated improved pit latrines.* UNDP, World Bank, Washington, D.C.

Mehl, J. A. (2008). *Pathogen destruction and aerobic decomposition in composting latrines: A study from rural Panama,* <http://cee.eng.usf.edu/peacecorps/Resources.htm> (Jan. 24, 2009).

Mihelcic, J. R., and Zimmerman, J. B. (2009). *Environmental engineering: Fundamentals, sustainability, design,* John Wiley & Sons, New York.

Pickford, J., ed. (1991). *The worth of water: Technical briefs on health, water and sanitation.* Intermediate Technology Publications, London.

SanPlat website. (2009). <http://www.sanplat.com/> (Jan. 24, 2009).

Schönning, C., and Stenström, T. A. (2004). *Guidelines for the safe use of urine and faeces in ecological sanitation systems,* Stockholm Environment Institute, Stockholm, Sweden.

Tawney, E. (2006). *Composting toilet construction manual,* <http://cee.eng.usf.edu/peacecorps/ Resources.htm> (Jan. 24, 2009).

Wegelin-Schuringa, M. (2000). *On-site sanitation: Building on local practice,* Occasional Paper Series No. 16. IRC International Water and Sanitation Centre, The Hague, Netherlands.

WHO/UNICEF. (2006). *Meeting the MDG drinking water and sanitation target: The urban and rural challenge of the decade,* WHO Press, Geneva, Switzerland, <http://www.who.int/water_sani tation_health/monitoring/jmpfinal.pdf> (June 5, 2007).

World Health Organization. (WHO). 2006. *Safe use of wastewater, excreta and greywater, Volume IV: Excreta and greywater use in agriculture.* WHO Press, Geneva, Switzerland.

))) Further Reading

Mara, D. D. (1985). *The design of pour flush latrines.* UNDP, World Bank, Washington, D.C.

McConville, J. R. (2003). *How to promote the use of latrines in developing countries,* <http://cee.eng .usf.edu/peacecorps/Resources.htm> (Jan. 24, 2009).

Moe, C. L., and Izurieta, R. (2003). *Longitudinal study of double vault urine diverting toilets and solar toilets in El Salvador,* 2nd International Symposium on Ecological Sanitation. Lübeck, Germany, April 7–11.

Ocwieja, S. (2009). *Life cycle approach for evaluating sustainability of biogas latrines,* <http://cee.eng .usf.edu/peacecorps/Resources.htm> (May 21, 2009).

Rybczynski, W., Chongrak, P., McGarry, M. (1982). *Appropriate technology for water supply and sanitation—Low-cost technology options for sanitation, a state of the art review and annotated bibliography.* World Bank, Washington, D.C.

Shaw, R., ed. (1999). *Running water: More technical briefs on health, water and sanitation.* Intermediate Technology Publications, London.

Strauss, M., and Blumenthal, U. J. (1990). *Use of human wastes in agriculture and aquaculture: Utilization practices and health perspectives, Duebendorf, Switzerland,* <http://www.eawag.ch/ organisation/abteilungen/sandec/publikationen/publications_wra/downloads_wra/human_ waste_use_ETC_SIDA_UA.pdf> (Jan. 14, 2009).

Winblad, U., and Kilama, W. (1985). *Sanitation without water.* MacMillan Publishers Ltd., London.

21

Wash Areas and Soak Pits

))) 21.1 Appropriate Improvements

Many areas, both rural and urban, have an immediate need for the removal of sewage and greywater but lack the space or funding for full septic, sewer, or other infrastructure installations. In these situations, wash area and soak pit projects can be an appropriate improvement, usually on an interim basis. In addition, in areas with poor soil quality, wastewater and greywater may not seep into the ground quickly enough to prevent pools of stagnant water. Therefore, designed and constructed systems are necessary to encourage infiltration. Sand, gravel, and rock can often be obtained locally, so the majority of project cost is for cement.

A *wash area* (Fig. 21-1) can be described as a sloped concrete slab that drains into a soak pit through a length of PVC pipe. Such areas are typically using for laundry. With a privacy superstructure added, they can also be used for bathing.

Soak pits, also referred to as *soakage pits*, *seepage pits*, *leach pits*, or *soakaways*, are pseudoseptic systems used in the developing world. They consist of a rock-filled hole near the area of drainage, with a layer each of sand and gravel at the bottom. Figure 21-2 shows a soak pit below grade. Most of the discharge seeps though the bottom of the pit, but it is also able to exit through the sides. Containing greywater underground while it has time to soak into the soil prevents surface pooling, where mosquitoes breed and other disease vectors gather. Soak pits can be integrated into the design of a composting latrine to collect urine, if there is no opportunity to use it as a fertilizer (see Chapter 20).

Figure 21-3 shows the plan view of a general wash area design integrated with a soak pit. In general, designs for wash areas and soak pits depend on the number of users and the frequency of use. An increase in population causes the surface area of the wash area and the volume of the soak pit to increase. Women should come together to decide whether a wash area and soak pit are needed in a community. They should help to determine the size, location, and layout of the project. If they do not agree on the design, they will likely not use the area.

Brooke Tyndell Ahrens, Jennifer McConville, and Meghan E. Housewright contributed to this chapter.

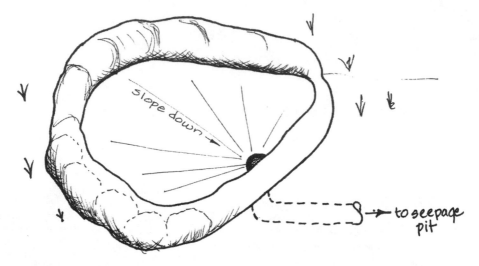

Figure 21-1. Wash Area with Greywater Flowing by Gravity Toward a Soak Pit (Not Shown).

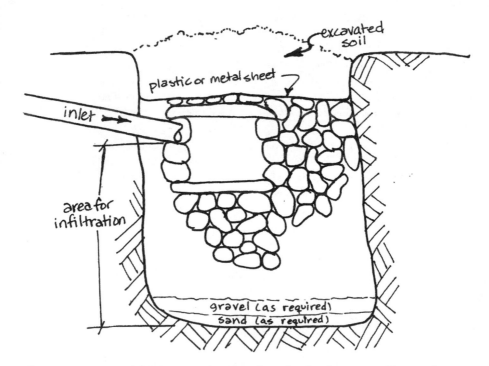

Figure 21-2. A Soak Pit Design Showing the Inlet for Greywater Flow and a Large Area Available for Infiltration to the Subsurface.

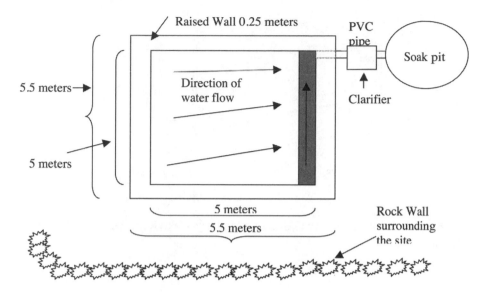

Figure 21-3. Plan View of Hypothetical Integrated Wash Area and Soak Pit Design for a Village Center.

Note: Wash area and soak pits are sized based on the number of users.

))) 21.2 Design Considerations

21.2.1 Wash Area Design

Wash areas differ in size, shape, and design because of the needs and preferences of the women using the wash area. There are no definitive guidelines on sizing a wash area. They may be 1 m × 1 m for a personal space, 2 m × 2 m for an area that serves several families, or 5 m × 5 m for use in a larger village. Figure 21-4 shows a cross section of a wash area with greywater moving left-to-right toward a soak pit (not shown).

21.2.2 Soak Pit Design

In the same way that an increase in users causes the surface area of a wash area to increase, it also increases the volume of a soak pit. Other factors that affect pit size are soil type and amount of water disposed of by each user. A soak pit serving a single water tap stand or a small family's latrine may only need to be 1.5 m deep and 1 m in diameter in sandy soil. However, if the soak pit drains the apron of a village pump in rocky, clayey soil, it may need to be 3 m deep and 2 m in diameter. Soak pits can be deeper, wider, or rectangular instead of circular, depending on local tradition and available space. A general rule of thumb is that soak pits should be designed so that the volume of the pit below the pipe exit can hold the amount of water discharged into the pit within 24 h.

Soak pits may be covered at the surface with local materials (sticks, leaves, mud), plastic, or a concrete slab. Figure 21-5 shows the components of a soak pit covered by a

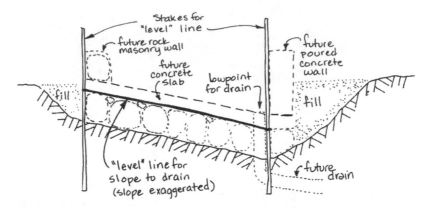

Figure 21-4. Wash Area with Greywater Moving Left to Right Toward a Soak Pit (Not Shown).

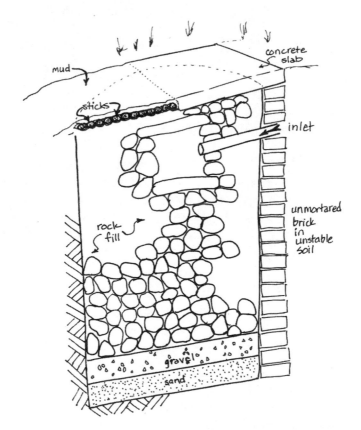

Figure 21-5. Components of a Soak Pit Covered by a Concrete Slab.

Note: Water enters through the inlet pipe in the upper right. The figure shows two options for covering the soak pit. The pit can be lined (right side) or unlined (left side), depending on soil stability.

concrete slab. If the soak pit is located in a village center and will have people and carts passing over it frequently, it should have a concrete slab cover. Additional issues to consider in soak pit design are listed in Box 21-1.

As shown in Fig. 21-6, a shallow infiltration trench can be constructed instead of a soak pit in areas with a high water table (where a deep hole is not practical or environmentally responsible) and where there is enough land surface area for trenches. Trenches for perforated pipe should be a minimum of 0.5 m deep and covered with plastic or geotextile fabric to prevent natural sedimentation from entering perforated pipe.

))) 21.3 Construction

21.3.1 Wash Area Construction

Wash areas are commonly constructed as composite slabs with raised edges or low walls to contain water. However, poured concrete can be used also. Floors should slope to the drain, which is covered by a screen to prevent debris from entering the soak pit. The wash

---

))) Box 21-1 Additional Considerations for Soak Pits

Adjacent construction—Soak pits or trenches should be located approximately 5 m from buildings. Avoid locating them under roadways and prevent subsequent construction over the pit or trench.

Soil infiltration and type—If sandy soil is present, the water will drain faster than in a clayey soil, but the walls of the pit may cave in if they are not supported. Clayey soils require a larger soak pit to compensate for slow water infiltration. Infiltration rates of porous soils range from 10 to 30 L/day per square meter of sidewall. Perform a soil infiltration test.

Water table or water source—Soak pits or trenches should be located a minimum of 30 m from wells or groundwater sources, and preferably downstream in terms of groundwater movement (typically downhill). The bottom of the pit or trench should be at least 1 m above the rainy season water table and at least 1 m above any impermeable layers. Do not use a pit or trench where the presence of contamination could result in pollution of a source of drinking water.

Solids, oils, and soaps—Water laden with solids, oils, and soaps can clog the soak pit. The wash area and drain require daily cleaning to reduce the amount of solids that enter the soak pit. Install a screen on the drain in the wash area to filter solids. Settling tanks or grease traps between the wash area and the soak pit reduce sediment and grease from entering the soak pit.

Single or multiple pits—Single pits or trenches need occasional cleaning or emptying. It may be advantageous to construct two smaller pits (or shorter trenches) and alternate between them so that one can be cleaned out while the other is in use.

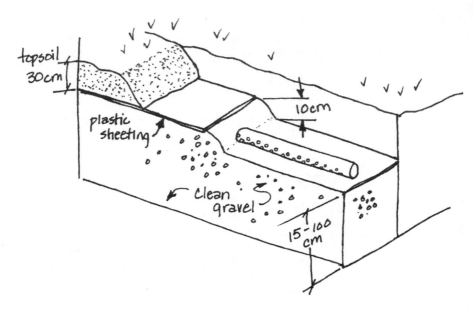

Figure 21-6. Shallow Infiltration Trenches Can Be Constructed in Areas with a High Water Table.

Source: Redrawn from Davis and Lambert (2002) with permission.

area floor finish should be smooth enough to enable cleaning, but it should have a broom finish to minimize slippage concerns. Refer to Chapter 7 for how to properly slope the wash area during construction.

21.3.2 Soak Pit Construction

Construct a soak pit by excavating to the calculated dimensions. (Refer to Chapter 7 regarding working in excavations.) If the soak pit area is sandy, the hole may need to be lined with bricks to support the sides (see Fig. 21-5). Place a layer of sand approximately 4 cm deep on the bottom of the pit. Place a 4-cm layer of gravel on top of the sand. Then fill the hole with large rocks. Place a large, flat rock below the pipe exit to encourage the greywater to spread throughout the soak pit. Maintain an open space at the pipe outlet, as shown in Figs. 21-2 and 21-5. In some instances, rocks may not be available, so the area filled with large rocks is often left open (this is referred sometimes as a dry pit). However, we recommend if possible, that broken tiles or brick be added to provide a surface for a biofilm to establish itself that will assist in breakdown of organic matter.

⟫⟫ 21.4 Maintenance

Remove shoes while using the wash area to reduce the amount of soil that enters the soak pit. Clean the screen on a regular basis, and sweep the surface daily. Remove trash from the area, which can quickly clog the drain and pipe.

The life of a soak pit is generally 3–5 years, after which it may need to be opened, cleaned, and refilled with clean sand and gravel, along with the original rocks (after scrubbing them). Wearing gloves prevents exposure to pathogens that may be present in the system.

))) Reference

Davis, J., and Lambert, R. (2002). *Engineering in emergencies*, ITDG Publishing, Rugby, Warwickshire, U.K.

))) Further Reading

Ahrens, B. T. *Step by step techniques for constructing a wash area and soak pit*, <http://cee.eng.usf.edu/peacecorps/Resources.htm> (Jan. 24, 2009).

Ahrens, B. T., and Mihelcic, J. R. (2006). "Making wastewater construction projects sustainable in urban, rural, and peri-urban areas," *J. Engineering for Sustainable Development: Energy, Environment, Health*, 1(1),13–32.

McConville, J. R., and Mihelcic, J. R. (2007). "Adapting life cycle thinking tools to evaluate project sustainability in international water and sanitation development work," *Environmental Engineering Science*, 24(7), 937–948.

22

Wastewater Lagoons

))) 22.1 Overview of Lagoons

Wastewater lagoons (sometimes referred to as *ponds*) are engineered to confine wastewater for treatment before it is discharged into a natural watercourse or reused. Sometimes they are referred to as *polishing, stabilization,* or *maturation lagoons* (or *ponds*), depending on the type of treatment used. Although they require more land area than mechanical treatment systems, their operational costs are low because the energy input is supplied by nature and they require no full-time staff.

Oxygen is provided by several methods: transfer of overlying air directly into the wastewater, mixing because of wind, and algal photosynthesis. Wastewater lagoons are well suited to tropical and subtropical climates because they function better in the presence of sunlight and at higher temperatures. They are effective in removing pathogens, and their discharge can be integrated into a plan of water reuse, especially for agriculture or aquaculture. In addition, synthetic covers can be placed on top of anaerobic lagoons to capture methane, which can then be used as source of energy.

Table 22-1 provides a description of the various types of wastewater lagoons. They are typically laid out in series in the following order: anaerobic lagoons, then facultative lagoons, and finally maturation lagoons. Not all three types are required for every situation. For example, it is not common to design for an anaerobic lagoon that only serves a community's need to treat municipal wastewater. The selection of the type of lagoons placed in series depends on wastewater characteristics and the required effluent quality. Anaerobic lagoons basically perform primary treatment, the facultative lagoon can be thought of as performing secondary treatment, and the maturation lagoon, tertiary treatment. Facultative lagoons can be easily incorporated into a water reuse plan that provides water for agriculture or aquaculture. Table 22-2 provides some general information on what types of lagoons are used for particular water reuse scenarios.

Table 22-1 shows that the various types of wastewater lagoons support different biological processes, influenced by the lagoon depth and whether the lagoon is mixed and oxygenated. They are designed in such a way as to remove primary wastewater constituents, including total suspended solids (TSS), biochemical oxygen demand (BOD), nutrients, and pathogens (see Table 22-3 for performance of facultative lagoons placed in series). Major mechanisms for pathogen removal in a lagoon include adsorption to particles, sedimentation, lack of food and nutrients, solar ultraviolet radiation, increase in temperature and pH, presence of microbial predators, straining and filtration in sediments, the presence

Table 22-1. Types of Lagoon Systems

Type of Lagoon	Description
Aerated	Not recommended because they require mechanical energy. However, solar and wind-powered aerators are available. Typically placed in series, in front of facultative lagoons (and not placed with an anaerobic lagoon). Aeration consists of either mechanical surface aerators or submerged, diffused aeration systems. Requires less area than a facultative lagoon and can operate effectively in colder climates. Weeds can grow because the lagoon is shallower than other types.
Anaerobic	Usually used to pretreat wastewaters, especially those of higher strength. Not commonly used with conventional domestic wastewater treatment. Functions much like a septic tank. Deep, nonaerated, and not mixed. Performance decreases at temperatures below 15 °C.
Facultative	Uses a combination of aerobic, anoxic, and anaerobic processes. Not mechanically mixed or aerated. Does not function well in colder climates. Also referred to as a stabilization, oxidation, photosynthetic, or aerobic–anaerobic lagoon.
Maturation	Also referred to as a *tertiary* or *polishing* lagoon. Primarily used to remove nutrients (N and P) and remaining pathogens.

Source: Adapted from Mihelcic and Zimmerman 2009.

of toxins and antibiotics excreted by some organisms, and natural die-off (Mihelcic and Zimmerman 2009).

))) 22.2 Design Considerations

If there is a high loading of grit, wastewater should be pre-treated with a manually cleaned bar screen and a horizontal grit chamber before routing it into the lagoon. Both of these pretreatment methods remove larger solids (Fig. 22-1). It is highly recommended that grit

Table 22-2. Types of Lagoons That Correspond to Particular Water Reuse Scenarios

Type of Reuse	General Rule on Types of Lagoons Required to Meet Water Quality Guidelines
Restricted irrigation—Includes irrigation of crops, except salads and vegetables eaten uncooked	Anaerobic lagoon followed by facultative lagoon
Unrestricted irrigation—Includes salads and vegetables eaten uncooked	Anaerobic lagoon followed by facultative lagoon and maturation lagoon
Wastewater-fed fish ponds	Anaerobic lagoon followed by facultative lagoon to maintain total nitrogen load into fish ponds of 4 kg N/ha-day. It is important to check pathogen and ammonia concentrations in the facultative lagoon effluent.

Note: Always refer to the most recent WHO guidelines on pathogen concentrations allowed in effluent for various uses and further guidelines for the safe use of treated wastewater. Peña Varón (2004) provides methods to determine the percent removal of human intestinal nematode eggs and *E. coli* as a function of lagoon retention time.

Table 22-3. Overall Removal Efficiency of Conventional Wastewater Constituents and Pathogens Treated by Facultative Lagoons Placed in Series

Wastewater Constituents and Pathogens	Removal Efficiency
BOD_5	70–95%
TSS	55–95%
Virus	2–4 log units
Bacteria	2–6 log units
Helminth eggs	2–4 log units
Protozoan cysts	2–4 log units

Note: Units of pathogen removal are in number of log units. Helminth eggs and protozoan cysts can be close to 100% removed because of long hydraulic retention times that maximize sedimentation.

Source: Data are from Oakley 2005a.

Figure 22-1. A Bar Screen Followed by a Horizontal Grit Chamber (a Sedimentation Tank for Large Particles) Can Be Used as Pretreatment When There Is a High Grit Load That Could Prematurely Fill the Lagoon. Both Are Manually Cleaned.

Source: Adapted with permission from Oakley (2005a).

chambers be installed to work on high sediment loads that can occur because of seasonal rainy seasons, lack of stormwater infrastructure, and topography.

Table 22-4 provides information for sizing the various types of lagoons. Lagoons are designed using guidelines for hydraulic detention time, organic loading rate, and depth. The organic loading rate is a measure of the mass of organic waste (measured in terms of BOD_5) added per day to a unit volume (or unit area) of a lagoon. The influent BOD_5 can be measured or estimated. Estimations are based on knowledge of wastewater influent characteristics found in environmental engineering textbooks. Be aware, though, that wastewater influent varies from community to community. We have observed a wide variety of ranges, and actual field data commonly do not follow what is observed in the United States or Europe. For example, wastewater may be of higher strength (i.e., higher BOD_5) because the waste stream is not diluted, either because of lower overall water usage or because a septic tank is used to pretreat the waste stream. On the other hand, BOD_5 could be lower because food waste is not usually routed to sewers, as it may be in parts of the world that have embraced garbage disposals.

Organic loading on a unit volume basis can be written as:

Organic loading (gm $BOD_5/m^3 \cdot day$) =

$$\frac{Flow\ (m^3/day) \times \left[\dfrac{organic\ concentration\ of\ incoming}{wastewater\ (mg\ BOD_5/L)}\right] \times \dfrac{1{,}000}{m^3} \times \dfrac{g}{1{,}000\ mg}}{lagoon\ volume\ (m^3)} \quad (22\text{-}1)$$

Equation 22-1 should be used with Table 22-4 to determine the appropriate lagoon volume (or lagoon area if organic loading is expressed in terms of per unit area). The area

Table 22-4. Design Considerations for Wastewater Lagoons

Type of Lagoon	Organic Loading (Units Vary and Are per Volume or Area)	Water Depth (m)	Hydraulic Detention Time (days)	BOD$_5$ Removal	Pathogen Removal
Anaerobic lagoon	100–350 g BOD$_5$/ m^3/day	2–5	20–50 Minimum of 1 day	60% at 20 °C 70% at 25 °C	Vibrio cholerae, helminth eggs
Facultative lagoon	15–80 kg BOD$_5$/ha/ day	1.2–2.4	20–180 Minimum of 4 days	70–90%	Vibrio cholerae, helminth eggs
Maturation lagoon		1–1.5	10–15 Minimum of 3 days	Small amount of BOD removal. Removes 80% of total N, 50% of total P	Viruses (especially rotavirus, astrovirus, norovirus), Bacteria (Salmonella spp., Shigella spp., pathogenic strains of E. coli.)

Sources: Data from Peña Varón 2004, the National Small Flows Clearinghouse, and Mihelcic and Zimmerman 2009.

of the lagoon can be determined by setting the depth based on the guidelines provided in Table 22-4, using Eq. 22-2.

$$\text{Lagoon area} = \frac{\text{Lagoon volume}}{\text{Table 22-4 depth}} \tag{22-2}$$

The hydraulic detention time (HRT) in days that must be satisfied equals

$$\text{HRT} = \frac{\text{Lagoon volume (m}^3)}{\text{Flow rate (m}^3/\text{day)}} \tag{22-3}$$

A lagoon can be round, square, or rectangular. Length-to-width ($L:W$) ratios are usually ≤ 3 and not <1. In all cases, the corners should be rounded to assist water movement.

Anaerobic lagoons are smaller than other lagoon types and receive a high organic loading. Because of the high organic loading, any oxygen in the wastewater is immediately consumed. Thus, the lagoons are void of oxygen. The primary removal mechanisms in anaerobic lagoons are sedimentation of solids and degradation of organic matter by anaerobic processes. They can produce odors (primarily hydrogen sulfide, H_2S, the smell of rotten eggs), especially if the influent sulfate concentration is greater than 500 mg SO_4^{2-}/L. It is thus best to locate this type of lagoon at least 500 m downwind from the nearest household. If the organic loading of an anaerobic lagoon is lower than the design guideline, the lagoon will not function properly because there will be oxygen present.

Figure 22-2 shows the various treatment zones found in a facultative lagoon and the physical, chemical, and biological processes that occur. An aerobic zone is located near the surface. It is aerated primarily by algal photosynthesis but also because of oxygen trans-

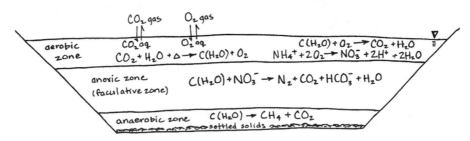

Figure 22-2. A Facultative Lagoon Consists of Three Zones for Wastewater Treatment.

Note: Oxygen is transferred to the upper aerobic zone via gaseous diffusion or algal photosynthesis. Here organic carbon (measured as BOD, written as $C(H_2O)$) is oxidized to CO_2, and ammonia nitrogen (written as NH_4^+) is oxidized to NO_3^-. In the middle anoxic zone, organic carbon oxidization is coupled to NO_3^- reduction via denitrification reactions. In the lower anaerobic zone, solids accumulate and fermentation reactions break down organic carbon to CH_4 and CO_2.

fer from the overlying air and wind-induced turbulence. Facultative lagoons should thus have a bright rich green color because of the presence of algae (Box 22-1). A turbid green pea color appearance (dull green) is not desired and indicates the presence of blue-green algae that block sunlight. The concentration of algae in a well-functioning lagoon is typically in the range of 500–1,000 µg chlorophyll-a per L. The presence of algae results in a diurnal variation in dissolved oxygen (mg O_2/L) and pH because with sunlight, algae can raise dissolved oxygen concentrations to as high as 20 mg/L and pH above 9.

As shown in Fig. 22-2, in the presence of oxygen, organic carbon (written as $C(H_2O)$, measured as BOD_5) is converted to carbon dioxide (CO_2), and ammonia nitrogen (NH_4^+–N or NH_3–N) is converted to nitrate (NO_3^-). In the process, biomass solids are produced (i.e., more algae and bacteria). An anaerobic zone forms on the bottom of the lagoon where the solids settle. This part of the lagoon supports anaerobic biological fermentation processes that convert $C(H_2O)$ into methane (CH_4) and CO_2. In between these two layers is an anoxic layer, which is also called the *facultative zone*. In this zone, denitrification reactions take place where NO_3^- is reduced to nitrogen gas (N_2) and is coupled to oxidation of $C(H_2O)$.

Typical organic loading values of facultative lagoons range from 15 to 80 kg/ha/day. One possible problem with facultative lagoons is that they cool off quickly and can experience a large reduction in biological activity during colder months (a problem for northern climates). In addition, algae may accumulate in the lagoon effluent, causing TSS to go well above 20–100 mg/L in a poorly designed lagoon. This problem is especially important when lagoons discharge into sensitive surface waters (e.g., saltwater bays) that contain ecologically critical reef systems. Figure 22-3 shows a typical layout of a lagoon system, with pretreatment, facultative lagoons, and maturation ponds. Maturation ponds typically follow facultative lagoons. They assist in further removal of pathogens and the nutrients nitrogen and phosphorus.

))) Box 22-1. Interpreting the Color of a Facultative Lagoon

Bright rich green—Indicates good conditions, with plenty of green algae. The green color is suspended throughout the water column.

Dull green or yellowish—Could mean that an undesirable type of algae (blue-green, filamentous algae) is becoming dominant in the lagoon. Blue-green algae are stringy and can clump, block sunlight, and cause hydraulic short-circuiting. They may exhibit a green pea soup color. They dominate lagoons when pH is low or when protozoa eat all the green algae. They can be physically removed.

Gray or black—Can indicate anaerobic conditions.

Tan, brown, or red—Can indicate soil in the water from bank erosion or the presence of algae with different pigmentation.

Source: From NSFC (n.d.).

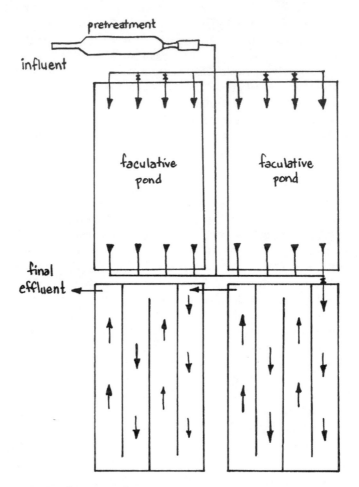

Figure 22-3. Typical Set-Up of a Lagoon System.

Note: Arrows indicate the movement of wastewater. A pretreatment step is followed by facultative lagoons and maturation lagoons. The facultative lagoons are designed in parallel so that one can be taken out of service every 5–10 years for solids removal. The maturation lagoons depicted here are designed with internal baffles that facilitate plug flow.

Source: Adapted with permission from Oakley (2005a).

))) 22.3 Construction

Lagoons can be excavated manually or with earth-moving equipment. They should be sited on flat topography. Their bottoms should be constructed as level as possible to facilitate continuous flow of wastewater, except near the inlet. If the soil has a permeability of greater than 10^{-7} m/s, a 250-mm compacted clay or synthetic liner should be placed under the lagoon. Internal embankments are typically constructed with a slope of $1:3$ (vertical to horizontal), and outer embankments are typically constructed with a slope $1:2$ (vertical to horizontal). Outer embankments are not always used; that is, the internal embankment can slope up to ground level. Native grass can be planted on the slope to minimize erosion. Stone riprap (i.e., large stones placed to reduce the erosion-causing energy of wave action) or another local material can be placed near the top water level on the internal embankment to prevent wind-induced wave erosion.

In terms of hydrology, the lagoon inlet and outlets should be located as far apart as possible and diagonally opposite one another to minimize hydraulic short-circuiting. Excessive wind can also cause hydraulic short-circuiting, so lagoon layout in terms of inlets and outlets should take into consideration local wind speed and direction. Inlets and outlets are usually designed with a scum box to remove scum. In addition to scum removal, outlets should also be designed to remove duckweed that may grow on the lagoon surface (Peña Varón 2004).

))) 22.4 Operation and Maintenance

Before start-up, all vegetation should be removed from the inside of the lagoon. Facultative and maturation lagoons are commissioned before the anaerobic lagoon is to prevent odor release from the facultative lagoons. During start-up, fill those lagoons with freshwater to permit development of algae and other bacteria. After the facultative and

Figure 22-4. Surface Plants, Scum, and Unwanted Algae Can Be Removed by Manual Cleaning with a Rake, Skimmer, or Net.

Note: Plants such as duckweed, water meal, and hyacinth can grow on the water surface. If allowed to grow, they will block penetration of sunlight, compete with algae for space, provide a breeding ground for mosquitoes, and minimize wind-induced aeration. Scum should be removed for similar reasons, and it may also clog inlets and outlets. Algae can multiply quickly and die off in cloudy weather or after abrupt temperature changes. Dead, matted algae on the surface can block penetration of sunlight and can also cause odors. Matted algae should be broken up with a boat or rake and dispersed. The grassy bank should be cut to discourage burrowing animals from destabilizing the bank and dikes.

Table 22-5. Relationship of Treatment to World Health Organization Guidelines for Reuse of Wastewater[a]

Category	Reuse Conditions	Exposed Group	Intestinal Helminths[b]	Fecal Coliforms[c]	Wastewater Treatment Expected to Achieve the Required Microbial Guideline
A	Irrigation of crops likely to be eaten uncooked, sports fields, public parks[d]	Workers, consumers, public	≤1	≤1,000	A series of lagoons designed to achieve the microbiological quality indicated, or equivalent treatment
B	Irrigation of cereal crops, industrial crops, fodder crops, pasture, and trees[e]	Workers	≤1	No standard recommended	Retention in lagoons for 8–10 days or equivalent helminth and fecal coliform removal
C	Localized irrigation of crops in Category B if exposure to workers and the public does not occur	None	Not applicable	Not applicable	Pretreatment as required by irrigation technology but not less than primary sedimentation

[a]In specific cases, local epidemiological, sociocultural, and environmental factors should be taken into account and the guidelines modified accordingly.

[b]*Ascaris* and *Trichuris* species and hookworms, given as arithmetic mean number of eggs per liter.

[c]Measured during the irrigation period, given as geometric mean number per 100 mL.

[d]A more stringent guideline limit (≤200 fecal coliforms/100 mL) is appropriate for public lawns, such as hotel lawns, with which the public may come into direct contact.

[e]In the case of fruit trees, irrigation should cease 2 weeks before fruit is picked (termed the *withholding period*), and fruits should not be picked off the ground. Sprinkler irrigation should not be used.

Source: From Oakley 2005a, courtesy of National Small Flows Clearinghouse.

maturation lagoons are commissioned, fill the anaerobic lagoon with untreated wastewater and inoculate them with sludge from a nearby anaerobic reactor, if available. Then ramp up the anaerobic lagoon to its design loading over a 2- to 4-week period (Peña Varón 2004). Periodically remove unwanted and dead algae, scum, and floating plants with a rake or skimmer (Fig. 22-4). Horizontal nets (similar to fishing nets) that span the width of the lagoon can be constructed to manually remove algae from a lagoon. These nets can be sewn with local materials from a hardware store with soda bottles filled with sand that hold the net down as two operators walk along opposite sides of the lagoon to skim the surface. Another option is to purchase ducks, which consume algae as food.

At a minimum, routine maintenance should include the following (Peña Varón 2004):

- removing screenings and grit from pretreatment structures;
- periodically cutting grass on embankments to discourage burrowing animals and removing intruding vines and other plants that could harm a lagoon liner;
- removing scum, unwanted algae types, and floating plants from the surface of facultative and maturation lagoons (especially important to allow light transfer to the lagoon for photosynthesis);
- examining lagoon inlets and outlets to remove any material blocking wastewater flow;
- repairing embankments that are damaged by burrowing animals such as rodents or rabbits; and
- removing sludge from the anaerobic lagoon every year or at a minimum every 2–5 years when the lagoon is one-third full of sludge

Facultative lagoons are designed to store sludge over their design life, but they may require solids removal every 5–10 years (or more frequently if a grit chamber is not used). Lagoons should always be designed and operated with consideration of reusing the treated wastewater. This is especially important in water-scarce regions. Table 22-5 shows the relationship of treatment to guidelines set by the World Health Organization (2006) for reuse of wastewater in the agriculture sector and for landscaping.

))) References

Mihelcic, J. R., and Zimmerman, J. B. (2009). *Environmental engineering: Fundamentals, sustainability, and design*, John Wiley and Sons, New York.

National Small Flows Clearinghouse (NSFC). (n.d.) <http://www.nesc.wvu.edu/nsfc/nsfc_index.htm> (Jan. 20, 2009).

Oakley, S. M. (2005a). "A case study of the use of wastewater stabilization lagoons in Honduras." *Small Flows Quarterly*, Spring, 6(2), 36–51.

Peña Varón, M. R. (2004). *Waste stabilization ponds for wastewater treatment*, IRC International Water and Sanitation Centre, Delft, Netherlands <http://www.irc.nl/content/download/13614/186208/file/WSPs.pdf> (Jan. 20, 2009).

World Health Organization (WHO). (2006). WHO guidelines for the safe use of wastewater, excreta and greywater, Vol. II: Wastewater use in agriculture, World Health Organization, Geneva, Switzerland, <http://whqlibdoc.who.int/publications/2006/9241546832_eng.pdf> (Oct. 1, 2008).

))) Further Reading

Crites, R., and Tchobanoglous, G. (1998). *Small and decentralized wastewater management systems.* McGraw-Hill, New York.

Nelson, K., Cisneros, B. J., Tchobanoglous, G., Darby, J. L. (2004). "Sludge accumulation, characteristics, and pathogen inactivation in four primary waste stabilization ponds in central Mexico," *Water Research,* 38; 111–127.

Oakley, S. M. (2005b). *Lagunas de estabilización en Honduras: Manual de diseño, construcción, operación y mantenimiento, monitoreo y sostenibilidad,* U.S. Agency for International Development—Honduras (USAID—Honduras), Red Regional de Agua y Saneamiento de Central América (RRAS-CA), Fondo Hodureño de Inversión Social (FHIS), Tegucigalpa, Honduras, June. (*Design and operations manual for wastewater stabilization ponds in Honduras,* published in Spanish).

U.S. Environmental Protection Agency (EPA). (n.d.). *Wastewater technology fact sheets,* Washington, D.C., <http://www.epa.gov/owm/mtb/mtbfact.htm> (Jan. 20, 2009).

23

Constructed Wetlands and Evapotranspiration Beds

〉〉〉 23.1 The Differences from Natural Wetlands

Constructed wetlands and evapotranspiration beds are different from natural wetlands because the ecosystems of plants and soil microbes are greatly simplified (plants are often monocultures); design and operation can be modified and controlled to meet higher pollutant capacity, climate flexibility, and treatment requirements; and they are typically lined with an impermeable liner so that contaminants are hydraulically separated from groundwater or surface water until treatment standards are reached.

A septic tank, oxidation pond, or other treatment system is placed before the natural system to reduce influent solids, except for the method of applying raw sewage to the wetland surface. To prevent premature clogging, the hydraulic retention time of the septic tank should be >1.5 days and the tank should be emptied routinely. Wetland effluent may undergo additional treatment or may be discharged to surface water or groundwater.

The two types of constructed wetlands are the free-water surface (FWS) wetland and the subsurface flow (SSF) wetland. Evapotranspiration (ET) beds are a type of SSF wetland where the effluent flow is zero. FWS wetlands (also referred to as *surface flow wetlands*) are similar to natural open-water wetlands in appearance and treatment mechanisms. Figure 23-1 shows an FWS wetland with its three zones of treatment.

In an FWS wetland, the majority of the surface area has aquatic plants rooted below the water surface. Water travels in deep sheet flow over the soil (or sand) and through the plant stems. The two zones with significant vegetation (Zones 1 and 3) typically have dissolved oxygen levels close to 0 mg/L. The area with no surface vegetation (Zone 2) is exposed to sunlight and open to the air to increase oxygen transfer. Zone 2 may have submerged aquatic plants to enhance the dissolved oxygen content. Wetland influent can be aerated by setting the inlet distributor above the water level so that the water cascades down.

SSF wetlands (also referred to as *reed beds* and *vegetated submerged beds*) use gravel as the aquatic plant rooting medium (Fig. 23-2). The water level is intentionally maintained below the gravel surface. Several hydraulic regimes may be used: horizontal flow, vertical

---

This chapter was written by Valerie J. Fuchs with contributions from Kelly L. Stanforth, Ed Stewart, Ryu Suzuki, and Glenn A. Vorhes.

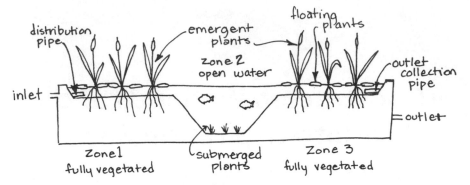

Figure 23-1. The Free-Water Surface (FWS) Wetland. Zones 1 and 3 Are Planted with Vegetation.

Note: Here, water travels in deep sheet flow over the soil and through the plant stems. Zone 2 has no surface vegetation, and water is exposed to sunlight and is open to air to increase the potential for oxygen transfer.

upflow, and vertical downflow. Horizontal flow is typical for continuous, gravity-fed systems (Fig. 23-3). Vertical-flow wetlands achieve good oxygen transfer for biological nitrification by using an aeration mechanism or by using control valves to alternately drain parallel beds, which allows void spaces to fill with air. The need for control systems makes vertical-flow wetlands mechanically more complicated, and design parameters are still not well under-stood. Therefore, vertical-flow wetlands will not be discussed further, and the term "SSF wetland" will be used here to represent a horizontal subsurface flow wetland.

Table 23-1 provides a comparison of FWS and SSF wetlands. FWS wetlands can be designed to provide long-term nitrogen removal because of the aerobic open-water zone that allows biological nitrification. However, this situation is only the case for wetlands

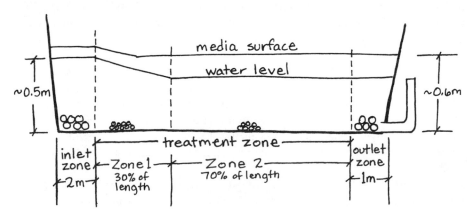

Figure 23-2. The Subsurface Flow (SSF) Wetland.

Note: Water flows from the inlet zone (left) toward the outlet zone.

Source: Adapted from EPA 2000.

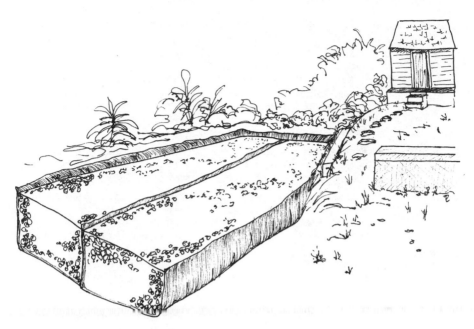

Figure 23-3. Example of an SSF Wetland Before Planting.

Note: The wetland has two parallel plastic-lined rock media beds, each with inside dimensions of 62.5 ft (19.1 m) length × 15.5 ft (4.7 m) width × 1.6 ft (0.5 m) depth of medium. Medium is irregularly shaped washed stone ordered as 0.5-in. river shingle. The bulk of the stone ranges in size from 0.25 to 1 in. (6–25 mm) diameter. The media porosity was measured to be 0.37 without plant roots. The black plastic liner running down the middle separates the two beds. The wetland will be planted with wild cane (*Gynerium sigattatum*).

that are shallow or well-mixed. In deeper water zones, good top-to-bottom mixing may not occur, so deeper water may not exhibit nitrification. SSF wetlands do not typically provide long-term nitrogen removal without plant harvesting or oxygenation of the water by means of cascading or mechanical aeration.

FWS and SSF wetlands should be lined if the surrounding soil is highly permeable. This procedure will ensure that untreated wastewater does not contact underlying groundwater before treatment. Bed liners may consist of compacted native clay, asphalt, concrete, or geosynthetic (plastic) membranes. Clay liners require careful compaction to ensure adequate thickness throughout the bed. Plastic liners are more difficult to install, but training local laborers to install and repair the plastic liner may ensure the best technology transfer and cost savings over time. The typical size for liners in the United States is 30 mil (1 mil equals one thousandth (10^{-3}) of an inch or 0.0254 mm). However, 10–20 mil liners are available in developing countries. If you are placing a plastic liner over a rocky wetland bed, place 2–5 cm of sand between the bed and liner to protect the plastic and reduce the possibility of punctures. Place another layer of sand on top of the plastic for an SSF if the wetland medium is sharp gravel. Any liner should overlap the earthen berm or wall surrounding the wetland to retain the wastewater within the wetland and to prevent surface water runoff from entering.

Table 23-1. Comparison of Free-Water Surface and Subsurface-Flow Wetlands

Characteristics	Free Water Surface	Subsurface Flow
Wastewater exposure	Open-water aerobic zone (Zone 2 in Fig. 23-1) enhances biological nitrification and provides wildlife habitat. Surface water allows mosquito breeding and requires isolation for public health safety.	Wastewater remains 2–4 in. below the media surface. There is no surface water to attract aquatic birds, little risk of human exposure, and no mosquito breeding.
Hydraulics	Not likely to overflow from an accumulation of solids at the inlet.	Surface flooding will occur at the inlet (see Fig. 23-2) if there is excessive accumulation of solids.
Bedding	The sand or sandy loam plant rooting medium has a lower cost than the gravel medium used in SSF wetlands.	The rock plant rooting media should be hard rock or stone-washed to minimize fines, with a diameter ranging between 6 and 38 mm (0.25–1.5 in.). This is more expensive than FWS wetland media. Larger rocks of 50–150 mm (2–6 in.) diameter may be used at the entry and exit zones to facilitate fluid dispersion. Because these stones are too large for the plants to use as stable rooting media, this size rock will prevent plant roots from clogging the inlet and outlet pipes (EPA 2000). However, the availability of good rock at an affordable cost may dictate the exact size used.
Effluent zone	FWS wetlands typically release effluent to a surface water body (e.g., riparian area, stream, river, lake), but they can also be designed to flow to a soil absorption or infiltration system.	SSF wetlands typically release effluent to a groundwater infiltration system, such as a leach field or other soil absorption system, but can also release to surface water.
Dimensions	Recommended aspect ratio (L : W) is 5 : 1 to 10 : 1. Depth of water may range from >6 cm in vegetated zones (Zones 1 and 3 in Fig. 23-1) to >1 m in open water zones (Zone 2).	Recommended aspect ratio is in the range of 0.25 : 1 to 1 : 1. Depth of gravel may be 35–70 cm.
Bed bottom	≤3% grade	≤0.5% grade

FWS wetlands typically have a much higher length-to-width ratio than SSF wetlands because filtration by SSF wetland media requires lower solids cross-sectional loading at the inlet. FWS wetlands are suited to treat oxidation pond effluent (Chapter 22) because the high algal solids exiting the pond do not cause the wetland inlet to clog and overflow its banks. The gravel and plant roots in an SSF wetland function as a filter and surface for attached biological growth treatment. Other primary effluents (besides that from a septic tank) may also be applied to an SSF wetland, as long as the wetland cross-sectional area is designed to handle the solids loading. In larger communities, a primary settling basin may handle the solids load before passing the clarified effluent to the SSF wetland for biochemical oxygen demand (BOD), nitrogen, phosphorus, and pathogen removal.

Solids accumulation eventually alters hydraulics and affects system performance. Over time, solids removal may be required. If this is a concern, the wetland may be designed with larger rock sizes or larger bed sizes to reduce the accumulated effect of solids buildup. It is difficult to provide an estimation of when solids accumulation will pass from a nuisance to a problem. The time period could be several or many years, depending on the solids loading and design. One study determined that the average long-term sediment accretion for a wide range of FWS wetlands in North America was approximately 6 mm/year, and a 5–10 mm/year accumulation rate was recommended for design (Keller and Knight 2004). Wetland sludge is a good fertilizer, but care should be taken to eliminate human exposure to pathogens. For wetlands larger than single-family applications, it is recommended that there be two or more parallel trains of wetlands so that one can be taken out of service and drained while the other continues to treat wastewater.

The bed, liner, and earthen berm may require occasional maintenance. The bed should be inspected periodically for clogging. If clogged, the bed can be drained and dried (organics will dry out and reduce clogging) or flooded to flush out accumulated sediments.

))) 23.2 Treatment of Wastewater Constituents

Constructed wetlands are a widely accepted technology for reducing pathogens, BOD, and total suspended solids (TSS). Wetlands can also be designed to treat nitrogen and phosphorus, with some limitations. Table 23-2 shows typical reduction of BOD, TSS, ammonia nitrogen, and phosphorus that can be expected in FWS and SSF wetlands.

FWS wetlands typically have a higher measured TSS influent loading in the field than SSF wetlands, possibly reflecting their use after oxidation ponds and the required

Table 23-2. Removal of BOD, TSS, Ammonia Nitrogen (NH_4^+), and Phosphorus (e.g., $H_2PO_4^-$) Observed in 170 FWS Wetlands and 1,329 SSF Wetlands Located in 19 Countries

	Free-Water Surface Wetland			Subsurface Flow Wetland		
	Influent Concentration (mg/L)	Effluent Concentration (mg/L)	% Removal	Influent Concentration (mg/L)	Effluent Concentration (mg/L)	% Removal
BOD	172	12	93	155	11	93
TSS	185	14	91	50	14	72
Ammonia nitrogen	32	3.8	88	48	2.9	94
Phosphorus	3.8	1.8	53	8	2.8	65

Note: Values are taken from the 50th percentile of the total number of FWS wetlands and SSF wetlands examined.

Source: Data from Wallace and Knight 2004.

pretreatment of TSS before treatment by an SSF wetland. SSF wetlands are used far more commonly for small-scale applications, whereas FWS wetlands are used for larger applications. For Table 23-2 data, the 50th percentile design flow for SSF wetlands was 2.6 m^3/day (approximately the wastewater flow of a 4-bedroom house in the United States), and the 50th percentile design flow for FWS wetlands was 389 m^3/day.

23.2.1 Pathogens

Pathogens are removed in surface wetlands primarily by sedimentation and predation. They also die if the hydraulic residence time is longer than the organism lifespan. UV exposure kills pathogens in FWS wetlands that are designed with open-water zones. Filtration is the primary removal mechanism in SSF wetlands.

The database of wetlands operating in the United States and Canada indicates that SSF wetlands and FWS wetlands with open zones provide a 2-log (99%) reduction in fecal coliforms. Fully vegetated FWS wetlands (i.e., no open-water Zone 2) demonstrated slightly poorer removal of pathogens with an average of 1–2 log (90–99%) reduction (EPA 2000). Other studies have shown similar results: SSF wetlands provide >99% reduction in bacteria and viruses, whereas FWS wetlands typically provide >90% reduction of bacteria and viruses (Chendorain et al. 1998; Vacca et al. 2005). Higher temperatures and increased sunlight found in tropical or arid climates of many developing countries should result in similar to better removal of pathogens. Waterfowl excreta in an FWS wetland may increase fecal coliforms in the effluent.

23.2.2 Biochemical Oxygen Demand and Total Suspended Solids

BOD and TSS are removed primarily by sedimentation, filtration, and consumption by microorganisms and plants. FWS wetlands in the United States and Canada that have open-water zones produce measured effluent concentrations below 30 mg/L BOD and 30 mg/L TSS when the areal mass BOD loading is maintained below 60 kg/ha-day (54 lb/acre-day), and the areal mass TSS loading is maintained below 50 kg/ha-day (45 lb/acre-day) (Table 23-3). The BOD (or TSS) loading equals $Q \times$ wetland influent BOD (or TSS) concentration.

Table 23-3. Maximum Wetland Areal Mass Loading Rates for BOD, TSS, and Inorganic Nitrogen and the Expected Effluent Concentration

Constituent	Free-Water Surface Wetland Loading	Subsurface Flow Wetland Loading	Effluent Concentration
BOD	60 kg/ha-day (54 lb/acre-day)	60 kg/ha-day (54 lb/acre-day)	30 mg/L
TSS	50 kg/ha-day (45 lb/acre-day)	200 kg/ha-day (178 lb/acre-day)	30 mg/L
Ammonia nitrogen (measured as TKN)	5 kg/ha-day (4.5 lb/acre-day)	NA	10 mg/L

Note: NA means not applicable.

Source: EPA 2000.

As also shown in Table 23-3, SSF wetlands in similar climates have been measured to produce effluents below 30 mg/L BOD and 30 mg/L TSS when maximum BOD loading (averaged over a month) remains below 60 kg/ha-day (54 lb/acre-day) and the TSS loading remains below 200 kg/ha-day (178 lb/acre-day) (EPA 2000).

The higher temperatures in tropical or arid climates of many developing countries will be advantageous to biological activity, which should improve BOD removal. SSF wetlands function in cold climates because the subsurface helps to regulate the temperature.

23.2.3 Nutrients (Nitrogen and Phosphorus)

Plants take up nitrate (NO_3^-), ammonia (NH_4^+), and phosphorus (found as polyphosphates, e.g., $H_2PO_4^-$) from water as they grow, and organic nitrogen and phosphorus return to the water when the plants die or slough off roots (Box 23-1). Rapidly growing young plants take up inorganic phosphorus, ammonia, and nitrate faster than mature plants. Fresh media and soil will adsorb ammonia and inorganic phosphorus until all the adsorption sites are saturated (Koottatep and Polprasert 1997; Reed et al. 2001). Therefore, inorganic nitrogen and phosphorus removal may be greater in the early years of a wetland and taper off when the media are saturated and the plant community matures. Shale or clay-based media may help phosphorus adsorption (Forbes et al. 2004). No routine cutting or harvesting is required to operate a wetland, but regular harvesting does increase nutrient removal (Koottatep and Polprasert 1997; EPA 2000).

))) Box 23-1. Biochemical Reactions

Chemoheterotrophs are organisms that use organic matter (i.e., $C(H_2O)$) as a carbon and energy source and, under aerobic conditions, consume oxygen in obtaining that energy:

$$C(H_2O) + O_2 \rightarrow CO_2 + H_2O + \Delta$$

Chemoautotrophs are organisms that use CO_2 as a carbon source and inorganic matter as an energy source, and they usually consume oxygen in obtaining that energy. An example of chemoautotrophy is nitrification, the microbial conversion of ammonia nitrogen to nitrate.

$$NH_4^+ + 2HCO_3^- + 2O_2 \rightarrow NO_3^- + 2CO_2 + 3H_2O + \Delta$$

For complete nitrogen removal, a variety of bacteria, including those of the genus *Pseudomonas*, can convert nitrate to nitrogen gas (N_2) (the process is termed *denitrification*). Assuming the biodegradable organic material (measured as carbonaceous biochemical oxygen demand, CBOD) can be written as $C_{10}H_{19}O_3N$, the removal of nitrate to nitrogen gas can be written as:

$$C_{10}H_{19}O_3N + 10NO_3^- \rightarrow 5N_2 \text{ (gas)} + 10CO_2 + 3H_2O + NH_3 + 10OH^-$$

Source: Adapted from Mihelcic and Zimmerman 2009.

Most of the heavily vegetated sections of FWS wetlands (Zones 1 and 3 in Fig. 23-1) and all the SSF wetland flow regimes have some anoxic water environments that support denitrification. In an SSF wetland, long-term nitrogen removal (via denitrification of aqueous nitrate to nitrogen gas) may be achieved by biological growth processes that take place on plant roots and in the gravel medium. In FWS wetlands, there may be sufficient oxygen at plant root tips to nitrify ammonia to nitrate, but most of the wastewater does not touch the root tips unless the medium is shallow enough that plants roots grow throughout the depth of the flow regime. Therefore, FWS wetlands with open areas that provide oxygen transfer are the only constructed wetlands that provide nitrification. In FWS wetlands, inserting islands or hummocks of plants and soil in the deep-water zones causes water movement and turbulence, allowing for better mixing and oxygen transfer.

Long-term phosphorus removal after the initial plant growth and media adsorption phase may be slight and may happen only though sedimentation. Wetland plants accumulate phosphorus during the plant growth phase, but this phosphorus is released again when the plants die. If plants are harvested regularly or soil is replaced when its adsorption capacity for phosphorus has decreased, there can be renewed phosphorus removal. In both FWS and SSF wetlands, the phosphorus in the sediments deposited on the bottom of the wetland or within the subsurface media remains in the wetland until the settled solids (i.e., sludge) are removed or disturbed.

))) 23.3 Design Considerations

This section lays out some aspects of wetland design, hydraulic design, wetland sizing, selection of plants, and consideration of disease vectors. This list (adapted from Crites and Tchobanoglous 1998) provides important steps in the design process of FWS and SSF wetlands:

1. Define treatment requirements.
2. Characterize wastewater.
3. Gather background information.
4. Evaluate the site.
5. Determine pretreatment level.
6. Select vegetation.
7. Determine vector control measures.
8. Determine design parameters.
9. Design system components.
10. Determine monitoring requirements.

23.3.1 Site Evaluation

A site evaluation provides information for the size and layout of the wetland design. It may also help to determine local vegetation that could be used. The site evaluation may also control the aspect ratio (AR) (length-to-width ratio) that was discussed in Table 23-1. Siting the wetland near the community or household producing the wastewater reduces the cost of piping, and placing it near agricultural users creates the potential for use of reclaimed water. In the absence of reliable wastewater characterization, septic tank and oxidation pond effluent can be estimated (Table 23-4).

Table 23-4. Typical Effluent Characteristics That May Be Used for Estimating Constructed Wetland Influent Strength

Constituent	Septic Tank Effluent (mg/L)	Oxidation Pond Effluent (mg/L)
BOD	129–147	11–35
Soluble BOD	100–118	7–17
COD	310–344	60–100
TSS	44–54	20–80
Volatile suspended solids	32–39	25–65
Total nitrogen	41–49	8–22
Ammonia, NH_3, and NH_4^+	28–34	0.6–16
Nitrate, NO_3^-	0–0.9	0.1–0.8
Total phosphorus	12–14	3–4
Orthophosphate	10–12	2–3
Fecal coliform (CFU/100 mL)	$10^{5.4}$–$10^{6.0}$	$10^{0.8}$–$10^{5.6}$

Source: EPA 2000.

Constructed wetlands should be outside or above river floodplains, and placement should provide gravity inflow and outflow. There must be additional room available for any required pretreatment. There must also be a location for wetland discharge, either on site through a leach field or into nearby surface water. The slope of the site may constrain the wetland type. If there is a 0–3% slope, FWS wetlands are recommended, whereas SSF wetlands should have a near 0% bottom slope. If the site is hilly or mountainous, small SSF wetlands may be possible, with each wetland placed in series down the slope (similar to a terrace approach).

Earthen berms, 15–25 cm high, are usually constructed around the perimeter of the wetland bed. Soil excavated to form the bed may be used but requires compaction. Compact the berm uniformly to minimize the chance that any part will wash away from runoff. The berm may be replaced by a wall or other structure designed to provide 15–25 cm of freeboard above the wetland, protecting it from surface runoff and providing storage during high rainfall or flooding. Freeboard is the distance from the water line to the top of the berm.

23.3.2 Plant Selection

Some typical wetland plants are listed in Table 23-5. The table indicates whether the plant will survive best in an FWS or an SSF wetland (based on tolerance for water depth, rooting depth, and need for soil base). Indigenous aquatic plants are recommended. They should be selected based on growth habit, tolerance to flooding, and tolerance to wastewater. Some mature aquatic plants can survive in deeper water than others, and some can better sprout new shoots in deeper water. Cattail and bulrush can successfully sprout new shoots in water 27 cm (10.5 in.) deep and probably deeper. However, the sprouts of some reeds

Table 23-5. Typical Plants Used in FWS and SSF Wetlands

Common Name	Scientific Name	FWS	SSF
Bulrush	*Scirpus spp.*	X	
Cattail	*Typha latifolia*	X	
Common arrowhead	*Sagittaria latifolia*	X	
Papyrus	*Cyperus papyrus L.*	X	
Torpedo grass	*Panicum repens L.*		X
Common reed	*Phragmites australis*		X
Rush	*Juncus spp.*		X
Sedge	*Carex spp.*		X
Yellow flag	*Iris pseudacorus*		X
Umbrella palm	*Cyperus alternifolius*		X
Wild cane	*Gynerium sigattatum*		X

Note: Indigenous plants should always be given preference. The final use of a plant (e.g., animal fodder, roof thatch) may also be a consideration in plant selection.

do not emerge through as little as 14 cm (5.5 in.) of water (Borst et al. 2002). Decorative or dwarf varieties such as soft rushes (*Juncus balticus, Juncus effuses*), umbrella palms (*Cyperus alternifolius, Cyperus papyrus*), woolgrass (*Scirpus cyperinus*), and graceful cattail (*Typha laxmanii*) have also been used to create attractive wetlands (Schellenberg 2001). Wild cane (*Gynerium sigattatum*) grows to more than 3–4 m high and has been successfully used in Jamaica (Stewart 2005).

Easy access to the plants should be a design feature if harvesting is planned. Harvesting emergent plants in a small SSF wetland is relatively easy. The labor involved in cutting vegetation in larger SSF and FWS wetlands will likely result in irregular or no harvesting, unless there is value in the cuttings (e.g., animal fodder, roof thatch). If plants are grown in an SSF wetland for human benefit (e.g., pumpkin, bamboo), it will be important to keep these plants away from the inlet end because of the potential for this area to experience surface flooding. Phragmites are extremely common in European constructed wetlands because of their high growth rate and market value. They are sold for use as roofing thatch and fiber for the paper industry. Harvesting the plants is also a removal mechanism for nutrients. Papyrus (*Cyperus papyrus L.*), a floating plant with fiber value, and torpedo grass (*Panicum repens L.*), a plant that has animal fodder value, have been used with positive results in Uganda (Denny 1997).

If plants are harvested for animal fodder, steps should be taken to restrict animals from the wetland and allow them to feed only on harvested plants. Harvested vegetation can also be dried and burned for fuel, chopped and composted for fertilizer, or chopped and used as mulch. The harvesting schedule may be once a year or based on labor availability, water quality requirements, plant growth rates, and harvest value. Floating plants, such as hyacinth, may be harvested from the bank with a pole fitted with a hook on the end, but care must be taken to avoid human contact with the wastewater-drenched plants. In SSF wetlands, plants should be cut in a way that allows new growth (typically leaving several centimeters of stalk).

Consideration should be given to whether animals (e.g., goats, cattle, waterfowl) or people may be attracted to the leaves, stalks, or fruit (seeds), and what consequences this may have. Small rodents may burrow into SSF wetland media or through the surrounding earthen berm to eat vegetation, causing leakage or reduced treatment. Unless the SSF is designed for animal habitat, a fence or barrier that extends below the ground surface can be installed to deter rodents and foraging animals.

FWS wetlands cannot be made mosquito-free, but breeding can be controlled to the same level as local wetlands, and the local community should be educated on this issue (Wallace 2005). The mosquito fish *(Gambusia affinis)* is widely used to control wetland mosquito populations and can survive in temperatures ranging from 0.5 to 42 °C, pH from 5 to 9.5, and dissolved oxygen levels as low as 0 mg/L (Knight et al. 2003). Because of the risk of introducing an invasive species, the mosquito fish may not be a prudent option. Plants bunched in hummocks rather than uniform thick stands allow predators (fish, dragonflies, etc.) better access to mosquito larvae. Periodic flooding may also flush stagnant areas used by breeding mosquitoes. Because SSF wetlands are designed to not have open-water areas, the habitat for mosquitoes is eliminated.

Emergent plant shoots or *rhizomes* (root sections) should be inserted into the soil or medium at a spacing of 75–100 cm before the wetland is filled with water in an FWS wetland, or with the water drawn down in an SSF wetland. The shoots should grow and spread on their own. For most species, there will need to be water level control when starting operation to allow new growth to rise above the water level. For FWS wetlands, loamy or sandy soils are preferable because emergent wetland plants adapt to them most easily. Sediment accretion above the soil continues to provide nutrients for plant growth.

23.3.3 Inlet and Discharge Requirements

The inlet pipes of FWS wetlands may be placed so that the influent flow can be visually observed and adjusted to ensure even distribution across the width of the bed. This goal may be achieved with a distribution pipe placed under the water surface of an open-water inlet zone, or the inlet distribution may cascade with open-air discharge. SSF wetland inlet pipes should be below the water surface and preferably deep enough so that plant roots cannot clog the inlet holes or slits. Inlet structures may be gated pipe, slotted pipe, or V-notched weirs that drain into an entry zone of large rock (to minimize clogging) to distribute the flow across the wetland inlet. Outlet structures may consist of perforated pipe submerged to the bottom of the wetland, or a gravel ditch, draining to an outlet pipe. Outlet pipes should be adjustable in height to adjust the water level in the wetland. Risers on the ends may be helpful for removing any obstructions or solids that may accumulate over time.

23.3.4 Further Design Considerations for Free-Water Surface Wetlands

Hydraulic Considerations and Organic Loading

Proper hydraulic design takes into account the maximum expected volumetric flow rate, maximum expected precipitation, bed geometry, construction materials available, site topography, bed volume loss from solids accumulation, and increase in flow resistance caused by accumulation of solids.

Inlet and outlet piping, wetland bed shape, solids accumulation, plant placement, and media all influence the path of water. Short-circuiting causes unexpected reduction in treatment efficiency. A wetland built as a plug flow reactor provides the most efficient treatment with the least amount of short-circuiting. This goal may be achieved by placing several wetland beds in series because the flow must mix when passing out of the previous wetland bed and into the next. The construction of a deep open-water section across the full width of an FWS wetland also improves mixing (see Zone 2 in Fig. 23-1). Wetland beds in series may suffer excessive buildup of solids at the inlet of the first bed. This problem may be addressed in an FWS wetland by increasing the depth of the entry zone.

The calculation of head loss along the length of an FWS wetland is sometimes not necessary because a typical FWS wetland with a recommended 5:1 to 10:1 AR may have a hydraulic gradient slope of only 1 cm in 100 m (EPA 2000). However, the head loss for the entire wetland length should be smaller than the difference between the inflow and outflow elevation.

Head loss (h_L) is determined based on the hydraulic gradient slope, water velocity, and Manning's n for roughness, as shown in Eqs. 23-1 to 23-3.

$$h_L = s \times L \tag{23-1}$$

$$s = \left(v \times \frac{n}{d^{2/3}} \right)^{0.5} \tag{23-2}$$

$$n = \frac{a}{d^{0.5}} \tag{23-3}$$

In the above equations, h_L is head loss (m); s is hydraulic gradient slope (m/m); L is wetland length (m); v is water velocity (m/s); n is Manning's roughness coefficient (unitless); d is water depth (m), and a is the resistance factor based on surface roughness. These values can be obtained from engineering design books referenced at the end of this chapter.

There is significant complexity in the biological, physical, and chemical mechanisms responsible for pollutant removal in constructed wetlands, so the proposed design models do not always fit all the collected data. Table 23-3 presents maximum mass loading rates for BOD, TSS, and ammonia nitrogen that should reliably keep a wetland's effluent below the listed effluent concentration. The use of maximum areal loading guidelines and recommended hydraulic retention times (HRTs) makes sizing for TSS, BOD, and nitrogen at times an iterative design process.

A typical FWS wetland designed for treating BOD and nitrogen has a heavily vegetated inlet zone to capture suspended solids and consume BOD, an open-air aerobic zone to improve biological nitrification, and a heavily vegetated outlet zone to denitrify (Box 23-2). There may be multiple vegetated and open zones to achieve the desired treatment objectives. The following information (EPA 2002) can help size an FWS wetland:

- The first vegetated zone (Zone 1 in Fig. 23-1) acts as an anaerobic settling chamber. The HRT should be 1–2 days.
- The HRT of the open zone (Zone 2 in Fig. 23-1) should be less than the amount of time required for algae to form. It depends on climate and temperature as

》》 Box 23-2. Five Steps to Design Free-Water Surface Wetlands

1. Determine the limiting effluent requirements for BOD, nitrogen, or pathogens.
2. Calculate the surface area using the Kadlec and Knight model (Eq. 23-4, below) for BOD, nitrogen (N), or pathogens (the largest area will control the design); increase by a 20% safety factor if you desire.*
3. Select the aspect ratio (AR) (see Table 23-1) based on site constraints. Calculate surface dimensions.
4. Check the head loss (Eq. 23-1) to ensure that it is smaller than the elevation difference between the inflow and outflow points. This amount allows continuous flow.
5. Design Zones 1 through 3 based on HRT, volume, flow rate, and calculated length and width.

*Be cautious when using unsubstantiated safety factors to increase the wetland size. Safety factors increase the materials, cost, and labor associated with wetland construction and operation.

well as nutrient limitations. In the United States and Canada, this time is typically 2–3 days.

- The HRT for the second vegetated zone (Zone 3 in Fig. 23-1) is typically 1 day to achieve denitrification.

For the three zones in total, a minimum HRT is 4 days. The vegetated and open zones can be determined as the last step in FWS design, based on this minimum combined HRT, the wetland design volume, and calculated area.

Kadlec and Knight Model for Sizing FWS and SSF Wetlands

The Kadlec and Knight model is commonly used to size wetlands (Example 23-1) (Kadlec and Knight 1996). It can be applied to both FWS and SSF wetlands. This method first determines whether the treated effluent needs to meet specific effluent constraints for BOD, coliforms, and/or nitrogen. The required wetland area (A) is determined separately based on each applicable effluent constraint, using Eq. 23-4. The largest of these calculated areas is then selected.

$$A = \left(\frac{0.0365 \times Q}{k_T} \right) \times \ln\left(\frac{Ci - C^*}{Ce - C^*} \right) \tag{23-4}$$

In Eq. 23-4, A is the wetland area required (hectares); Q is the volumetric flow rate (m³/day); k_T equals a rate constant for BOD, nitrogen, or pathogen removal at a specific temperature T (m/day); Ci is the influent concentration of BOD, nitrogen, or pathogens; C^* is the background natural concentration of BOD, nitrogen, or pathogens; and Ce is the effluent target concentration of BOD, nitrogen, or pathogens. Units for concentrations are mg/L for BOD and nitrogen, or coliforms per 100 mL. Typical values for the constants in Eq. 23-4 are presented in Table 23-6.

Table 23-6. Typical Values of Area-Based Constants Used for Kadlec and Knight Wetland Kinetics Model (Presented in Eq. 23-4)

Parameter	FWS k_{20} (m/year)	SSF k_{20} (m/year)	FWS θ (unitless)	SSF θ (unitless)	FWS C*	SSF C*
BOD	34	117	1.00	1.06	$3.5 + 0.053\,C_i$	3.0
Total nitrogen (N)	22	10	1.05	1.05	1.50	1.5
Pathogens (measured as fecal coliforms)	75	100	1.00	1.003	300	200

Note: C* values are measured in mg/L or coliforms/100 mL and can be estimated in the case where actual field measurements are not available.

Rate constants are usually reported at a standard temperature, such as 20 °C. They can be adjusted to a different temperature using Eq. 23-5.

$$k_T = k_{20}\,\theta^{T-20} \tag{23-5}$$

In Eq. 23-5, k_{20} is the rate constant reported at 20 °C (or some other temperature); θ is a temperature correction factor (unitless); and T is the water temperature of interest (°C).

Two types of rate constants are used in constructed wetland design: volume-based and area-based. If the depth of the water does not change, the two types can be related (Crites and Tchobanoglous 1998). Volume-based removal constants (k_V) have units of time^{-1}. Equation 23-4 uses an area-based rate constant (k_A), with units of length divided by time. The area- and volume-based constants are related by depth of water in the wetland:

$$r_{BOD} = -k_V(C_i) = -k_A(A/V)(C_i) = -(k_A/H)(C_i) \tag{23-6}$$

In Eq. 23-6, r is the rate of loss of concentration per unit time per unit volume, k_V is a volume-based rate coefficient for constituent removal (time^{-1}), C_i is the wetland constituent concentration (mass/volume), k_A is an area-based rate coefficient for constituent removal (length/time), A is the surface area of the wetland (length^2), V is the volume of the wetland (length^3), and H is the water depth (length).

⟫⟫⟫ Example 23-1. Using the Kadlec and Knight Model to Design an FWS Wetland

A town of 7,000 people generates a wastewater flow of 530 m^3/day through an existing oxidation pond. The pond effluent contains 15 mg/L BOD, 40 mg/L TSS, and 15 mg/L ammonia nitrogen. Regional authorities have recently passed a 10 mg/L BOD effluent requirement for all new wastewater facilities. There is no nitrogen limit. The wastewater temperature in the coldest month is 9 °C. Assume an AR of 7:1 (see Table 23-1) and an average water depth (d) of 0.5 m.

Solution

The design follows the five steps listed in Box 23-2.

Step 1

First, determine the limiting effluent requirements. The design does not have an effluent requirement for nitrogen or pathogens. Therefore, BOD loading controls the wetland area. The effluent requirement for BOD is 10 mg/L.

Step 2

The required surface area can be determined using the Kadlec and Knight model (Eq. 23-4) with the constants provided in Table 23-6. The problem statement provided C_e (10 mg/L), C_i (15 mg/L), and Q (530 m^3/day). k_{20} will need to be adjusted for the cooler temperature of 9 °C (using Eq. 23-5). Also, the BOD background concentration (C^*) must be determined.

$$k_T = k_{20}\theta^{T-20}$$

$$k_{9\,°C} = 34 \text{ m/year} \times 1.00^{9-20} = 35 \text{ m/year}$$

According to Table 23-6, the BOD background concentration (C^*) is based on the given influent ($C_i = 15$ mg/L) and the following approximation:

$$C^* = 3.5 + 0.053 \times C_i = 4.3 \text{ mg/L}$$

The area can then be determined from Eq. 23-4.

$$A = \left(\frac{0.0365 \times Q}{k_T} \right) \times \ln\left(\frac{Ci - C^*}{Ce - C^*} \right)$$

$$A = \left(\frac{0.0365 \times 530 \text{ m}^3/\text{d}}{34 \text{ m/yr}} \right) \times \ln\left(\frac{15 \text{ mg/L} - 4.3 \text{ mg/L}}{10 \text{ mg/L} - 4.3 \text{ mg/L}} \right)$$

$$= 0.36 \text{ hectares}$$

A 20% safety factor (incorporated as the term 1.2 in the following equation) will be added to the calculated area:

$$A \times SF = 0.36 \text{ ha} \times 1.2 = 0.432 \text{ ha} = 4,320 \text{ m}^2$$

If the wetland had also needed to be sized based on nitrogen and pathogens, the largest area obtained would have been selected.

Step 3

Using the given AR of 7:1 ($L: W$), the surface width and length can be determined.

$$A = W \times L = W \times 7W$$

$$W = \left(\frac{A}{7} \right)^{0.5} = \left(\frac{4,320 \text{ m}^2}{7} \right)^{0.5} = 25 \text{ m}$$

$$L = AR \times W = 7 \times 25 \text{ m} = 175 \text{ m}$$

Step 4

The head loss needs to be checked to ensure adequate head between the inflow and out-flow points (using Eq. 23-1). The velocity in the wetland equals

$$v = \frac{Q}{D \times W} = \frac{530 \text{ m}^3/\text{day}}{0.5 \text{ m} \times 25 \text{ m}} = 42.4 \text{m/day} \times \frac{\text{day}}{24 \text{ hr}} \times \frac{\text{hr}}{60 \text{ min}} \times \frac{\text{min}}{60 \text{ sec}} = 0.00049 \text{ m/sec}$$

The Manning's n is determined from Eq. 23-3 based on a resistance factor (a) of 0.487 for sparse vegetation and $d \geq 0.5$ m (values are available in Crites and Tchobanoglous 1998).

$$n = \frac{a}{d^{0.5}} = \frac{0.487}{0.5 \text{m}^{0.5}} = 0.689$$

The hydraulic gradient (slope) is determined from Eq. 23-2.

$$s = \left(v \times \frac{n}{d^{2/3}} \right)^{0.5} = \left(0.00049 \text{ m/s} \times \frac{0.689}{0.5 \text{ m}^{2/3}} \right)^{0.5} = 0.023 \text{ m/m}$$

The head loss along the 175-m length of the wetland is determined using Eq. 23-1.

$$h_L = s \times L = 0.023 \text{ m/m} \times 175 \text{ m} = 4.02 \text{ m}$$

This head loss is much larger than the wetland water depth (d) of 0.5 m. The designer must go back through the design and adjust the assumptions until the head loss is rea-sonable (i.e., smaller than the water depth or the difference between inflow and outflow elevations). One assumption that was not made at the beginning of the problem was the bed slope. If the slope of the bed is assumed to be 3% (0.03 m/m), the difference between inflow and outflow elevations (h) equals

$$h = 0.03 \text{ m/m} \times 175 \text{ m} = 5.25 \text{ m}$$

In this case, the calculated head loss (4.05 m) is smaller than the difference between inflow and outflow elevations. Thus, our design is acceptable.

Step 5

The specific depths for Zones 1–3 now need to be designed. First, determine the hydraulic residence time (HRT):

$$\text{HRT} = V/Q = L \times W \times d/Q = 175 \text{ m} \times 25 \text{ m} \times 0.5 \text{ m}/530 \text{ m}^3/\text{day} = 4.13 \text{ day}$$

This value is slightly greater than the recommended minimum HRT of 4 days for an FWS wetland (discussed earlier in this chapter), so it is acceptable.

The specific zone volumes can be determined based on the HRTs

$$\text{recommended HRT for Zone 1} = 1 \text{ day}$$

$$\text{recommended HRT for Zone 3} = 1 \text{ day}$$

$$\text{HRT for Zone 2} = 4.13 - 1 - 1 = 2.12 \text{ day}$$

$$\text{volume for Zone 1} = \text{HRT Zone 1} \times Q = 1 \text{ day} \times 530 \text{ m}^3/\text{day} = 530 \text{ m}^3$$

volume for Zone 2 = HRT Zone 2 × Q = 2.13 day × 530 m³/day = 1,129 m³

volume for Zone 3 = HRT Zone 3 × Q = 1 day × 530 m³/day = 530 m³

Now use algebra to calculate the zone lengths and depths. Assume the entire FWS has a constant width, calculated previously as 25 m, and assume that Zones 1 and 3 have equal lengths and depths.

$$L1 = L3$$

$$D1 = D3$$

$$L1 + L2 + L3 = 175 \text{ m} \qquad \text{(determined previously)}$$

$$(D1 + D2 + D3)/3 = 0.5 \text{ m} \qquad \text{(assumed average depth from problem statement)}$$

$$L1 \times D1 = L3 \times D3 = \text{volume}_{1,3}/W = 530 \text{ m}^3/25 \text{ m}$$

$$L2 \times D2 = \text{volume}_2/W = 1,129 \text{ m}^3/25 \text{ m}$$

With the six equations and six unknowns above, the linear system can be solved for

$$D1 = D3 = 0.36 \text{ m}$$

$$D2 = 0.78 \text{ m}$$

$$L1 = L3 = 59 \text{ m}$$

$$L2 = 58 \text{ m}$$

This example does not account for the effects of evapotranspiration or precipitation on the flow rate, although the 20% safety factor may deal with excessive inputs. Local weather and precipitation records can be investigated to determine evapotranspiration and precipitation rates in the area.   **}}}**

23.3.5 Hydraulic Considerations for Subsurface Flow Wetlands

As described in the FWS hydraulic section, proper hydraulic design should account for the wastewater flow rate, precipitation entering the bed, geometry of the wetland bed, available construction materials, site topography, and solids accumulation. Plant roots do not seek the submerged inlet distribution pipe of an SSF wetland as long as the bed water surface is above the inlet pipe, ensuring that the plant roots are satisfied and do not creep to find water. A shallow distribution pipe requires rocks to be placed at the inlet that are too large (>5 cm diameter) for the plant roots to grasp. Adjustable outlets, such as PVC elbows (which can be turned up or down), are recommended for controlling the water level. It is conservative to assume the evapotranspiration rate and the infiltration rate are zero. Several centimeters of freeboard should be added above the expected maximum water level during the heaviest rainstorm at maximum predicted influent flow rate.

The potential effects from short-circuiting can be reduced by correct design for solids loading, even plant placement throughout the bed, distributed inlet and outlet piping, and installation of several parallel and series wetland trains. An SSF wetland may be built with parallel wetland beds to divide the inlet loading and to allow a bed to be taken out of service for maintenance. Practitioners suggest that SSF wetlands with low solids loading and properly designed media size should not clog or require solids removal for many years (Crites and Tchobanoglous 1998; Reed et al. 2001).

The height of the water is greater at the inlet of an SSF wetland because of the greater resistance to flow caused by solids accumulation. The media depth may be increased at the inlet end to account for the higher water level. Most of the BOD and TSS treatment in SSF wetlands occurs in the first few meters. Studies have indicated that long aspect ratios do not necessarily promote better treatment in SSF wetlands, so aspect ratios should not exceed 1:1 (EPA 2000). For a wetland with greater width than length, there should be no need to calculate the minor head loss expected along the length of the wetland bed.

Because most TSS accumulates at the inlet end of an SSF wetland bed, one recommendation to prevent media clogging is to design SSF wetlands for ≤348 lb/acre-day (390 kg/ha-day) TSS loading over the entry zone alone (Crites and Tchobanoglous 1998). The TSS loading is determined as

$$\text{TSS Loading} = \frac{\text{TSS mass loading/day}}{\text{entry zone cross-sectional area}} \tag{23-7}$$

where the TSS mass loading/day equals $Q \times \text{TSS}_{\text{influent}}$, Q is the total flow of wastewater (m³/day), and $\text{TSS}_{\text{influent}}$ is the average concentration of TSS in the influent to the wetland (mg TSS/L).

In some situations, it might make sense to shift some of the space and other resource requirements to the primary treatment unit and away from the wetland. A lower TSS in the wetland influent may thus be accomplished with a higher level of primary treatment.

Darcy's law represents flow through porous media. It is used in SSF wetland design to determine the head loss through the media based on an assumed hydraulic conductivity. In simple terms, *hydraulic conductivity* is the measure of how easily the soil or media can transmit water and is described as the volume of water that can pass through an area of permeable media over a measured time under a unit hydraulic gradient (e.g., m³/m²-day). It may also be shown as a velocity or the speed at which the water can move in the soil under a unit hydraulic gradient (e.g., cm/day).

Darcy's law is written as

$$Q = K \times Ac \times s \tag{23-8}$$

where Q is the flow rate (m³/day); K is the hydraulic conductivity (m³/m²-day); Ac is the cross-sectional area (m²); and s equals the hydraulic gradient slope (m/m). The cross-sectional area (Ac) equals the width of the wetland (W) multiplied by the average depth of water (d). The hydraulic gradient slope (s) equals the change in water height (h_L) (i.e., the head loss) divided by the length of travel through the wetland (L). Table 23-7 provides some values of porosity and hydraulic conductivity as a function of subsurface media.

Because of solids accumulation at the inlet of SSF wetlands, it is recommended that the first 30% of the length of an SSF wetland bed be assumed to have a hydraulic conductivity (K) of 1% of that of clean media, and the remaining 70% of the bed should be assumed to have a K value that is 10% of the clean K value (EPA 2000). Gravel rang-

Table 23-7. Porosity and Hydraulic Conductivity for Different Materials

Material	Porosity (Unitless)	Hydraulic Conductivity (m/day)
Coarse gravel	0.28	150
Medium gravel	0.32	270
Coarse sand	0.39	45
Medium sand	0.39	12
Fine sand	0.43	2.5
Silt	0.46	0.08
Clay	0.42	0.0002
Medium-grained sandstone	0.37	3.1
Fine-grained standstone	0.33	0.2
Limestone	0.30	0.94
Basalt	17	0.01
Weathered granite	45	1.4

Source: From Todd 1980.

ing from 6 to 25 mm (0.25–1 in.) diameter with a porosity of 0.38 corresponds to a K in the range of 1,000–50,000 m^3/m^2-day (Reed 2001). If it is assumed that a clean K equals 10,000 m/day, then the first 30% of the bed would have a K of 100 m/day and the remainder of the bed would have a K of 1,000 m/day (Box 23-3, Example 23-2).

))) Box 23-3. Seven Steps to Design Subsurface Flow (SSF) Wetlands

1. Determine the limiting effluent requirements for BOD, TSS, nitrogen, and pathogens.
2. Calculate the surface area using the Kadlec and Knight model (Eq. 23-4). If desired, increase by a 20% safety factor. Again, because of the increase in materials, labor, and potential effects from the safety factor increase, apply safety factors with care.
3. Using Darcy's law (Eq. 23-8), calculate the cross-sectional area required to hydraulically accept the flow. Reduce the hydraulic conductivity by 90% to account for clogging and root growth that occurs over time.
4. Select an appropriate aspect ratio (AR) (Table 23-1), constrained by site conditions, and calculate surface dimensions.
5. Calculate the water depth by dividing the calculated cross-sectional area (determined in Step 3) by the width (determined in Step 4).
6. Add the water depth to the chosen freeboard to obtain the required media depth (d_m).
7. Check the head loss to ensure adequate head between the inflow and outflow points (Eq. 23-1). Ensure that the head loss is less than the calculated water depth (determined in Step 5).

))) **Example 23-2.** Using the Kadlec and Knight Model to Design an SSF Wetland

A community of 2,000 people generates 190 m³/day (50,000 gal/day) of septic tank effluent. The septic tank effluent characteristics are 130 mg/L BOD, 50 mg/L TSS, and 30 mg/L nitrogen (as NH_3–N). An effluent BOD standard of 10 mg/L would greatly improve water quality in the riparian area where the wetland empties, but no treatment objectives are required for TSS, pathogens, and nitrogen. The available gravel medium has a K of 30,000 ft/day (9,144 m/day), with a porosity of 0.4. Decaying plant matter provides a background BOD concentration (C^*) of 5 mg/L. Assume a wastewater temperature in the wetland of 9 °C and an area-based, first-order k_{20} of 117 m/year (Table 23-6). Assume an AR of 0.5 : 1, a bed slope of 0.5% (0.005 m/m), and a freeboard of 0.2 m.

Solution
The design follows the seven steps listed in Box 23-3.

Step 1
The only effluent requirement imposed is for BOD, 10 mg/L.

Step 2
Adjust k_{20} for the cooler temperature of 9 °C (using Eq. 23-5).

$$k_T = k_{20}\theta^{T-20}$$

$$k_{9\,°C} = 117 \text{ m/year} \times 1.06^{9-20} = 61.6 \text{ m/year}$$

The required surface area is determined using the Kadlec and Knight model (Eq. 23-4).

$$A = \left(\frac{0.0365 \times Q}{k_T}\right) \times \ln\left(\frac{Ci - C^*}{Ce - C^*}\right)$$

$$= \left(\frac{0.0365 \times 190 \text{ m}^3/\text{d}}{61.6 \text{ m/yr}}\right) \times \ln\left(\frac{130 \text{ mg/L} - 5 \text{ mg/L}}{10 \text{ mg/L} - 5 \text{ mg/L}}\right)$$

$$= 0.35 \text{ ha}$$

The calculated area is then increased by the 20% safety factor (1.2 in the following expression).

$$A \times SF = 0.36 \times 1.2 = 0.43 \text{ ha} = 4,300 \text{ m}^2$$

Step 3
The cross-sectional area can be determined using Darcy's law, rearranged to solve for the area (Ac in Eq. 23-8). The hydraulic conductivity is reduced by 90% (K multiplied by 0.1) to account for clogging and root growth.

$$Ac = \frac{Q}{K \times S} = \frac{190 \text{ m}^3/\text{day}}{9,144 \text{ m/day} \times 0.1 \times 0.005 \text{ m/m}} = 41.55 \text{ m}^2$$

Step 4

Using the given AR (0.5 : 1), the surface width and length can be determined as follows:

$$A = W \times L = W = \times 0.5W$$

$$W = \left(\frac{4,300 \text{ m}^2}{0.5} \right)^{0.5} = 92.7 \text{ m}$$

$$L = AR \times W = 0.5 \times 92.7 \text{ m} = 46.4 \text{ m}$$

Step 5

The water depth is determined based on the cross-sectional area and width.

$$d = Ac/W = 41.55 \text{ m}^2/92.7 \text{ m} = 0.44 \text{ m}$$

Step 6

The media depth is determined based on the water depth and freeboard. The freeboard was provided as 0.2 m.

$$d_m = d + \text{freeboard} - 0.44 \text{ m} + 0.2 \text{ m} = 0.64 \text{ m}$$

Step 7

The head loss is checked (Eq. 23-1) to ensure adequate head between the inflow and out-flow points. The bed slope was given as 0.5%. The length (L) was calculated previously as 46.4 m.

$$h_L = s \times L = 0.005 \text{ m/m} \times 46.4 \text{ m} = 0.23 \text{ m}$$

Because this value is less than the water depth (0.55 m), there should be adequate head to maintain flow between the inlet and the outlet.   **)))**

23.3.6 Design Considerations for an Evapotranspiration Bed

Evapotranspiration (ET) beds require a septic tank for pretreatment of solids and a sand bed with a distribution network that is sealed by an impermeable liner. The basic principle of the ET system is to distribute septic tank effluent through the base of the sand bed, allowing water to travel to the upper portion of the bed through capillary action. Gravel is often installed under the sand bed to provide short-term effluent storage. At the surface, water evaporates directly to the atmosphere or becomes lost through plant transpiration. Nutrients are assimilated by the shrub and herbaceous plants. Table 23-8 lists the components of an evapotranspiration system along with some details related to design and construction.

ET systems are an accepted technology for arid and semiarid regions (Bernhardt 1973; EPA 2002). They are feasible when evaporation exceeds annual rainfall by at least 620 mm (24 in.) (EPA 2002). Installation should be reserved for sites where geological conditions prevent the use of a leach field or some other type of subsurface disposal and all other on-site treatment options are explored and deemed inappropriate or unfeasible (Salvato 1992).

Table 23-8. Components of an Evapotranspiration Bed System

Component	Comments
Septic tank	Critical for removing solids, oil, and grease before influent enters the sand and gravel bed.
Distribution network and gravel storage	Septic effluent drains by gravity into a distribution box, which evenly distributes effluent to the distribution laterals within the sand bed. The laterals can be perforated plastic pipe that are spaced at 90–120 mm (3–4 ft). Accurate placement and slope of the distribution network (box and laterals) is critical because gravity is the single force moving effluent through the sand bed. If a gravel storage bed is used, it is placed under the sand bed, is typically 150–300 mm (6–12 in.) long, and provides storage of septic tank effluent during short intervals when precipitation or wastewater inflow exceed evapotranspiration (Bernhart 1973). The distribution network is placed on top of the gravel bed. Gravel should be clean and free of small particles before installation and should have 20 mm (0.8 in.) mean diameter. EPA (2002) states that storing wastewater too deep in a gravel storage area limits the positive effects of capillary action because the lack of wicking properties of gravel may restrict the rise of wastewater to the ground surface for evaporation.
Sand bed	Septic tank effluent is distributed through the lower portion of the ET sand bed, where the wastewater then travels to the upper portion of the bed through capillary action. Studies indicate that a uniform sand with a median diameter equivalent to 0.10 mm promotes capillary rise to above 1 m. Pore sizes <0.10 mm retard capillary rise and promote bed clogging. Sand particles too large will provide insufficient capillary rise (Bernhart 1973; Salvato 1983). Additionally, the topsoil layer on the sand layer, if composed of a sand, silt, clay, and loam soil mixture and vegetation, promotes wicking action by maintaining sufficient moisture content within the bed.
Impermeable liner	To prevent vertical seepage, use a hardpan clay liner or impermeable plastic liner. EPA (2002) recommends a 20-mil polyethylene liner. Two layers of 10-mil liner or comparable material may be substituted.
Topsoil and vegetation	Proper sloping of the topsoil cover can impede 30% of total precipitation from seeping into the bed (Bernhardt 1973; Salvato 1992). Other studies recommend assuming 100% infiltration of precipitation (Bennett and Linstedt 1978) in design calculations.
Vegetation	Vegetation must be tolerant of both extreme wet and dry conditions and excess nutrients and salts. Restrict vegetation to herbaceous and shrub varieties, not trees or aggressive species, where roots may penetrate the perforated pipe or liner. Native plants adapted for these extreme environments are most effective. In the initial years, plants may require some watering, especially if the ET system is not used daily. After they are established, less care will be required. Include numerous plant species in the design, and quickly replace any varieties that do not survive with ones that are successful.

ET systems sometimes require a large footprint to accommodate the required sand bed. Several items should be considered when locating the facility:

- a convenient location for public access,
- a large enough area to accommodate the sand bed,
- a site with the most wind and sunlight to maximize evapotranspiration, and
- an area that does not hold water during heavy rains (i.e., natural depressions).

Installation of a perimeter fence around the sand bed may be helpful if the facility is constructed near a heavily populated area.

A simple water balance study correlating evapotranspiration, precipitation, and wastewater loading rate provides a sufficient estimate of ET bed size requirements (Example 23-3). This method is similar to that used by others (Bernhardt 1973; Salvato 1992; EPA 2002).

Heavy rainfall is accounted for by providing sufficient storage. Precipitation data used should sufficiently represent the wettest months and include enough data to accurately capture the wettest potential years. Bennett and Linstedt (1978) suggest evaporation and precipitation data from the critical year, or a 10-year span is sufficient, whereas Salvato (1983) recommends using the monthly pan evaporation minus the wettest year of a 10-year period.

Wastewater inflow varies depending on the number of people served, the type of facilities (i.e., toilets, sinks, showers) connected to the system, and the expected daily use. Some ET systems may be designed to treat only greywater. A good practice when designing ET systems is to select low-flow technology and to incorporate issues of water conservation into hygiene education.

Evapotranspiration is the sum of evaporation and plant transpiration. It is influenced by numerous factors, including soil pore size and distribution, soil moisture, wind, relative humidity, and vegetation type. Transpiration alone is an unreliable parameter for ET design because the life expectancy of the vegetated surface is unpredictable (Bennett and Linstedt 1978). Overall, pan evaporation estimates should be sufficient for determining the evapotranspiration rate for ET beds. Mara (1976) recommends that evapotranspiration can be estimated as 80% of Class A evaporation pan data. However, other resources recommend a range of 70–80% (Kadlec and Knight 1996).

))) Example 23-3. Designing an Evapotranspiration Bed

This example takes place in a location where the climate has two major seasons: rainy and dry. The rainy season begins in May or June and extends through November or December. The heaviest rains, typically in September and October, provide 30–40% of the total annual rainfall. This heavy rainfall must be accounted for during the design of an ET bed by providing sufficient storage.

Design an ET bed system that will serve 70 people using a public sanitation facility with 4 toilet units that require 7.5 L (2 gal) per flush. Assume that each person uses the facility once per day. (Most people are fishing during the day.) Water from a hand-washing facility is discharged to a soak pit. No gravel storage bed is included in the design.

The first table below provides averaged evaporation (E) and precipitation (P) data obtained over 30 years from 10 hydrological basins located near Kingston, Jamaica. First estimate the size of the ET bed. Then, perform calculations to confirm that there is sufficient monthly storage to accommodate the rainy season.

Fill in the remainder of the first table by determining the evapotranspiration rate (ET) and allowable hydraulic loading rate (LR). The LR is the average volume of water that can be removed in a month from the system for a given bed area.

For simplicity, assume the evapotranspiration (ET) is

$$ET = E - 0.7 \times P \qquad (23\text{-}9)$$

The allowable loading rate that can be removed by evapotranspiration is determined as

$$LR = ET \times \frac{ft}{12\,in.} \times \frac{7.48\,gal}{ft^3} \qquad (23\text{-}10)$$

The ET bed has zero outflow, so 100% of the inflow must be removed by evapotranspiration (ET).

Parameter	Jan	Feb	Mar	Apr	May	June	July	Aug	Sept	Oct	Nov	Dec	Average
Evaporation (E) (in./month)	4.5	4.7	6.6	6.6	7.1	7.7	8.3	7.1	6.1	5.6	4.5	4.9	6.1
Precipitation (P) (in./month)	1.5	1.8	1.5	2.6	5.5	4.1	2.9	3.6	7.7	8.0	2.9	1.7	3.7
Evapotranspiration (ET) (in./month)	3.5	3.5	5.5	4.8	3.2	4.8	6.3	4.6	0.7	0.0	2.5	3.7	3.6
Loading Rate (LR) (gal/ft²-month)	2.2	2.2	3.4	3.0	2.0	3.0	3.9	2.8	0.4	0.0	1.6	2.3	2.2

As previously stated, EPA (2002) recommends the use of ET systems at sites where the evaporation exceeds annual rainfall by at least 620 mm (24 in.). Here, annual precipitation is 44 in. (1,100 mm), and annual evaporation is 74 in. (1,800 mm), a difference of 700 mm. Thus, the ET system is feasible.

The area of the ET bed can be determined by dividing the wastewater flow by the average allowable loading rate (LR) determined in the above table (2.2 gal/ft²-month).

The wastewater flow that requires treatment after passing through a septic tank is

$$70\text{ people} \times 2\text{ gal/day} \times 30\text{ day/month} = 4{,}200\text{ gal/month}$$

The required ET bed area is

$$A = \frac{Q}{LR} = \frac{4{,}200\text{ gal/month}}{2.2\text{ gal/ft}^2\text{-month}} = 1{,}900\text{ ft}^2 \qquad (23\text{-}11)$$

The available storage in the sand (V) can be determined from the calculated area (A), the bed depth, and the porosity of the media. Assume the bed depth (d) is set at 2 ft and the sand porosity (η) is 0.40.

$$V = A \times d \times \eta \times \frac{7.481\text{ gal}}{ft^3} = 11{,}400\text{ gal} \qquad (23\text{-}12)$$

We now need to confirm that there is adequate monthly storage. This confirmation requires another table that examines the available storage every month. LR is taken from the table above. The ET here is determined from

$$ET \text{ (gal/month)} = LR \text{ (gal/ft}^2\text{month)} \times A \text{ (ft}^2\text{)} \tag{23-13}$$

The monthly surplus is the volume of the bed storage that is taken up by wastewater that is not evapotranspirated. It is the difference between ET (gal/month) and the monthly input of wastewater (Q). It is cumulative, as shown in the table below.

Parameter	Jan	Feb	Mar	Apr	May	Jun	Jul	Aug	Sept	Oct	Nov	Dec
Loading rate (LR) (gal/ft²-month)	2.2	2.2	3.4	3.0	2.0	3.0	3.9	2.8	0.44	0.0	1.6	2.3
Evapotranspiration (ET) (gal/month)	4,620	4,620	7,140	6,300	4,200	6,300	8,190	5,880	933	0	3,360	4,830
Monthly input of wastewater (Q) (gal/month)	4,200	4,200	4,200	4,200	4,200	4,200	4,200	4,200	4,200	4,200	4,200	4,200
Cumulative monthly surplus (gal/month)	0	0	0	0	0	0	0	0	3,267	7,467	8,307	8,937
Bed capacity used (%)	0	0	0	0	0	0	0	0	29	66	73	78

At the end of the calendar year, 71% of the bed volume is taken up by moisture. During the first 8 months of the new year, the bed has the potential to remove 13,650 more gallons of water through ET than the wastewater that is put into it, so the surplus from the previous year's rainy season will be removed before the next rainy season begins. In some instances, for safety, another storage or treatment technology is partnered with the ET bed to store or treat overflow during the rainy season.

The final design is summarized in the table below, which encompasses the results of the calculations and some additional information on ET bed design.

Parameter	Value
Aspect ratio (AR)	3 : 1
Slope of topsoil	1–2%
Spacing of distribution pipes	4 ft
Average loading rate (LR)	2.2 gal/ft²-month
Available storage by sand (V)	11,400 gal
Area (A)	1,900 ft²
Length (L)	25 ft
Width (W)	75 ft
Depth (d)	2 ft

⟩⟩⟩ References

Bennett, E., and Linstedt, D. K. (1978). *Sewage disposal by evapotranspiration*, EPA-600-2-78-163. EPA Office of Research and Development, Municipal Research Laboratory, Cincinnati, Ohio.

Bernhardt, A. (1973). *Treatment and disposal of wastewater from homes by soil infiltration and evapotranspiration.* University of Toronto Press, Toronto, Canada.

Borst, M., Riscassi, A. L., Estime, L., and Fassman, E. L. (2002). "Free-water depth as a management tool for constructed wetlands." *J. Aquatic Plant Management.* 40, 43–45.

Chendorain, M., Yates, M., and Villegas, F. (1998). "The fate and transport of viruses through surface water constructed wetlands." *J. Environ. Quality*, 27, 1451–1458.

Crites, R., and Tchobanoglous, G. (1998). *Small and decentralized wastewater management systems.* McGraw-Hill, New York.

Denny, P. (1997). "Implementation of constructed wetlands in developing countries." *Water Sci. and Technol.*, 35(5), 27–34.

Environmental Protection Agency (EPA). (2000). *Manual: Constructed wetlands treatment of municipal wastewaters*, EPA 625/R–99/010, EPA Office of Research and Development, Cincinnati, Ohio.

Environmental Protection Agency (EPA). (2002). *Onsite wastewater treatment manual.* EPA/625/R–00/008, U.S. Office of Water Program Operations Washington, D.C., Feb.

Forbes, M. G., Dickson, K. R., Golden, T. D., Hudak, P., and Doyle, R. D. (2004). "Dissolved phosphorus retention of light-weight expanded shale and masonry sand used in subsurface flow treatment wetlands." *Environ. Sci. and Technol.*, 38(3), 892–898.

Kadlec, R. H., and Knight, R. L. (1996). *Treatment wetlands.* Lewis Publishers. Boca Raton, Fla.

Keller, C., and Knight, B. (2004). *Sediment accretion and ageing in treatment wetlands.* Memorandum to HDR from Wetland Solutions, Gainesville, Fla.

Knight, R. L., Walton, W. E., O'Meara, G. F., Reisen, W. K., and Wass, R. (2003). "Strategies for effective mosquito control in constructed treatment wetlands." *Ecological Engineering*, 21, 211–232.

Koottatep, T., and Polprasert, C. (1997). "Role of plant uptake on nitrogen removal in constructed wetlands located in the tropics." *Water Sci. and Technol.*, 36(12), 1–8.

Mara, D. (1976). *Sewage treatment in hot climates.* John Wiley & Sons, Chichester, U.K.

Mihelcic, J. R., and Zimmerman, J. B. (2009). *Environmental engineering: Fundamentals, sustainability, and design*, John Wiley and Sons, New York.

Reed, S. C., Aulenbach, D., Bavor, H., Bouwer, H., Crites, R., Kinshella, P., Middlebrooks, J., Otis, R., Parten, S., Polprasert, C., Reimold, R., Shober, R., Smith, R., Tchobanoglous, G., Wallace, A., Watson, J., and Zimmerman, M. (2001). *Natural systems for wastewater treatment*, WEF Manual of Practice No. FD–16, Water Environment Federation, Alexandria, Va.

Salvato, J. A. (1983). "Rational design of evapotranspiration bed." *J. Environ. Engrg.*, 109(3), 646–660.

Salvato, J. A. (1992). *Environmental engineering and sanitation*, 4th ed., John Wiley & Sons, New York.

Schellenberg, D. (2001). *Constructed wetlands in East Texas, design, permitting, construction and operations*, Vols. 1 and 4, Pineywoods Resource Conservation and Development Council, Nacogdoches, Tex.

Stewart, E. (2005). *Evaluation of septic tank and subsurface flow wetland for Jamaican public school wastewater treatment*, <http://cee.eng.usf.edu/peacecorps/Resources.htm> (Jan. 23, 2009).

Todd, D. K. (1980). *Groundwater hydrology*, 2nd ed., John Wiley & Sons, New York.

Vacca, G., Wand, H., Nikolausz, M., Kuschk, P., and Kastner, M. (2005). "Effect of plants and filter materials on bacteria removal in pilot-scale constructed wetlands." *Water Res.* 39, 1361–1373.

Wallace, S. D. (2005). "Constructed wetland systems: Design approaches," PowerPoint presentation. In M. A. Gross and N. E. Deal, eds., *University curriculum development for decentralized wastewater management.* National Decentralized Water Resources Capacity Development

Project. University of Arkansas, Fayetteville, Ark. <http://www.onsiteconsortium.org/files/Constructed_Wetlands_Design_Methods.pdf> (Jan. 23, 2009).

Wallace, S., and Knight, R. L. (2004). "Water Environment Research Foundation (WERF) wetland database." *9th International Conference on Wetland Systems for Water Pollution Control.* Avignon, France, September 26–30. <http://www.nawe-pa.com/docs/Technical%20Paper%20-%20IWA%202004%20WERF%20database.pdf> (Jan. 31, 2008).

⟫⟫ Further Reading

Siliva, H. (1990). *Evapo-transpiration method of sewage disposal.* Ministry of Health/Environmental Control Division, Kingston, Jamaica.

Stanforth, K. L. (2005). A sanitation plan for a fishing village in Jamaica that incorporates evapotranspiration bed technology, <http://cee.eng.usf.edu/peacecorps/Resources.htm> (Jan. 23, 2009).

Solid Waste

Solid Waste

24

Solid Waste Management

))) 24.1 Definition

Solid waste is a material that is abandoned, recycled, or considered inherently wastelike. Hazardous waste is typically considered a subset of solid waste. *Municipal solid waste* (commonly referred to as trash or garbage) consists of items such as food scraps, packaging, bottles, cans, yard waste, furniture, clothing, newspapers, appliances, paint, and batteries.

Chapter 21 of Agenda 21 (the Rio Declaration on Environment and Development) outlines the environmentally sound management of solid wastes through the following hierarchy (UNDESA 2005): (1) minimizing wastes; (2) maximizing environmentally sound waste reuse and recycling; (3) promoting environmentally sound waste disposal and treatment; and (4) extending waste service coverage.

Sustainable solid waste management requires consideration of issues of health, the environment, energy, and economic opportunities for participants. For example, microenterprises can be readily incorporated into a solid waste management plan. Projects can also initiate not only a solid waste management plan, but also health and hygiene education.

))) 24.2 Health Issues

In most developing countries, solid waste is disposed of in open dumps, and scavengers sort through the waste looking for materials of value. Solid waste contains a high percentage of organics, which attracts flies, rodents, and scavenging animals. Disease vectors such as rats, mice, maggots, cockroaches, and mosquitoes are found in collection areas. As one example, the Andes mosquito, which transmits dengue fever and yellow fever, lays eggs in improperly disposed of cans and tires. The general population does not always understand the important connection among these vectors and diarrheal diseases, malaria, and other diseases. Figure 24-1 shows an F-diagram developed for solid waste, analogous to the F-diagram for fecal–oral transmission of disease shown in Chapter 2 (Fig. 2-2).

The lack of adequate solid waste disposal has many undesirable consequences. Waste is commonly processed by open burning, which discharges air pollutants into a

---

Jennifer L. Post, Alexis M. Troschinetz, Brooke Tyndell Ahrens, and Emily L. Owens contributed to this chapter.

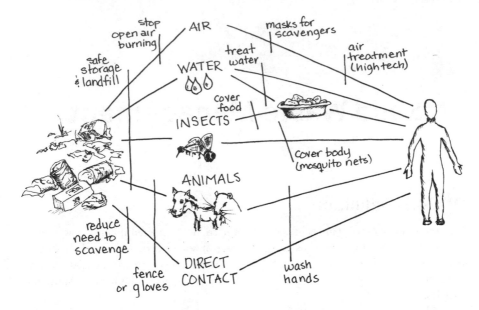

Figure 24-1. The F-Diagram for Solid Waste.

Note: A combination of engineering activities and hygiene education can effectively act as barriers to routes of disease transmission. Three engineering barriers are safe storage, preventing the need for open air burning, and proper disposal (e.g., landfilling). Reducing waste is also effective.

community. This solution results in environmental and public health problems. The work of transporting (and possibly burning) trash is often carried out by already vulnerable segments of a population, such as children or certain socioeconomic classes. Also, discarded solid waste blocks storm-water drains, resulting in flooding. Some people suggest that the presence of litter can reduce the morale of a community.

))) 24.3 Generation and Characterization of Solid Waste

To properly design a municipal solid waste management plan, it is important to understand the generation rate, discard rate, and composition of solid waste. This information can be obtained by conducting surveys and waste characterization studies that represent the community.

Some issues to consider when developing such a study are seasonal production of waste, seasonal employment, seasonal agricultural activities, the presence of nearby markets for solid waste, the income of residents, and household factors related to cleaning and food preparation. The study needs to identify the locations of major producers of certain types of solid waste (e.g., markets that produce large quantities of organic waste), existing disposal sites, how the quantity and type of waste generated may change in the future, and the presence of any hazardous or medical wastes that are currently generated (or may be generated in the future). It is also important to learn what perceptions individuals, households, and the community have toward solid waste, as well as the specific components of a municipal solid waste management plan.

Generation rates are much higher in the United States and Europe, compared to the developing world. Material can be separated into discarded material and materials that are recovered for recycling or energy generation. Generation rates are thus greater than the discard rate (which will require disposal). Generation rates for developed and developing countries are compared in Fig. 24-2. The lower end of solid waste generation rates is

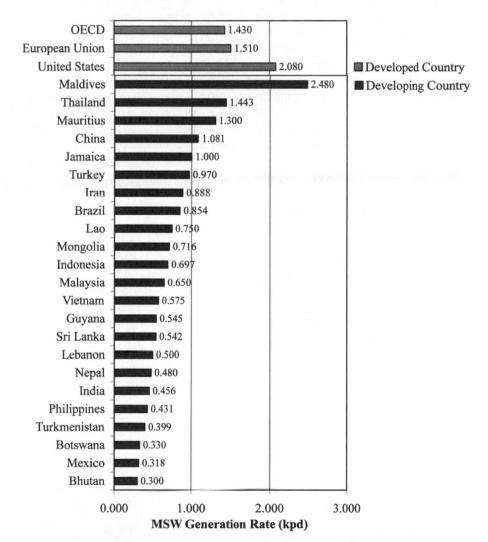

Figure 24-2. Municipal Solid Waste Generation Rates (kg per Capita per Day) for 23 Developing Countries, Compared to the United States, the European Union, and Organisation for Economic Co-operation and Development Countries.

Source: Reprinted with permission from Troschinetz and Mihelcic (2009).

0.3 kg/capita-day, but they can approach and exceed 1 kg/capita-day. The high generation rate for the Maldives is partially related to understanding their economic activity; tourism is the largest industry, accounting for 28% of GDP (and >60% of foreign exchange receipts), and most staple foods are now imported.

A wide variety of factors affect waste generation rates. Historically as a country's economy developed, its waste generation rate increased. The quantity of waste generated is directly affected by several factors. In addition to being affected by household income, waste generation is also related to the number of people in a household. A greater number of people in a household results in less waste generation on a per-person basis. Socioeconomic development and the degree of industrialization influence waste generation rates by generally affecting income and consumption patterns. Climate and seasonal changes affect the amount of organic material generated from preparing fresh foods. There are also social and religious customs that influence waste generation (and composition).

Table 24-1 shows the range of municipal solid waste composition for 19 developing countries (Troschinetz and Mihelcic 2009). There is a significant percentage of organic material (as high as 80%). Packaging (paper and cardboard) makes up a much smaller amount of the waste stream compared to the United States or the European Union. Small island nations in the Pacific have a relatively higher percent of metal cans. This difference has implications for each stage of the municipal solid waste management system.

A general rule of thumb is that high-income households generate more inorganic material from packaging waste, whereas low-income households produce more organic material from preparing food from base ingredients. However, some high-income households may generate the same amount of organic material because they can afford servants to prepare fresh, unpackaged food. In some parts of the world, women sweep the area around their homes in the morning and evening and place the material in a trash can or corner. This material consists mainly of organics (i.e., grass, leaves) and perhaps even paper and plastic.

Figure 24-3 shows the difference between municipal solid waste composition in developed countries (United States and the European Union) versus less developed coun-

Table 24-1. Range of Composition of Municipal Solid Waste Generated by Households in 19 Developing Countries

Municipal Solid Waste Composition	Percentage by Weight
Paper and paperboard	5.6–25.0
Textiles	0.0–6.4
Plastics	3.3–30.0
Glass	0.0–12.5
Metals	0.7–12.5
Organic material	17.0–80.0
Other waste	0.0–40.6

Note: Organic material is usually much higher in developing world solid waste streams, whereas packaging is much less.

Source: Troschinetz and Mihelcic 2009.

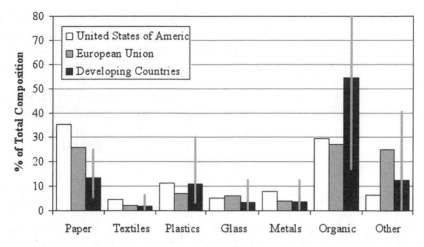

Figure 24-3. Comparison of Municipal Solid Waste Composition of Developed Countries (United States and European Union) Versus the Average of 19 Developing Countries.

Note: Vertical bars provide the range of composition of each material type for the developing countries only.

Source: Reprinted with permission from Troschinetz and Mihelcic (2009).

tries (LDC). On average, developed countries' waste streams have half as much organic material, twice the portion of paper and cardboard, and similar fractions of glass and plastic.

⟩⟩⟩ 24.4 Components of a Solid Waste Management Plan

24.4.1 Pollution Prevention

Pollution prevention is focused on increasing the efficiency of a process to reduce the amount of waste generated. It should always be a preferred method to manage any waste. A pollution prevention hierarchy shows that source reduction is always the favored method over recycling and treatment. Disposal is thus the least preferred alternative. According to the U.S. Environmental Protection Agency (EPA), source reduction prevents waste either by redesigning products or by otherwise changing social patterns of consumption, use, and waste generation.

Public understanding of the importance of reduction, recycling, and reuse is integral to the success of solid waste management programs. The incorporation of tireless educational campaigns into source reduction and recycling programs has proven to be a key factor in increasing diversion rates and thereby lowering collection, processing, and disposal costs. In essence, communities that invest the most in education on source reduction and recycling appear to attain better results and lower costs (Wells 1994).

Incentives for source reduction and recycling can also save money on disposal. The introduction of packaging taxes and other economic incentives for reducing waste and encouraging recycling in many countries has had a larger effect on recycling and source reduction when compared to just increasing public awareness and organizing recycling programs alone (World Bank 2003).

In most parts of the world, people already reuse cans, bags, and other plastics and metals. Discarded wastes may be used for containers in the reselling of products (i.e., juice, spices) or recycled into shelters or toys and art (e.g., metal milk cans hammered into toy trucks). The importance of scavenging cannot be overemphasized in a successful solid waste management plan, and it is discussed later in this chapter.

24.4.2 On-Site Storage

Figure 24-1 showed that storage is a key barrier used to prevent disease transmission. Unfortunately, it is estimated that 10–20% of municipal solid waste in the developing world ends up as litter, so it is very common to observe large quantities of litter next to homes and alongside roads (Fig. 24-4).

Figure 24-4. Greywater and Trash Are Often Commingled Outside Homes, Providing Breeding Grounds for Disease.

On-site storage is usually the responsibility of the homeowner, tenant, or business. Figure 24-5 shows three aboveground on-site storage options. In low-income areas, any convenient container is used, including fabricated metal drums; woven, plastic, or metal buckets; bulk containers such as rice sacks; small plastic bags; milk crates; and cardboard boxes. The metal drums may be whole or cut in half. Sometimes holes are placed along the sides of the drums to provide ventilation and drainage.

Plastic bags, if used alone and not collected frequently, are vulnerable to scavenging animals (e.g., dogs, pigs, rats) that break open the bags and scatter the waste. Containers such as milk crates, uncovered buckets, and cardboard boxes also attract pests because of the odors of decomposition. The best containers are those that have a lid, are vermin-proof, are able to contain all the waste generated between times of collection, and can be safely handled by waste collectors. An aboveground wooden (or metal) platform can be constructed to ensure that animals do not get access to the waste.

Larger generators, such as businesses or neighborhoods, can store their discarded waste in large trash containers (e.g., Dumpster brand) or an area that contains several metal drums. These areas should be secured to ensure that animals cannot scavenge the waste and that it does not blow away, causing a litter nuisance to nearby community members. Typically, it is worthwhile to place these "new" collection sites in the same location

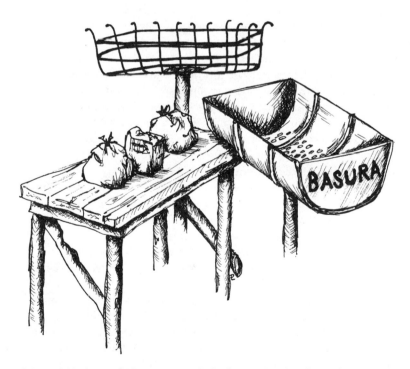

Figure 24-5. A Variety of Aboveground Platforms Made of Local Materials That Store Solid Waste Before Collection.

Note: The platforms keep the waste off the ground and away from rodents, roaming dogs, and pigs.

where community members have traditionally disposed of their waste. For example, if residents are used to transporting waste from their house to an open dump or area where waste was burned, it may be easiest to place a secured community collection system in the same location. This development will avoid requiring behavioral changes in a community that has been transporting their waste to a particular location for years.

Individuals are more likely to use a community collection system if it is located close to the point of generation. Sphere Standards (see http://www.sphereproject.org/) recommend that in emergency situations, no dwelling be farther than 15 m from a collection point. A general rule of thumb is to provide 100 L of collection volume per 10 families, though this figure will depend on the specific discard rate of a community (Oxfam 2007). A 100-L container can be manufactured from a 50-gal drum cut in half and can have legs welded to the bottom (Fig. 24-5).

24.4.3 Collection and Transport

Decisions on collection include type of service (e.g., curbside pickup or pickup behind the home), frequency of collection, type of collection vehicle, crew size, and collection route design. Once the waste is collected, it must be transported to a processing site, a disposal site, or both. Regular public collection usually occurs more often for urban households than for rural households. Table 24-2 shows that in Jamaica's St. Ann Parish, 63% of urban households had access to regular collection, whereas only 37% of rural households did. The table also shows that in areas where collection is not regular, households process their own waste by open burning, dumping (in their yard or into surface water), or burying.

If collection exists, it is typically performed by a government agency (e.g., a municipality) or a privately owned and operated group. Disposal sites are often located closer to the garbage generators than they are in the developed world. Therefore, waste is generally transported directly to a disposal site by the collector. To provide readers with an idea of scale, Sikasso (Mali), a city of 150,000, has 10 collection areas located on its periphery. In some cases, the solid waste is brought to several collection areas on the periphery of a city, where the waste is then transferred to a common disposal site.

Collection is often done with open-bed trucks (which can spread litter), although it is also common to see a handcart or donkey cart used where the disposal (or transfer site) is closer to a community. Municipalities may have trucks with a several tons of capacity that can handle hilly areas and cover large distances. If vehicular collection is used, it may be more economic if the disposal site is located within 30 min of the point of collection; otherwise, a transfer station may be used that allows waste to be transferred to larger vehicles.

As an example of nonvehicular collection, men with wheelbarrows or donkey carts may collect trash from a household each day and take it to a collection area on the periphery of the city. In this situation, the collectors typically do not empty a residence's trash unless a trash can or bags are used for on-site storage (which is the responsibility of the homeowner to obtain). Those without trash cans may send their children to carry their trash to collection areas or hire local children to perform the same duty. Handcarts may not be appropriate for hilly communities; they have a shorter range (approximately 1 km) and may only be able to accommodate a load of approximately 0.5 m^3. In some cases, the collector operates a large truck (and perhaps a tractor) and can take large piles of trash at

Table 24-2. Disposal Methods by Urban and Rural Households in the Parish of St. Ann, Jamaica

Disposal Method	Total Households	Total Households Using a Particular Disposal Method (Percent)	Total Households That Are Urban (Percent)	Total Households That Are Rural (Percent)
Total	45,380		28	72
Regular public collection	13,929	30.7	63	37
Irregular public collection	654	1.44	48	52
Private collection	316	0.70	13	87
Burn	26,236	57.8	12	88
Bury	643	1.42	7	93
Dumping (all types)	2,947	6.50	9	91
Seas, rivers, and ponds	222	0.49	9	91
In own yard	1,764	3.89	6	94
At municipal site	506	1.12	20	80
Other	455	1.00	11	89
Other	196	0.43	14	86
Not reported	458	1.01	34	66

Source: Statistical Institute of Jamaica 2001.

once from central areas. In terms of cost, a private enterprise in Mali (in 2004) would offer trash pickup for about 1.000 fCFA a month (approximately US$2).

The frequency of collection is typically determined by the amount of waste generated, the type of waste, and the climate. It may be determined by the availability (and reliability) of a motorized collection vehicle and the limited hours that a disposal site is open. For example, a municipality may own only one truck that frequently breaks down and is also needed for other municipal projects.

The primarily organic wastes from markets are typically collected immediately after the market closes. These materials are swept into piles and placed in collection containers or trucks by hand or with the use of a tractor for larger urban markets. Because of their high organic content and high volume, market wastes are easy to target for compost programs directly at the point of generation.

24.4.4 Processing

Processing solid waste is intended to improve disposal options, recover valuable resources, and prepare materials for recovery as new products or energy. Processing includes volume reduction, size reduction, component separation, and resource recovery. For medical wastes, processing is not recommended. Instead, medical wastes should be disposed of in a way that prevents humans from coming into contact with the waste.

Municipal solid waste should be seen as a useful, sustainable energy source, particularly for small island developing countries. This notion is especially important because these nations may be 100% dependent on imported petroleum. Municipal solid waste can in fact become part of a larger energy plan that includes solar, wind, and biofuels from local plant materials.

Figure 24-6 shows the common sight of open burning of municipal solid waste. Residents often burn large community trash piles when they become too large. Burning of solid waste is a method of volume and size reduction. However, open burning typically takes place near living areas, and children are often present. It can create health problems of air pollution, and the burning attracts disease vectors (see Fig. 24-1). The importance of developing a well-organized reduction, reuse, collection, and transfer system is clear.

Figure 24-7 depicts a typical brick (or block) incinerator that can reduce the volume of solid waste. Figure 24-8 shows a metal drum incinerator. The metal drum incinerator can be constructed from 200-L (50-gal) metal drums that have a fire chamber located below metal grating that separates the waste from the fire. A cover is placed on top of the drum, and a chimney is added. An incinerator constructed from a metal drum should only be viewed as a temporary solution (e.g., for the emergency disposal of medical wastes).

Figure 24-6. Open Burning of Municipal Solid Waste, Which Is a Method of Volume and Size Reduction.

Note: Open burning is not recommended because of health considerations. The figure shows the close proximity of homes and children.

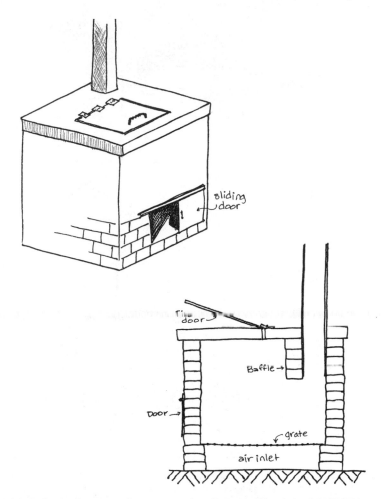

Figure 24-7. An Incinerator Constructed to Process Municipal Solid Waste.

Note: Care should be taken to separate electronic waste, plastics, and batteries, which may contaminate the ash. The resulting ash can be used as a desiccant in compost latrines or can be applied to agricultural fields.

Source: Redrawn from Davis and Lambert (2002) with permission.

24.4.5 Disposal

No matter what source reduction, recycling plan, and processing options are developed, there is always some residue of which we must dispose. In much of the developed world, the practice of open dumping of refuse has been (or is being) replaced by the sanitary landfill. This evolutionary process is taking place throughout the world. A sophisticated sanitary landfill requires a high level of planning, design, monitoring, and operation. Thus, a community can only adopt a sanitary landfill if the local economy can afford a high level of economic expenditures and can ensure the presence of trained staff. However, hybrid

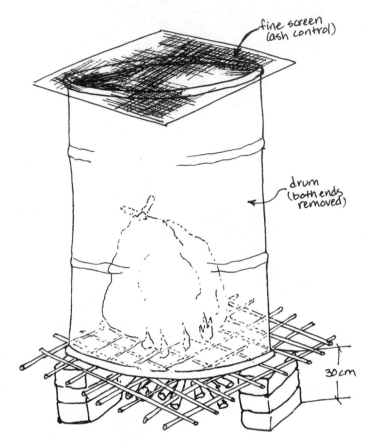

Figure 24-8. Metal Drum Incinerators Are Often Used in Emergency Situations.

Note: In the United States, similar burn barrels are common in rural areas. Because of the different composition of solid waste in the United States, informal backyard burning of solid waste is now the major source of U.S. dioxin emissions (EPA 2006).

landfills and improved dumps can be designed and operated that are appropriate for a particular community situation. For example, dumps can be better managed to prevent exposure to disease vectors by adopting use of cover material, as discussed below. Soil cover can be used in a dump to prevent contact of the deposited waste material with disease vectors, such as rodents and flies.

The three goals of landfilling are to compact the waste as much as possible, place it in an environmentally safe location that will not negatively affect groundwater or surface water resources, and ensure that rodents, flies, and human scavengers do not contact the waste material (Fig. 24-1). In a modern sanitary landfill, the refuse is dumped, compacted, and covered with soil on a daily basis. Although this process is more expensive than a dump, it reduces the nuisances and health hazards associated with decaying and burning refuse. Table 24-3 provides examples of wastes that can pose problems during landfilling.

Table 24-3. Examples of Wastes That Can Pose Problems during Landfilling

Waste	Problems Encountered in the Landfill
Night soil (fecal wastes) and sludge	Ideal upper moisture limit is 70%, so waste may have to be dewatered first.
Industrial and chemical wastes	Potential groundwater contamination; health problems of scavengers because of explosivity, inhalation of aerosols and vapors, and contact with the skin.
Tires	Too many tires may result in unpredictable settlement of the landfill. Tires can be reused in other applications. If landfilled, they should be shredded first. Tires should never be placed near the top of landfills (they may surface as the more dense material below them settles).
Bulky wastes and construction debris	Includes furniture, appliances, baskets, and cages. This material has low density, so it should be broken up by hand or mechanically before being landfilled to ensure even settlement. It may also collect water, which can serve as a breeding ground for insect vectors.
Asbestos	Dust containing asbestos fibers poses severe human health risk. Asbestos wastes should be first sealed in a bag and then deposited below municipal solid waste.
Electronic waste	Contains heavy metals, which can leach and contaminate soil and a groundwater resource.

Source: Adapted from Rushbrook and Pugh 1999.

Scavengers can be involved in a managed waste disposal process (Box 24-1). For example, the collected waste material can be brought to a central location, where it is first tipped onto a managed site (on the ground or a concrete pad). Scavengers could then go through the material before placing it back in a vehicle for final disposal. The scavengers would be provided with gloves and protective footwear; soap and water; education about health and safety issues associated with the contents of municipal solid waste; and appropriate immunizations as part of their compensation for their important role in solid waste management.

))) 24.5 Construction, Operation, and Closure of Landfills

24.5.1 Site Selection

Site selection criteria should include good drainage; availability of suitable soil for daily and final cover; visual isolation; access to major transportation routes; and location away from wetlands, airports, and floodplains. No cultural or religious sites should be present, and the landfill should never be located in an area that has important biodiversity.

The economic livelihood of scavengers cannot be disrupted. It is thus important in the planning process to involve the scavenging community in other economic opportunities, a

))) Box 24-1. Role of Scavenging

It is common for scavengers, the "informal sector," to participate in solid waste management activities in developing countries. This situation occurs primarily because of inadequate municipal services, which creates a large need for informal waste collection and economic opportunities. In China, scavengers collect 8–10% of waste before municipal collection because individual collectors are more aggressive than the municipal collectors and they make part of their living through the collection of the recyclables (Wang and Nie 2001).

> When scavenging is supported—ending that exploitation and discrimination—it represents a perfect illustration of sustainable development that can be achieved in the Third World: jobs are created, poverty is reduced, raw material costs for industry are lowered (while improving competitiveness), resources are conserved, pollution is reduced, and the environment is protected (Medina 2000).

Unfortunately, the majority of scavengers have inadequate access to medical care, work in unsafe conditions, have limited job stability, and receive few educational or other employment opportunities.

Scavenging is a form of processing. It is similar to the hand separation processing used in the developed world in addition to mechanical systems, such as magnetic separation and air classification. In a solid waste management plan, engineers must work with the scavenging community to ensure that their important contribution to solid waste management is maintained, while recognizing the need to improve their access to immunizations, health care, and education.

fair relocation process, or perhaps in the process of upgrading an existing dump into a more controlled sanitary landfill. The construction of a new landfill might result in the creation of a new scavenging community that will appear near the landfill site. The development of water, sanitation, and housing associated with this community should become part of the overall disposal site planning process.

Rainfall frequency affects the seasonal accessibility of transportation routes to the disposal site. Precipitation needs to be managed to prevent washout of the solid waste disposal site. A landfill should not be located where it would reduce the temporary water storage capacity of a floodplain, nor should it be located in a 10-year floodplain. If the landfill is located in the 100-year floodplain, the engineer should ensure that it does not restrict the flow of water during the flooding event.

Landfills should never be built or expanded in wetlands. They should also not be located in areas that are susceptible to destabilizing events. These events include debris flows resulting from heavy rainfall; fast-forming sinkholes caused by excessive groundwater withdrawal; rock falls set off by geological faults or explosives; and the sudden liquification of soil after a long period of repeated wetting and drying. Fortunately, destabilizing events are usually historical, so a general walk-through of potential sites can visually show evidence of such events. It is also best not to site a landfill within 0.5 km of a fault line or other geological fracture. Because landfills attract scavenging birds, they should not be sited less than 1.6 km from an airport.

The site committee should also ensure that groundwater resources are not contaminated. Considerations include the following:

- there should not be downgradient use of groundwater for human or animal consumption;
- there should not be the presence of underlying geological formations (e.g., karst limestone and fractures) that could provide a conduit for groundwater contamination;
- the base of excavation of the landfill should be above the seasonal and long-term groundwater table; and
- the soils between the base of excavation and the groundwater table should be impermeable (e.g., clays) with a hydraulic conductivity of less than 10^{-6} cm/s.

))) Example 24-1. Estimating the Area Required for a Landfill

The volumetric site capacity that is required at the disposal site can be quickly estimated as follows:

$$\text{Site capacity (m}^3) =$$

$$\frac{\text{Waste discard rate (kg/capita-day)} \times \text{population} \times \text{design life (yr)} \times \dfrac{365 \text{ days}}{\text{yr}}}{\text{Compacted waste density (kg/m}^3)} \quad (24\text{-}1)$$

In Eq. 24-1, the design life is typically 10–20 years, and the compacted waste density is a function of whether machinery is used at the landfill site to compact the waste. Loose garbage collected in the developing world can have a density of 250–600 kg/m^3. For individual components of the solid waste stream, the density of loose food wastes is 350–400 kg/m^3, of garden waste is 64–80 kg/m^3, of newspaper is 80–100 kg/m^3, and of plastic containers is 32–48 kg/m^3 (Mihelcic and Zimmerman 2009). The future population should also be forecasted. Equation 24-1 shows that the life of a landfill can be extended by decreasing the waste discard rate or by increasing the waste density.

Note that the volumetric site capacity determined from Eq. 24-1 is already increased by 10–20% to account for the placement of daily and final cover soil on the deposited waste material. The primary purpose of the daily cover material is to prevent contact of the deposited waste material with disease vectors such as rodents and flies. It also minimizes oxygen penetration that could support a fire. Final cover soil not only prevents further contact of disease vectors with humans but also serves as a barrier to precipitation to reduce the amount of leachate generated in the landfill.

The area required for storage of the landfilled solid waste can be determined from

$$\text{Required site area (m}^2) = \frac{\text{Site capacity (m}^2) \times (1.1 \text{ to } 1.2)}{\text{Height of landfill (m)}} \quad (24\text{-}2)$$

The height of the landfill (m) is typically controlled by (Rushbrook and Pugh 1999):

- the availability of a natural void in the topography (in this case, the void (which might be a quarry, ravine, or other natural or anthropogenic depression) is filled to the surface);
- an aesthetic limit above the ground surface so that it does not stand out in a flat plain; and
- the geotechnical load-bearing constraints of the solid waste and the soil that limit the height and slope of an aboveground landfill. **⟩⟩⟩**

Landfills typically rise in a rectangular shape above the ground level because of the way waste is brought into the landfill on a daily and weekly basis. It is common to size the rectangle so the length is 2 times longer than the width and to incorporate a side slope of 3 (horizontal) to 1 (vertical). A buffer of approximately 20 m is placed on each side of the rectangle to allow for vehicle access and to ensure safety in case the slope becomes unstable.

24.5.2 Site Investigation

After a suitable site is identified, a site investigation is performed. The site investigation includes performing a topographic survey for surface contours and features; a hydrologic survey that looks at how the local hydrology will affect drainage requirements; and a hydrogeologic survey that will determine underlying geological formations and soil types, the seasonal high depth to the groundwater table, the direction of groundwater flow, and if possible, the current quality of the groundwater.

The topographic survey of the site is important to estimate the amount of available soil. Because 10–20% of the landfill volume may be taken up by daily and final cover soil, it is important to think not only about where this soil will be obtained, but also where it will be stored. The level of detail associated with the hydrologic and hydrogeologic survey will depend on the amount of resources and type of equipment available. In the absence of a detailed survey, the site can be visually inspected for rock outcroppings and depressions to make informed judgments. Groundwater flow can sometimes be determined by location of downgradient springs and streams.

The existing quality of groundwater can be determined easily if a well is present by measurement of total dissolved solids or conductivity. The equipment for both measurements is relatively easy to use and may be commonly available at a local laboratory or educational institution. Later samples can be obtained to determine if leachate emitted from the landfill is contaminating groundwater because leachate is high in dissolved solids and metals (along with organic material).

24.5.3 Design and Operation

The detailed design should contain a set of drawings that communicate how the site will be developed over the 10–20 year lifetime of the landfill. The design needs to communicate how the site will accept solid waste and how the waste will be landfilled. Also important is how the landfill will be restored at the end of its useful life and whether there will be any monitoring to ensure that the site is not creating an aesthetic problem or health hazard.

Issues of cover soil, leachate production, and gas production are also incorporated into the design and operation of a landfill. Some microbial activity will take place in a landfill. However, landfills are designed to minimize the infiltration of water and the penetration of oxygen. Thus, the biological processes that take place over time are primarily *anaerobic* (i.e., no oxygen is present).

Cover Soil

Daily application of cover soil (approximately 15–20 cm) is a standard part of operations. Covering waste regularly with soil reduces the exposure of waste to vector elements, such as rodents, flies, and mosquitoes (Fig. 24-1). A layer of soil also reduces fire risk by limiting the exposure of waste to sparks, flames or lightning, and the oxygen to feed a fire. Organic compost can serve as a cost-effective cover material in urban areas where cover soil is difficult to obtain and where there is a large amount of organic waste from local markets.

Leachate Production

The solid waste placed in the landfill does contain moisture, ranging from 10% to 60%. Water that infiltrates into the landfill is taken up by the solid waste like a sponge until the moisture-carrying capacity of the waste is reached. After that point, any precipitation that penetrates the solid waste causes an equal volume of water to be displaced. The displaced water travels vertically until it reaches groundwater. The displaced water that percolates through the landfill is called *leachate*.

Table 24-4 shows an example composition of leachate, which is highly concentrated with pollutants that can reduce the quality of surface or groundwater. Most of this analysis has been performed in the developed world. We are currently performing research on leachate composition in the South Pacific. Landfills are designed to minimize movement of water into the mass of refuse and thus to minimize the production of leachate. This goal is accomplished by sloping the top of the landfill (several percent) to maximize run-off of rainwater (while also preventing slope instability). In addition, if available, a less permeable cover made of clay can be placed on top of the completed landfill. This less-permeable cap is then covered by a layer of soil that supports plant growth to maximize evapotranspiration. The soil that lies between the bottom of the landfill and the underlying groundwater can slow movement of leachate if it has low permeability and can buffer contaminants in the leachate through biological, chemical, and physical processes. Arid areas pose less of a threat to leachate development because of low rates of precipitation.

Table 24-4. Composition of Landfill Leachate

Parameter	Range (mg/L)	Average (mg/L)
Chemical oxygen demand	4,500–8,310	7,090
Total organic carbon	169–2,820	1,270
Total suspended solids	130–189	160
NH_3–N		120
PO_4^{3-}–P		296

Source: Adapted from Mihelcic 1999.

Gas Production

Because landfills are designed to minimize the influx of oxygen and moisture, anaerobic biological processes occur that lead to the production of carbon dioxide and methane. The rate of decomposition in a landfill usually reaches a peak at about 2 years, then slows down and continues for 25 years or longer. Besides being a nuisance, methane production at a landfill can lead to fire and explosions if the gas migrates through the subsurface into a neighboring depression that is natural or anthropogenic. Methane concentrations as high as 40% have been detected at lateral distances up to 120 m from the edge of a landfill. Landfills are designed to either vent gases to the atmosphere where the methane may be burned, or in larger cases, to capture the gases for use as a source of energy. Gas movement is controlled by constructing vents and barriers. Gas can be channeled through coarse layers placed in the landfill. Figure 24-9 shows an example of a steel bar-encased vent pipe. It penetrates into the disposed waste to radially collect and then vertically vent landfill gas.

Disposal Methods

Figure 24-10 shows how an old hillside dump can be restored into a landfill. All or part of the waste has been historically deposited along the slope of the hillside. A layer of coarse material (e.g., gravel) is placed at the base of the waste, and a layer of less permeable clay material is placed along the face that contacts the waste and coarse material. The purpose of the less permeable clay material is to prevent the infiltration of rainwater. The clay material is sloped to maximize runoff of rainwater. The coarse material not only serves to stabilize the steep slope of the waste material but also collects any gas that is produced from the waste pile, where it can be vented to the atmosphere. The topsoil cover can be planted with vegetation to further minimize infiltration of rainwater by maximizing evapotranspiration.

Figure 24-11 shows how a pit of any size can use the principles of a sanitary landfill. This work can be accomplished by a household, several households, or a community. The daily amount of refuse is covered with a layer of daily cover soil (in this case, 10 cm) and then after the pit is filled up, a layer of final cover is added and sloped.

The *depression method* occurs at sites where natural features such as canyons, ravines, dry borrow pits, and quarries are available that can be filled in. Careful understanding of the site's hydrology is important. For example, canyons are filled from the inlet to the outlet to prevent backing up of water behind the deposited refuse. Each day's worth of collected refuse that is brought to the disposal site is entombed in a *cell*. Each cell is covered daily with soil (Fig. 24-12). When the landfill has exhausted its life, a final cover is placed on top of the landfill. Topsoil is replaced on the site, and the site is landscaped; groundwater may be continuously monitored; leachate is continuously collected and treated; and gases are continuously collected and vented.

The *trench method* is most suitable in locations where the depth to the groundwater table does not prevent one from digging a trench in the ground. In this method, a trench is excavated by manual labor or with a bulldozer. Refuse is then placed in the trench in layers that are compacted. The operation continues for the day until the desired daily height is reached. Again, daily cover is placed over the refuse to produce a cell. The

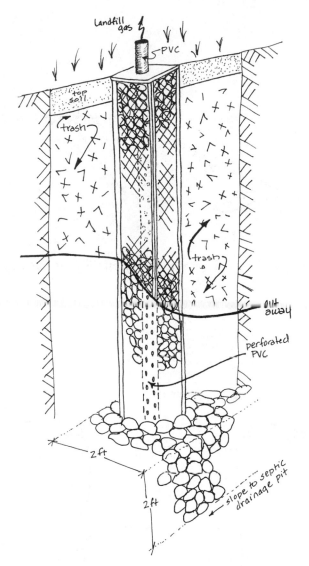

Figure 24-9. Drainage for Leachate and a Gas Collection Chimney Can Be Readily Incorporated into a Landfill.

Note: Rocks are placed in a steel cage to protect a perforated piece of PVC pipe in the middle. Waste is then placed around the cage.

ramp method is similar to the trench method except that solid waste is deposited along a sloped site.

The *area method* is used when the site topography does not allow below-grade disposal. Typically, an earthen berm is constructed and refuse is placed in thin layers against this berm and compacted. In a day, the compacted waste may reach a height of several

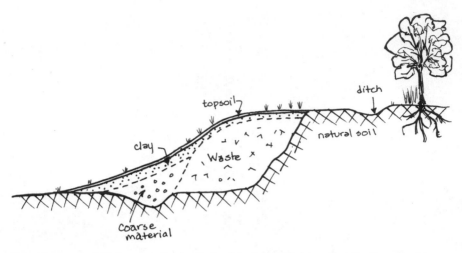

Figure 24-10. Old Dumps, Typically Located on Hillsides, Can Be Restored into More Secure Landfills.

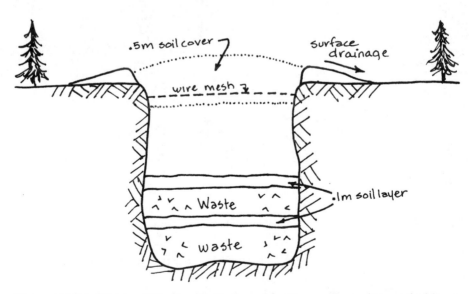

Figure 24-11. Disposal Pits Can Be Designed for One or Several Households, or a Community, to Dispose of Solid Waste.

Note: A layer of daily cover soil (in this case, 10 cm) is placed over the waste to prevent transmission of disease. A final layer of soil, sloped to maximize runoff, is placed on top of the pit after it is filled up. An optional wire mesh can be used in smaller pits where there is concern about animals burrowing for food.

Source: Redrawn from Davis and Lambert (2002) with permission.

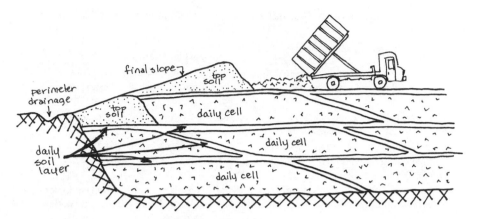

Figure 24-12. Cells of Solid Waste Are Covered with Layers of Soil Cover on a Daily Basis.

Note: A final sloped soil cover is added to minimize infiltration of rainwater.

meters. At the end of the day, daily soil cover is applied as a barrier to disease vectors. When the final design height is reached, a final soil cover is placed on top of the material and sloped.

A properly constructed and operated landfill does not have to require large inputs of expensive mechanical equipment and fossil fuels to run the machinery. Figure 24-13 depicts how a small community landfill can be operated manually. One difference is that

Figure 24-13. A Manually Operated Landfill That Has Most Aspects of a Sanitary Landfill (Cells, Daily and Final Cover, Drainage) Without the Presence of Mechanical Equipment.

Table 24-5. Comparison of Manual and Mechanical Placement of Solid Waste in a Landfill

Manual Placement	Mechanical Placement
Works for smaller sites handling up to 50 tons (4.54 × 104 kg) of solid waste per day.	No limit to size of landfill. Size is based on site conditions, population, expected population growth, and discard rate.
Compaction density is more variable. Loose solid waste can have density that ranges from 90 to 600 kg/m³. Depending on the waste composition, compaction densities may only reach 100–270 kg/m³.	Compaction densities can reach 480–770 kg/m³ after initial compaction and then increase to 700–1,100 kg/m³ after overburden is placed on the waste.
Lower in cost, does not require fossil fuels. Creates more employment.	Higher in cost, requires expensive equipment and fossil fuels. Requires fewer and better-trained workers.

Source: Adapted from Rushbrook and Pugh 1999; Mihelcic and Zimmerman 2009.

compaction densities do not approach that of a landfill that has heavy mechanical equipment. Table 24-5 indicates some differences between manual and mechanical placement of solid waste into a landfill. Manually operated landfills are appropriate for smaller communities that handle less than 50 tons (4.54 × 10^4 kg) of solid waste per day and where costs associated with mechanical operation are a limitation.

24.5.4 Landfill Completion and Closure

When a landfill has reached its capacity, it is ready for closure. A final cover is installed to further minimize rainwater infiltration while maximizing runoff and evapotranspiration. The final cover should be designed and constructed to have hydraulic conductivity no greater than 1×10^{-5} cm/s. The final cover (as high as 1 m) can be constructed of an infiltration layer composed of a minimum of 46 cm of earthen material to minimize the flow of water into the closed landfill. The cover must also contain an erosion layer to prevent the disintegration of the cover. The erosion layer must be composed of a minimum of 15 cm of earthen material capable of sustaining plant growth. For larger community landfills, plants should not have deep root systems that will penetrate the infiltration layer. Maintenance will require that trees and plants with deep root systems be removed on an annual basis. The landfill will continue to produce leachate and landfill gas well beyond its useful life, so a plan must be put in place to ensure that the landfill is monitored in the future.

⟫⟫ 24.6 Disposal of Medical Wastes and Animal Carcasses

Medical wastes and animal carcasses should not come in contact with human or animal scavengers. *Medical wastes* consist of *sharps* (e.g., needles, syringes) and *pathological waste* that has contacted body fluids. Sharps can be incinerated, although it is more common to dispose of them via some type of burial method. These wastes should be buried deeply so that they cannot be reached by animal or human scavengers. Sharps and pathological wastes should always be transported in sealed containers, like a bucket with a secure lid. They can also be buried in sealed concrete vaults. If the pathological waste is nonmetallic, it should be disposed of via incineration to ensure that a high enough temperature is reached to destroy pathogens. Figures 24-7 and 24-8 provide layouts for brick or metal drum incinerators. Figure 24-14 shows an example of an incinerator used to process medical wastes.

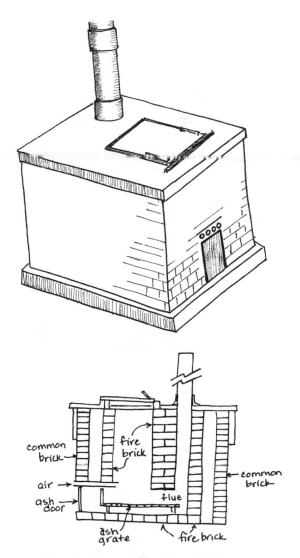

Figure 24-14. A Brick (or Block) Incinerator Built to Process Medical Wastes.

Source: Redrawn from Davis and Lambert (2002) with permission.

Animal carcasses (and animal waste materials) may contain disease and may find their way back into the food supply. Scavengers should not contact these materials, and these materials should not be reused. In emergency situations, animal carcasses can be treated with lime and a concentrated bleach solution (as purchased in the store) before being buried. The lime increases the pH of the waste material, which not only increases the decomposition process but also deters contact with rodents and flies. In cases where a sanitary landfill is used, the pretreatment of animal carcasses should not be necessary. However, care must be taken to ensure that the waste does not reenter the food supply, either through that of the family of the scavenger or by the waste material being resold to a food stand.

⟫⟫ 24.7 Recycling

Recycling has been a viable economic activity that has existed for long periods of time, but in a form often unrecognizable to citizens of affluent countries, who are accustomed to only separating their bottles and newspapers at home. As an economic activity, recycling is thus easier to develop in poorer communities precisely because it is labor-intensive.

Other factors besides economics are known to affect the success of recycling in the developing world. Table 24-6 summarizes 12 factors that relate to barriers and incentives to recycling in developing countries (Troschinetz and Mihelcic 2009). Activities that require the greatest collaborative inputs include waste collection and segregation, municipal solid waste management plans, and local recycled-material markets.

⟫⟫ 24.8 Composting

Composting is a process where organic waste is broken down by biochemical processes in the presence of oxygen to produce a stable humuslike product. Because of the large amount of organic waste found in the developing world's municipal waste stream (Table 24-1 and Fig. 24-3), composting is often believed to be a simple solution to reduce the discard rate. However, composting does produce a by-product that will require usage. One major drawback of composting is that there may be no demand for the product or no economic reasons to use it. For situations where there is no agricultural demand for compost, Box 24-2 provides an example of how it still has an economic advantage as landfill cover.

Table 24-7 provides examples of common materials that are recommended and not recommended for composting (Example 24-2). For a compost pile to function properly, it must be aerated, and the initial carbon to nitrogen (C:N) ratio should be in the range of 20:1 to 40:1 (by mass). Table 24-8 provides the C:N ratio for common materials added to a compost pile. Many materials are either too carbon-rich or too carbon-poor. In this case, a mixture of waste materials can be combined in calculated mass proportions to obtain the optimal initial C:N ratio. After the decomposition process has been completed, the C:N ratio will decrease to 15 or 20 because part of the carbon is converted to carbon dioxide (Box 24-3).

Table 24-6. Descriptions of 12 Factors Influencing Recycling as an Element of Sustainable Municipal Solid Waste Management (MSWM) in Developing Countries

Title	Description
Government policy	Presence of regulations, enforcement of laws, and use of incentive schemes
Government finances	Cost of operations, budget allocation to MSWM, stability and reliability of funds
Waste characterization	Assessment of generation and recovery rates, and composition of waste stream
Waste collection and segregation	Presence and efficiency of formal or informal collection and separation by scavengers, municipality, or private contractors
Household education	Extent of knowledge of waste management methods and understanding linkages among human behavior, waste handling, health, sanitation, and the environment within households
Household economics	Individuals' incomes influence waste handling behavior (reusing, recycling, illegal dumping), presence of waste collection and disposal fees, and willingness to pay by residents
MSWM administration	Presence and effectiveness of private and/or public management of waste (collection, recovery, disposal)
MSWM personnel education	Extent of trained laborers and skilled professionals in MSWM positions
MSWM plan	Presence and effectiveness of an integrative, comprehensive, long-term MSWM strategy
Local recycled-material market	Existence and profitability of market systems relying on recycled-material throughput, involvement of small businesses, middlemen, and large industries and exporters
Technological and human resources	Availability and effective use of technology and/or human workforce and the safety considerations of each
Land availability	Land attributes such as terrain, ownership, and development dictating MSWM

Source: Troschinetz and Mihelcic 2009.

))) Box 24-2. Use of Compost at a Landfill

Pendley (2005) demonstrated the economic advantages of using compost as daily cover in a landfill. The quality of the compost used in this way is not critical. Compost can also be used for erosion control at a landfill site and in land reclamation projects associated with landfill final cover and strip mining. The direct economic benefit of using compost as a landfill cover is derived from avoidance of the cost of hauling cover material from an off-site location. Some indirect economic benefits include the fact that less landfill space is required, market costs associated with promoting the product are eliminated, and quality control costs associated with using the compost for agriculture are reduced.

Organic material that is composted increases in bulk density (from 60 to 80 kg/m^3 to several hundred (and as high as 500) kg/m^3). Thus, even when compost is simply landfilled, a reduction in landfill space associated with disposal of organic wastes can be realized, which will extend the landfill's life. In addition, the composting process (which is aerobic) produces carbon dioxide, whereas organic carbon placed directly into a contained landfill (which is anaerobic) produces methane along with carbon dioxide. The global warming potential of a pound of methane is 25 times higher than that of a pound of carbon dioxide, so composting is positive in terms of climate change.

Table 24-7. Examples of Materials Recommended (and Not Recommended) for Composting

Materials Recommended for Composting	Materials Not Recommended for Composting
Yard and park waste	Meats
Plant-based kitchen waste and other green matter	Dairy foods
Waste residue of the food industry	Fats, oils, and grease
Sewage sludge	Pet excrement
	Fish scraps
	Diseased plants
	Bones

))) Example 24-2. Determining Proper Ingredients for Successful Composting

A poultry manure has a moisture content of 70% and is 6.3% N (on a dry mass basis). The manure is to be composted with readily available oat straw. The oat straw has a moisture content of 20% and is 1.1% N (on a dry mass basis) The desired C:N for the mixture is 30. The C:N ratio for poultry manure is 15, and the ratio for oat straw is 48. Determine the kg of oat straw required per kilogram of manure to provide the C:N ratio that will result in optimal composting.

Solution

Assume 1 kg of poultry manure dry mass. Let $X =$ kg of oat straw on a dry mass basis. The mass of carbon and nitrogen obtained from each material in the mixture is given by

Dry mass nitrogen from poultry manure $= 1 \text{ kg} \times (1 - 0.7) \times 0.063 = 0.0189 \text{ kg}$

Dry mass carbon from poultry manure $= 1 \text{ kg} \times (1 - 0.7) \times 0.063 \times 15 = 0.2835 \text{ kg}$

Dry mass nitrogen from oat straw $= X \times (1 - 0.2) \times 0.011 = 0.0088 \times X$

Dry mass carbon from oat straw $= X \times (1 - 0.2) \times 0.011 \times 48 = 0.4224 \times X$

The overall $C : N$ ratio is given by

$$30 = \frac{(\text{mass carbon from poultry manure} + \text{mass carbon from oat straw})}{(\text{mass nitrogen from poultry manure} + \text{mass nitrogen from oat straw})}$$

$$30 = \frac{(0.2835 + 0.4224 \times X)}{(0.0189 + 0.0088 \times X)}$$

Solving for X, we find that $X = 1.8$ kg. Thus, for every 1 kg of poultry manure, we must add 1.8 kg of oat straw to obtain an optimal $C : N$ ratio of 30. The poultry manure is a better source of nitrogen, and the oat straw provides a better source of carbon.

Source: From Mihelcic and Zimmerman 2009.

As a rule of thumb, when the moisture content of a compost pile drops below 40% or 60%, moisture will become limiting. Composting is thus possible under seasonal rainy conditions found in many parts of the world. An acceptable moisture content can be determined if you are able to squeeze a few drops of water from a handful of composting material. Smaller particle size is also better, which is why it is best to chop or grind large waste material before adding it to the pile. In addition, the pH should be near neutral (6–8).

Initially, mesophilic microorganisms actively metabolize and reproduce in the compost pile. If the conditions are optimal, the microbes will generate heat in the pile. Above 45 °C, thermophilic microorganisms will become predominant, and temperatures can approach 70 °C. As the carbon energy source diminishes, so too does the activity and the heat generation of the microorganisms. The temperature of the compost pile will decrease over time to a point that the mesophilic organisms become active again. A final maturation stage occurs when the biological activity and decomposition slow down, with temperature approaching ambient levels.

Mature compost will not have evidence of most of the initial organic materials that were added to the pile and will have a dark appearance and an earthy, humus-type smell. Compost is mature when the oxygen demand (i.e., the microbial activity) approaches zero. One simple test is to place a sample of compost in a plastic bag and keep it away from direct sunlight. If the bag does not swell or show condensation on the inside, the compost can be considered mature. The dry weight reduction because of loss of organic carbon will be one-half to one-quarter, and the final $C : N$ ratio will be approximately 15–20 (perhaps from an initial high of 30 or 40).

Table 24-8. Carbon-to-Nitrogen Mass Ratios for Common Organic Materials Placed in a Composting Pile

Material	C : N Ratio
Human feces	6–10
Urine	0.8
Cow manure	18
Poultry manure	15
Horse manure	25
Kitchen waste	25
Sawdust	200–500
Grass	12–15
Straw, wheat	128–150
Municipal wastewater treatment plant sludge	6

A compost pile needs to be aerated and can be easily turned with manual labor. It is best to turn the material near the edge of the pile (which is in contact with oxygen) to the center of the pile. Water may be added before, during, or after the turning. Heat is allowed to escape so that the microbial activity does not suppress itself, and the pile is left alone until the next turning. Passive aeration can also occur, whereby slotted piping is placed inside the pile to allow air to be continuously mixed into the pile.

>>> **Box 24-3. Three Simple Methods to Construct a Small-Scale Composting System**

- **Windrow method.** In the absence of a concrete pad, windrows can be constructed of bamboo (or other wooden material), chicken wire, and tie wire (Fig. 24-A). Lay bamboo logs approximately 3 m long parallel to each other and 1 m apart. Lay chicken wire on top and secure it to the bamboo with tie wire. Chop the organic waste by hand into small pieces and place it on top of the chicken wire. Transparent plastic sheeting can be laid on top of the windrow to retain heat and moisture.
- **Elevated barrel.** Puncture a metal barrel with several holes in the bottom and sides, then elevate it using hollow blocks or rocks (Fig. 24-B, a). Chop organic waste into small pieces and add it to the barrel. The barrel top can be covered with the transparent plastic sheeting to retain moisture and heat. If equipment is available, a door can be cut out near the bottom of the barrel to access matured compost.
- **Compost pit.** Manually dig a pit with square dimensions of 1 m and depth of 35–100 cm (Fig. 24-B, b). Chop organic waste into small pieces and add it to the pit. The whole pit can be covered with the clear plastic sheeting to retain moisture and heat.

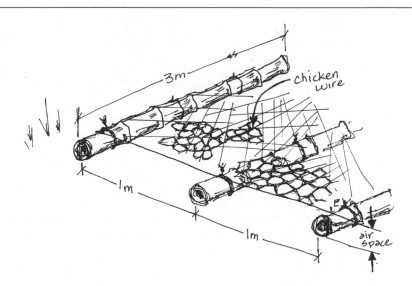

Figure 24-A. Windrows for Composting Can Be Constructed of Bamboo (or Other Wooden Material), Chicken Wire, and Tie Wire.

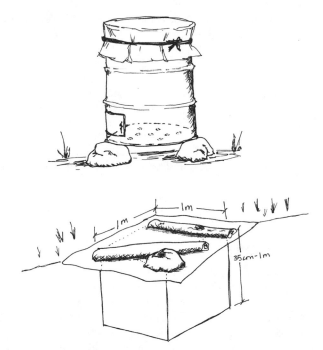

Figure 24-B. (a) An Elevated Metal Barrel Can Be Used for Composting by Puncturing Several Holes in the Bottom. (b) A Compost Pit with Square Dimensions of 1 m and Depth of 35–100 cm Can Be Manually Dug. Organic Waste Is Chopped into Small Pieces and Added to the Pit.

⟫⟫ References

Davis, J., and Lambert, R. (2002). *Engineering in emergencies*, ITDG Publishing, Rugby, Warwickshire, U.K.

Environmental Protection Agency (EPA). (2006). *An inventory of sources and environmental releases of dioxin-like compounds in the United States for the years 1987, 1995, and 2000*, EPA/600/P–03/002f, Environmental Protection Agency, Washington, D.C.

Medina, M. (2000). "Scavenger cooperatives in Asia and Latin America," *Resources, conservation and recycling*, 31(1), 51–69.

Mihelcic, J. R. (1999). *Fundamentals of environmental engineering*, John Wiley and Sons, New York.

Mihelcic, J. R., and Zimmerman, J. B. (2009). *Environmental engineering: Fundamentals, sustainability, design*, John Wiley & Sons, New York.

Oxfam. (2007). *Guidelines for solid waste management in emergencies*, An Oxfam Technical Manual, Oxfam Public Health Engineering, Oxford, U.K.

Pendley, P. S. (2005). *Feasibility and action plan for composting operation incorporating appropriate technology at Riverton disposal site, Kingston, Jamaica*, <http://cee.eng.usf.edu/peacecorps/Resources.htm> (Jan. 23, 2009).

Rushbrook, P., and Pugh, M. (1999). *Solid waste landfills in middle- and lower-income countries: A technical guide to planning, design, and operation*, World Bank Technical Paper No. 426, World Bank, Washington, D.C.

Statistical Institute of Jamaica. (2001). *Population census 2001—Jamaica Vol. 1 Country Report*. Statistical Institute of Jamaica, Kingston, Jamaica.

Troschinetz, A. M., and Mihelcic, J. R. (2009). "Factors influencing sustainable municipal solid waste management in developing countries: 12 barriers and incentives to recycling," *Waste Management*, 29(2), 915–923.

United Nations Department of Economic and Social Affairs (UNDESA). (2005). *Agenda 21—Chapter 21, Environmentally sound management of solid wastes and sewage-related issues*, Division for Sustainable Development, United Nations Department of Economic and Social Affairs, New York, <http://www.un.org/esa/sustdev/documents/agenda21/index.htm> (Jan. 23, 2009).

Wang, H., and Nie, Y. (2001). "Municipal solid waste characteristics and management in China," *J. Air and Waste Management Association*, 51, 250–263.

Wells, C. (1994). "The Brazilian recycling commitment: Helping stimulate recycling in a developing country," *Industry and Environment*, United Nations Environment Programme, Paris, France, 17(2), 14–17.

World Bank (2003). "Waste reduction, reuse, and recycling," *Thailand Environment Monitor 2003*, The World Bank Group, Washington, D.C., 13–18.

⟫⟫ Further Reading

Owens, E. L. (2008). *Material flow analysis for Kayangel State, Republic of Palau: Solid waste management on a small Pacific island*, <http://cee.eng.usf.edu/peacecorps/Resources.htm> (Jan. 24, 2009).

Post, J. L. (2007). *Solid waste management in Jamaica: An investigation into waste reduction strategies*, <http://cee.eng.usf.edu/peacecorps/Resources.htm> (May 21, 2009).

Post, J. L., and Mihelcic, J. R. (2009). "Waste reduction strategies for improved management of household solid waste in Jamaica," *International J. of Environment and Waste Management*, Vol. 3, in press.

Prüss, A., Giroult, E., and Rushbrook, P. (1999). *Safe management of waste from health care activities*, World Health Organization, Geneva, Switzerland.

United Nations Environment Programme (UNEP). (1996). *International source book on environmentally sound technologies (ESTs) for municipal solid waste management (MSWM).* Division of Technology, Industry and Economics, United Nations Environment Programme, Osaka, Japan, <http://www.unep.or.jp/Ietc/ESTdir/Pub/MSW/index.asp> (Jan. 23, 2009).

United Nations Environment Programme (UNEP). (2005). *Solid waste management.* Division of Technology, Industry, and Economics, United Nations Environment Programme, Osaka, Japan, <http://www.unep.or.jp/ietc/Publications/spc/Solid_Waste_Management/index.asp> (Jan. 23, 2009).

Wilson, D. C., Whiteman, A. D., and Tormin, A. C. (2001). *Strategic planning guide for municipal solid waste management,* published by the World Bank DFID on behalf of the Collaborative Working Group on SWM in Middle- and Low-Income Countries, Washington, D.C., <http://www.worldbank.org/urban/solid_wm/erm/start_up.pdf> (Jan. 23, 2009).

World Health Organization (WHO) website. (n.d.). *Needle-removing and recycling of the plastic from auto-disable (AD) syringes,* Waste Management—WHO Policy and Activities, Immunization Safety, World Health Organization, <http://www.who.int/immunization_safety/waste_management/update/en/index3.html> (Jan. 23, 2009).

World Health Organization (WHO). (1993). *Information management for municipal solid waste management services,* Environmental Health Centre (EHC), Western Pacific Regional Office, World Health Organization, Geneva, Switzerland.

Air Quality

25

Indoor Air

⟩⟩⟩ 25.1 Indoor Air Quality and Health

25.1.1 A Global Perspective

More than 70% of global person-hours are estimated to be spent inside a built structure, and more than 50% of a person's air intake over their life occurs inside the home (Sundell 2004). Some 33% of total person-hours take place in rural indoor settings of the developing world (Smith 1993). Indoor air quality is not affected solely by activities that take place in the home. Ambient air pollution also eventually affects the indoor environment (Smith and Mehta 2003). Accordingly, indoor air quality is the dominant route of human exposure to inhalation of air pollutants, even though sources of outdoor ambient air pollution emit higher absolute quantities of pollutants.

Table 25-1 lists the most common indoor air pollutants that affect human health, along with their sources. Indoor air pollution is implicated in the deaths of more than 2 million people every year and is responsible for 3.7% of disability adjusted life years (DALYs) contributing to the burden of disease. This statistic places indoor air pollution close to the lack of clean water and appropriate sanitation in terms of its effect on human health (refer back to Table 2-1).

Health concerns related to indoor air quality in the developed world include allergies; asthma; and airway infections because of high levels of biological contaminants, dust, and low ventilation rates. Cancer is also linked to indoor air pollution because of exposure to environmental tobacco smoke (ETS) and radon gas. As developing nations increase their economic wealth, issues such as ETS are becoming more prevalent, adding to the existing burden of disease associated with unprocessed solid fuels and inefficient cooking technologies (Bruce et al. 2002).

25.1.2 A Developing World Perspective

The majority of air pollution in the developing world is emitted from fuel combustion. This pollution originates primarily from unprocessed biomass fuels, which emit carbon monoxide and particulate matter (PM), as well as other hazardous materials and gases.

Although additional indoor air pollutants affect human health in rural settings, particulate matter is the most significant because of its ability to affect the upper airways of the respiration system. Particulate matter is measured as either PM_{10} (or particles less

---

This chapter was written by Matthew D. Babcock with contributions from Kurtis G. Paterson.

Table 25-1. Common Indoor Air Pollutants and Their Sources

Indoor Air Pollutant	Source(s)
Carbon monoxide (CO)	Incomplete combustion, environmental tobacco smoke
Sulfur oxides	Incomplete combustion
Nitrogen oxides	Incomplete combustion
Particulate matter (PM)	Incomplete combustion
Ozone	Photocopiers, laser printers, air cleaners
Volatile organic compounds (VOCs)	Off-gassing of furniture, fuel oil vapors, solvents, paints
Biological contaminants	Mold spores, animal dander, insect remains
Radon gas	Soil

Note: Incomplete combustion is associated with indoor cooking and heating in the developing world.

than 10 μm diameter) or $PM_{2.5}$ (particles less than 2.5 μm diameter). The smaller $PM_{2.5}$ particles penetrate deeper into the lower lungs.

World Health Organization (WHO) air quality guidelines set acceptable PM_{10} concentrations at 50 μg/m^3 (WHO 2005). In contrast, 24-h mean concentrations for PM_{10} found in developing world homes range from 300 to more than 3,000 μg/m^3. During use of a cook stove, indoor air PM_{10} concentrations can reach as high as 20,000–30,000 μg/m^3 (Bruce et al. 2002).

In the developing world, up to 90% of rural households use unprocessed solid fuels for their daily cooking and heating needs. This use can expose inhabitants to high levels of pollutants for 3–7 hours per day (Bruce et al. 2002). Thus, several percent of the total DALYs in developing countries from use of solid fuel occur because of acute respiratory infections (ARIs), chronic obstructive pulmonary disease (COPD), tuberculosis, asthma, lung cancer, ischemic heart disease, and blindness.

ARI and COPD are the main disease categories connected to indoor air pollution in the developing world. In children under the age of 5, ARIs caused by exposure to indoor air pollution are the largest category of both deaths (64%) and DALYs (81%) (Smith and Mehta 2003).

A significant fraction of the global burden of disease is associated with COPD, which makes it one of leading causes of death in the world, much of which is attributed to indoor smoke. More striking is the fact that 40–45% of the burden of COPD experienced by women in the developing world is caused by indoor air pollution from solid fuel use (Smith-Sivertsen et al. 2004). Other health effects related to high concentrations of indoor air pollutants found inside homes of the developing world are middle ear infection and low birth weight.

Because of the mostly chronic nature of diseases such as COPD and ARIs, rural populations in the developing world are less familiar with their causes and the long-term effects of indoor air quality in general. Interventions and studies designed to improve indoor air quality, although on the rise, are not as widespread as those designed to improve conditions related to water and sanitation.

⟫⟫ 25.2 Assessing Indoor Air Quality

Many environmental, technical, and behavioral parameters influence the air quality of a given interior space and the health of the inhabitants. Figure 25-1 lists several of these parameters and their relationships. There are three main methods for collecting indoor air assessment information related to the parameters shown in Fig. 25-1: household and community surveys, physical inspections, and air quality monitoring equipment.

Air quality monitoring equipment helps to obtain accurate data on the concentrations of particulate matter and carbon monoxide, but it is expensive. Access to such equipment may be limited. Most indoor air quality assessments are therefore likely to be performed through a combination of surveys and physical inspections of cooking and living spaces.

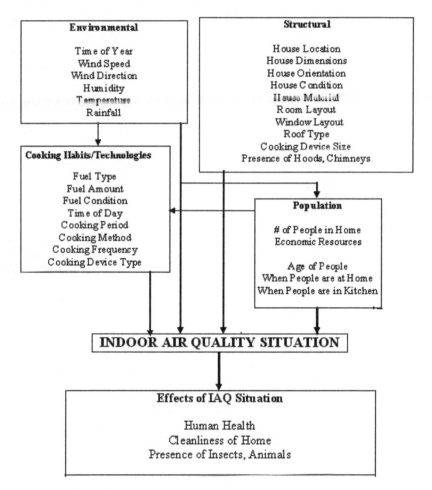

Figure 25-1. Parameters That Affect Indoor Air Quality.

Source: Babcock (2006).

Household surveys should include questions on family size, family health, and the perception of existing indoor air quality to judge the household's knowledge of and interest in improving their indoor air quality. It is important to interview both men and women. Women are usually exposed to more indoor air pollution and involved more in cooking, and men are usually involved in the construction, modification, or both of the cooking technology and cooking space.

The surveys should also request information on the type of food prepared, cooking methods and times, fuel types and quantities used, maintenance, and specifics of the cooking and heating technology. This information will provide an overall picture of the current indoor environment and related occupant behavior. Similar surveys can be performed with community organizations and for other spaces, such as school kitchens and meeting houses, where large amounts of cooking occur.

Information obtained from surveys will be useful later for the creation of educational units aimed at increasing awareness and demand for air quality improvement measures. An occupant's perception of poor indoor air quality determines whether he or she will be motivated to change the situation. Because poor indoor air quality is a daily occurrence, it may be viewed as a low priority that does not require immediate correction.

Physical inspections involve measuring cooking space and household dimensions, as well as observing cooking and heating processes and the resulting airborne emissions. Physical inspections help to cross-check the information obtained through the survey. They are also necessary for the design of new cooking technologies or the modification of an existing cooking space.

))) 25.3 Indoor Air Pollution and the Intervention Pathway

Indoor air quality interventions that reduce exposure can be identified along each stage of a generalized pathway (e.g., fuel substitution, reducing emissions with stove technology, improving ventilation, changing behavior). After completing the indoor air quality assessments, a specific pathway can be sketched out to depict the local situation. This pathway can then be used as a guide to determine where different types of interventions (technological, personal, or social) could be implemented to improve indoor air quality and to prevent the incidences of diseases related to indoor air quality.

In general, it is more desirable to intervene early in the generalized pathway. However, this goal is not always possible because of economic, social, or technological constraints. An accurate indoor air quality assessment highlights where proposed changes can be most effective and efficient.

))) 25.4 Solid Fuels and Fuel Substitution

25.4.1 Traditional Solid Biomass Fuels

Solid biomass fuels include wood, dung, crop residues, and charcoal. When incompletely burned, these solid fuels can release extremely high levels of particulate matter and carbon monoxide in addition to other pollutants.

Even though they are polluting, solid fuels are widely used because they are inexpensive and more accessible than more processed fuels. Solid biomass fuels are also renewable if managed sustainably. Use of a particular solid fuel is related to geography and access. Where neither wood nor charcoal is available, crop residues and animal dung will be used, both of which can be the most polluting fuel types.

Although local solid fuels may be more accessible than fuels such as kerosene and liquid petroleum gas (LPG), the use of solid fuels requires more labor and time. The reliance on solid fuels also puts pressure on local natural resources, especially forests. Unsustainable harvesting of fuel wood is a major cause of deforestation in many areas of the world. Because deforestation can adversely affect water supply, the use of solid fuels can have health effects above and beyond those already linked to indoor air pollution.

25.4.2 Fuel Substitution by the Energy Ladder

As the social and economic status of households improves, there tends to be a transition to fuel types that are less polluting in terms of their indoor environment. This concept has been termed the energy ladder principle (Fig. 25-2). Not every household follows the exact pattern depicted in the energy ladder. Households often continue to use solid fuels even when they have the economic ability to move up the energy ladder, either because

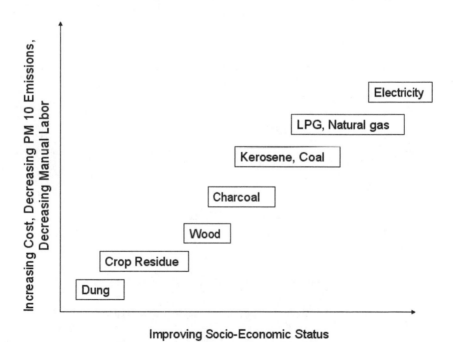

Figure 25-2. The Energy Ladder, Showing How Fuel Type Can Change as a Household's Social and Economic Status Increases.

Source: Adapted from Smith et al. (1994).

the solid fuels are more accessible or because their newly acquired resources are used for other household needs.

Many households use LPG or kerosene stoves in combination with a standard three-stone fire because of economics (it is expensive to continually purchase LPG or kerosene fuel); access (fuels like LPG and kerosene are usually not produced locally); and convenience (cooking in larger pots might be easier on a three-stone fire).

Using fuels further up the energy ladder decreases the amount of particulate matter and carbon monoxide that is emitted to the indoor environment, although they do emit some particulate matter and greenhouse gases. Figure 25-3 provides a comparison of the PM_{10} and greenhouse gas emissions of different types of fuel. The use of LPG- and kerosene-based stoves may also be seen as a status symbol. Although this development may help to speed the adoption of those technologies when financial resources allow, it can also create resistance to the adoption of other improved cook stoves.

Biogas emits less particulate matter (Fig. 25-3) and can be obtained from crop residues, manure, and even latrines. However, biogas systems are perceived as being more expensive to implement, and they require a level of coordination and technical experience that sometimes restricts them to larger projects.

Some rural areas have access to community electricity, usually generated by diesel engines. Although the use of electric stoves and heaters decreases the level of indoor air pollution, electricity generated in this fashion is not sustainable for many reasons. Perhaps

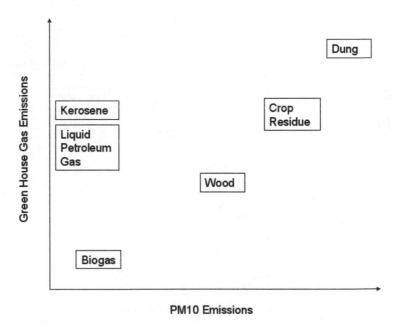

Figure 25-3. Emissions of Greenhouse Gases and Particulate Matter for Different Fuel Types.

Source: Adapted from McGranahan and Murray (2003).

the cleanest method of providing fuel for cooking and heating is the local generation and distribution of electricity based on renewable wind or solar energy, which is beyond the scope of this book.

))) 25.5 The Traditional Three-Stone Fire

Wood and other solid fuels can be burned efficiently using advanced stoves and appropriate cooking skills. Unfortunately, traditional cooking technologies often do not efficiently burn solid fuels and do not remove pollutants emitted from the cooking and heating space. The most common traditional cooking technology uses one of many variations of the three-stone open fire (Fig. 25-4). This system consists of a cooking pot placed on top of three large stones (or logs) arranged in a small circle. A small open fire is constructed in the center of the three stones. The stones are placed directly on a dirt or concrete floor or on a box of packed earth and ash if located in a wooden structure. If logs are used instead of stones, the ends of the logs will burn and can act as a steady heat source.

Table 25-2 lists some advantages and disadvantages associated with a three-stone fire. In general, the advantages are outweighed by the disadvantages in terms of health and the inefficient use of fuel. One small improvement to the three-stone system is to put out an open fire before it dies on its own, thus conserving fuel and preventing unnecessary air pollution.

Local weather conditions and planning around the weather also can affect the quality of the fuel source. For example, poor timing of the collection of a solid fuel can lead to the use of wet fuels, which emit more particulates. Particulate emissions can be reduced by storing small sets of damp fuel wood close to the stove to preheat and dry the wood. Although this approach is limited, preheating or predrying fuel can help to improve combustion and limit indoor air pollution.

Figure 25-4. A Three-Stone Fire.

Note: Thick logs or large bricks can be substituted for the stones.

Table 25-2. Advantages and Disadvantages of a Three-Stone Cooking Fire

Advantages	Disadvantages
Simple and inexpensive to construct and use	Much of heat generated is lost to surrounding air (especially with high wind)
Good for temporary, outdoor use in farmland or work site	Inefficient burning of solid fuel
Much of heat generated is transferred to the cooking pot (with low wind)	Smoke and other combustion products are released directly into cooking space
Fire is provided with sufficient oxygen	No on/off switch (i.e., fuel continues burning when not being used)
Easy to use with different pot sizes	Only one cooking pot can be heated at a time
	Can be physically unstable, lead to spilling of contents and/or burning of user

))) 25.6 Improved Stoves

The goals of improved cooking technology are to improve cooking efficiency and remove indoor air pollution from inhabited areas. Improving cooking efficiency involves achieving more complete combustion of the fuel, insulating the interior of the stove, cooking multiple pots on one stove, and directing more of the heat to the cooking pot. When fuel is used more efficiently, less fuel is used and fewer particulates, carbon monoxide, and other pollutants are released.

Stove entrances can be designed to be small to force the user to add less fuel at a time. This restriction allows the fuel to burn more completely before new fuel is added. Most improved stoves are also front loading and require the user to monitor their fuel more closely than with an open fire. Removing pollution from the cooking and living area requires a chimney. Stoves with chimneys are sometimes simpler to construct than more complex stoves that focus solely on fuel and cooking efficiency.

The selection of an appropriate stove technology requires careful coordination with the community and individual households to determine what is most appropriate based on economic, health, and cooking factors. Economic factors and the presence of local natural resources often play the largest role in determining the type of stove selected.

Two basic types of improved stoves are presented in this chapter, the Lorena stove and the Rocket stove. Both are low cost, simple to construct, and readily adaptable. Table 25-3 summarizes the differences between the two designs. Many other stove designs are available that are based on these types. For example, advanced stove designs such as the Rocket-Lorena and the Improved Plancha stove (discussed in Section 25.6.3) incorporate the beneficial aspects of both the Lorena and Rocket stoves. Readers interested in more detailed descriptions of types of improved cook stoves can find such information from the Partnership for Clean Indoor Air (Box 25-1).

25.6.1 Lorena Stoves

The Lorena stove, widely promoted in Central and South America, is named after its materials: *lodo* (Spanish for "mud") and *arena* (Spanish for "sand"). The Lorena stove (Fig. 25-5) consists of a relatively massive body made of a mud and sand mix, with a fuel

Table 25-3. Comparison of Lorena and Rocket improved stoves

Basic Improved Stoves	Advantages	Disadvantages
Lorena toves	Removes pollutants from the cooking space and can cook multiple pots at the same time	Low cooking and fuel efficiency because of required materials and massiveness of the design
Rocket stoves	Increases fuel efficiency and cooking efficiency, which creates less pollution	Requires good insulating materials and cooks only one pot at a time Does not have a chimney integrated into the design that would vent to the outside

access opening, multiple openings on the top surface for cooking pots, a chimney, and a tunnel carved into the body to connect everything (Fig. 25-5). It is simple to construct, and everything, including the chimney section, can be manufactured from local materials. Air pollutants are trapped within the stove and exit through the chimney, which also provides a slight draft that theoretically improves the burning efficiency. The mass of the stove and types of materials used in its construction keeps the surfaces relatively cool, preventing burns and the overheating of the surrounding area.

Lorena stoves improve indoor air quality. However, although originally promoted as being able to increase fuel and cooking efficiency, recent studies suggest that they do not increase overall efficiency. This problem occurs because the sand–clay mix is not insulating and much of the heat generated by the cook fire is absorbed into the large mass of the stove instead of being transferred to the pot. Because basic Lorena-type stoves do not increase fuel and cooking efficiency to a great extent, they may not be appropriate for areas where the emphasis is on fuel conservation in addition to improving indoor air quality.

Materials

The most important aspect of constructing a Lorena stove is to identify the correct sand–mud mix. While the word "mud" is used in the stove's name, the material of importance is soil with high clay content. The sand provides the mass of the stove, and the clay holds the sand together. Any type of sand works well, though angular grains are better. Any small rocks should be removed by sifting with a screen.

))) Box 25-1. The Partnership for Clean Indoor Air

The Partnership for Clean Indoor Air was created at the 2002 World Summit on Sustainable Development and consists of more than 170 nongovernmental, private sector, educational, governmental, and multilateral agencies with the mission of reducing exposure to indoor air pollution from energy use. The PCIA website (www. PCIAonline.org) provides extensive resources on cook stoves, including *Design Principles for Wood Burning Cook Stoves* (Bryden et al. 2006). This document provides details on stove theory, testing, safety, and design.

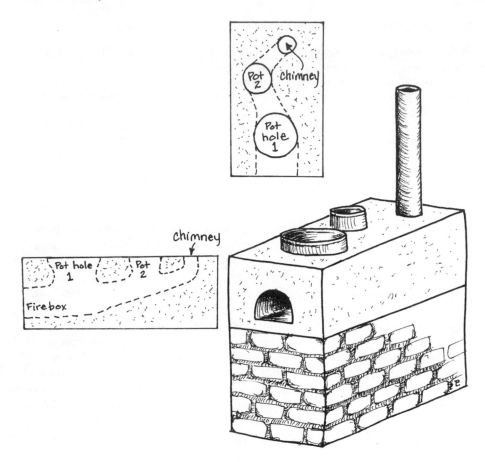

Figure 25-5. The Lorena Stove, Showing the Location of the Pothole Above the Firebox, a Secondary Pothole, and the Chimney. A Plan (Top) and Section (Left) View of the Lorena Stove.

Note: The stove user determines the location and number of potholes. Pothole size, which is based on cooking pot size, is also household-dependent.

The clay-containing soil is usually harder to find than the sand, but community members may know of the whereabouts of such soil if clay is used for other purposes. Clay can be found where reddish, slick, and shiny soils are located. A clay source can also be identified by observing cutaways found near road construction or those that exist naturally in broken terrain. The clay soil should also be free of small rocks and any organic material because their presence would make it more difficult to carve the stove. Avoiding organic material will require sifting the soil with a 1/4-in. screen and also excavating the soil from below the organic soil horizon. Table 25-4 explains several tests that can be used to evaluate a soil's clay content.

In addition to these tests, the sample should be fired to check for crumbling (i.e., flaking). A sample of soil can be test-fired by creating a small ball approximately 5 cm in

Table 25-4. Tests to Estimate Clay Content in a Soil Sample

Test	Procedure	Results	Analysis
Feel	Check the feel of a slightly wetted sample of soil.	Sticky and greasy Gritty Powder, residue	Contains clay Contains sand Contains silt
Shine	Make a small cake of the sample and rub it with a fingernail or the flat side of a knife.	Surface shines Surface remains dull	Contains clay Contains silt and/or sand
Thread	Make a small lump (15-mm diameter). Roll out the lump into a thinner thread of 3-mm diameter. If the sample breaks before 3 mm, then add more water. If it can be rolled thinner than 3 mm, then relump it until it just breaks apart at 3 mm. Reform ball of sample and apply pressure with fingers.	Ball deforms under large amount of pressure without cracking and crumbling Ball deforms, cracks, or crumbles under small amount of pressure Cannot be made into a ball	Contains clay Contains little clay Too much sand and/or silt
Ribbon	Roll a sample thread 15 mm in diameter and 10 cm long. The thread should not be sticky but should be able to be rolled to a 3-mm diameter without breaking. Place thread in the palm of one hand and hold the end between the thumb and forefinger. Flatten and advance the thread between thumb and forefinger. Form largest ribbon possible before breaking.	Long ribbon, 20–25 cm Short ribbon, 5–10 cm No ribbon	Contains large amount of clay Contains medium to small amount of clay Contains little or no clay

Source: Adapted from Evans and Boutette 1981.

diameter and letting the ball dry in the sun. After drying, the sample is placed on hot coals for 1 h, after which it is removed and allowed to cool. The sample should not crumble after cooling or when rubbed with the hand. If the sample crumbles easily, a better soil will need to be identified. Sample cracking is expected, however, until sand is added to form the Lorena mixture.

After identifying appropriate sources of sand and of clay-containing soil, a proper mix must be determined. Too much clay causes the stove body to crack, too much sand results in low stability and crumbling, and too much silt results in a dilute mix and flaking.

When making mixes of different sand:clay ratios, crush all lumps of clay and homogenize the mix as much as possible. Table 25-5 provides details for two tests to determine the correct sand:clay ratio. For each test, make multiple sample mixes of different sand:clay ratios.

A more thorough test is to make multiple test blocks. Do this procedure whenever encountering a new source of sand or clay. The test blocks should have different ratios of sand:clay and should be at least 80 cm long. After drying the blocks in the sun, visually

Table 25-5. Quick Tests to Determine the Correct Sand:Clay Ratio for Lorena Mixes

Test ame	Procedure	Results	Analysis
Ball	Make a sample ball of about 5 cm. Apply pressure to sample with thumb and forefinger.	Little to no deformation	Sample is acceptable
		Large amount of deformation	Sample requires more clay
Palm	Make a small sample that is wet enough to shine. Flatten the sample between two palms. Holding the sample in one palm, gradually turn over that palm and relax the grip on the sample.	Sample cleanly falls off of hand	Sample is acceptable
		Sample sticks to hand or leaves behind a lot of material	Sample requires more sand

Source: Adapted from Evans and Boutette 1981.

inspect the cracks. Investigate the surface strength by scraping the block with a sharp object to check for crumbling. Good mix designs have a hard surface and no deep cracking. Adding manure, hay, or grass to the mix may improve the eventual strength of the mix, but will make carving the stove more difficult

Add the correct amount of water to the mixture. One way to test the appropriate water content in a Lorena mix is to make a packed ball approximately 5 cm in diameter. Throw the ball about 1 m in the air and catch it with a flat palm. If the ball cracks after landing, there is insufficient water. If it deforms, there is too much water.

The other two components of a Lorena stove are the stove base and the chimney. Large stones or bricks can be used for the base if they are available, or a reinforced table can be constructed of wood. Another possibility is to create a box of soil that supports the stove. The walls of this box are usually wood or concrete blocks. Although the chimney can be constructed from carefully placed bamboo and adobe sections, it is typically made from metal tubing or folded sheet metal that is sealed well. Metal cans and food containers can be carefully attached together to form a chimney structure.

Construction

Stove construction requires time and physical labor, but little technical expertise. Here are the steps required in construction:

- Construction of the stove base.
- Construction of a wood or sheet metal form into which the stove body material is poured.
- Creation of the Lorena mix and packing it into the form.
- Setting of the mix.
- Carving of the stove.
- Addition of the chimney structure.

The base should be constructed in the desired location with the intended orientation of the stove in mind. The purpose of the base (typically 45–60 cm) is to raise the stove to a more comfortable height for the user. Keeping in mind that the body of the stove will have a height of approximately 30 cm, the desired height of the base should be determined by the user. The area of the base should be at least as large as the proposed stove body to provide appropriate support and prevent cracking. If possible given the selected base materials, toe space should be included under the base so that the user has to bend less.

After completing the base, a wood or sheet metal form (approximately 30 cm high) is usually constructed for the stove body. A form is not required but is useful to prevent slumping and cracking of the Lorena mix. The downside to a form is that the size and shape of the stove are often different for each household, so different forms must be made.

After the form is set and secured, make the chosen Lorena mix and place it into the form. Place the mix in stages. Pour out layers and then flatten each one with a piece of wood or your hand. Before adding the next layer, ensure that the previous layer is hard enough, though not necessarily dry. The mixture is ready for another layer if a finger lightly pushed into the mix does not go in more than 5 mm. If a layer has dried too much before the next layer is added, dampen the surface. Carefully place a piece of steel rebar inside the stove during the pouring of the Lorena mix to provide additional strength to the entrance ceiling. After filling the form to the top, level out the surface as much as possible.

Leave the Lorena mix to set until it is difficult to push a finger more than 1 cm into the mix. Then remove the form and smooth and even the side surfaces. It is important to allow the mix to dry sufficiently before beginning to carve out the holes that will hold the cooking pots (referred to as *potholes*) and the tunnel. The firmness of each section of the stove should be tested before carving. This drying process may take longer in wetter or colder climates.

Wet all carving instruments, usually a set of spoons and knives, so that the tools do not tug at the surrounding Lorena mix. Before carving, lay out the pothole pattern on the stove's surface, as well as the exit hole to the chimney. Determine the size of the potholes by the size of the household's cooking pots. Depending on the household or community use, the stove might be designed with one, two, three, or more potholes. All holes (including the chimney hole) that are carved out of the top surface of the stove should be at least 10 cm away from any edges and at least 5 cm away from each other. The firewood feed entrance should also be laid out on the appropriate side of the Lorena block. It should be located at least 5 cm above the bottom of the stove body, and the top of the entrance should be at least 10 cm below the top surface.

To begin carving the stove, dig downward through the center of the potholes. After digging these smaller holes partway, begin digging out the firewood entrance. The firewood entrance should be taller than it is wide for better structural strength. After carving inward, expand the firewood entrance to complete the firebox. Locate the firebox beneath the first pothole or slightly before it. Round the inside walls of the firebox to allow easier fire building. If possible, add a small metal or ceramic shelf to the fire feeding area. This shelf will support the fuel while allowing air to flow underneath.

As the Lorena mix dries, make more openings and expand others to the intended design. Wet the edge of the potholes and lightly turn the cooking pot inside the holes to judge the tightness. The pot should be able to rest deeply and snugly in the pothole for more efficient heat transfer. Slope the potholes inward (like a cone) so that multiple sized pots can be used in the same hole.

The tunnel that connects the potholes to each other and to the firebox should ideally be carved with a slight, continuous upward slope from the firebox to the chimney exit. During this carving step, proceed smoothly from one section to the next, trying to minimize hard corners. The tunnel should have a diameter the size of an adult's fist. A consistent size will ensure a steady draft.

After the stove is fully carved, apply a mixture of wood ash, charcoal, and water to the stove interior and the outside of the stove. Sink the chimney somewhat into the stove body and support it on nails pushed into the stove or with another holding device. Dig a small ditch underneath the area where the air enters the chimney shaft to collect debris that accumulates under the chimney. The top of the chimney should not be open, but instead should have a small roof.

25.6.2 Rocket Stoves

Rocket stoves are designed to improve fuel use and combustion by directing more heat toward the cooking pot. The most basic Rocket stoves (Fig. 25-6) generally consist of a short (approximately 30–40 cm high) cylindrical or rectangular L-shaped combustion chamber that is made of, and surrounded with, materials that have insulating properties. The cooking fire is made in the rear of the horizontal section, and fuel is added through the front opening. The cooking pot is placed on top of the vertical section, and is either sunk into the vertical section or is surrounded by a metal "skirt," which extends the top of the vertical section around the cooking pot. The purpose of sinking the cooking pot,

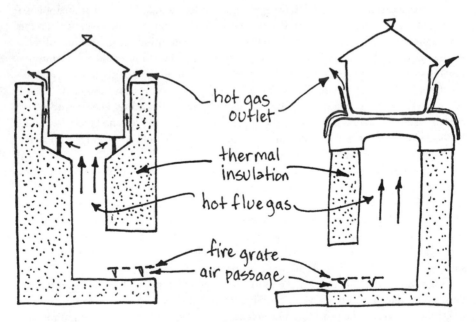

Figure 25-6. The Basic Rocket Stove with a Sunken Pot (Left) and a Pot Inserted in a Metal Skirt Located on Top of the Stove (Right).

Sources: Left, Adapted from Ministry of Energy and Mineral Development et al. (2004) and right, adapted from Still and Winiarski (2001).

or of extending a metal skirt, is to force the heated gas around the surface of the cooking pot, which in turn increases the efficiency of the stove. A firewood shelf is added to the combustion chamber entrance to allow better air flow underneath the fire.

Rocket stoves increase the overall efficiency of burning solid fuel and thus decrease the amount of air pollution. However, they do not have a chimney or other method of ventilating the air pollution from the home, so they might be considered as an alternative for outdoor cooking. Ventilation in the cooking area is especially a problem in tightly enclosed areas. Furthermore, the basic Rocket stove is also only designed to cook one pot at a time.

Materials

The primary L-shaped combustion chamber section of a Rocket stove is shown in Fig. 25-6. It can be constructed from brick, simple clay mixes, or ceramics, as long as the materials are strong, light, and insulating. Sheet metal can also be used to create the combustion chamber. Simple clay mixes that do not require firing are made by mixing clay with sawdust, pumice, wood ash, or chopped grass in a volume ratio of 1 : 1. The clay acts to keep the mix together and provides strength. The organic materials keep the mix light and allow air holes to form, which provide insulation. The final mix is more insulating than a Lorena-type mix, which uses sand and absorbs large amounts of the heat away from the cooking pot. Water is added to the mixture until it is moldable, and it is homogenized by further mixing. More advanced ceramics that are kiln-fired can also be used. They will usually consist of clay, cement, and organic materials.

Construction

Simple Rocket stoves can be constructed with the combustion chamber made from the same material as the rest of the stove (e.g., a stove made entirely of bricks) or with the combustion chamber constructed from a different material (e.g., a clay–sawdust mix structure built over a ceramic or metal combustion chamber). Figure 25-7 shows Rocket stoves of slightly different shapes made from the same material. The one on the right has a metal skirt placed on top.

The stove base is approximately 2 cm thick and can be constructed from a variety of local materials, including a clay–sawdust mixture. Figure 25-7 depicts a square base. One layer of bricks (or stones) that will serve as the outside of the stove is placed on top of the base a few centimeters from the edge (as shown for the Fig. 25-7 stoves) or along the edge of the circle. The width of the finished stove can be estimated by adding approximately 24 cm to the cooking pot diameter. A gap of width J is left in the row of bricks (or stones) where the combustion chamber entrance will be located. Table 25-6 shows how the dimensions of the combustion chamber are based on the diameter and capacity of the cooking pot. Figure 25-8 shows the relationship of the L-shaped combustion area dimensions listed in Table 25-6.

Obtain a form that is easy to remove and can occupy the L-shape of the combustion chamber and lay it out. Table 25-6 provides the diameter of forms that can be used for different pot sizes. The forms serve as placeholders for the L-shape while the clay–sawdust mixture is built up around them. Banana tree stems or similar wood species make excellent forms.

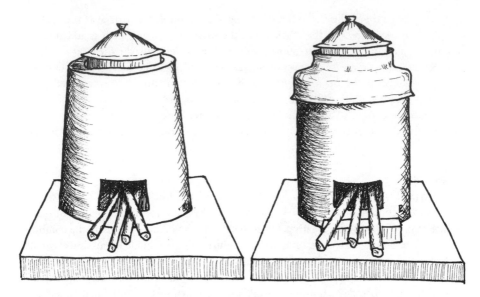

Figure 25-7. Rocket Stove Designs Without (Left) and with Metal Skirt (Right).

Note: Stoves can be constructed from brick, clay mixtures, and ceramics.

When the combustion chamber forms are in place, place a clay–sawdust mixture around the form. Build up the combustion area to the height of the vertical section, H, of the combustion chamber and out toward the perimeter of the stove base.

Place the desired cooking pot (wetted to allow for easier removal) on top of the vertical portion of the combustion chamber form and fill the area around the pot with

Table 25-6. Dimensions for a Rocket Stove Combustion Chamber Based on the Cooking Pot Diameter and Volume

Pot diameter, D (cm)	Pot volume (L)	J (cm)	$K = 1.5 \times J$ (cm)	$H = K + J$ (cm)	Form Diameter (cm)	Chamber Area (cm²)	Chamber Dimensions (cm)
≤ 20	≤ 2.7	11	16.5	27.5	12.4	121	11 × 11
21–27	2.7–7.5	12	18.0	30.0	13.5	144	12 × 12
28–30	7.5–9.8	13	19.5	32.5	14.7	169	13 × 13
31–35	9.8–15.7	14	21.0	35.0	15.8	196	14 × 14
36–40	15.7–24	15	22.5	37.5	17.0	225	15 × 15
41–50	24–35	16	24.0	40.0	18.0	256	16 × 16
46–50	35–47	18	27.0	45.0	20.3	324	18 × 18

Note: J is the height and width of the fuel opening, H is the height of the combustion chamber, and K is the distance from the top of the fuel opening to the top of the combustion chamber.

Source: Data from Ministry of Energy and Mineral Development, Republic of Uganda, and German Technical Cooperation (2004).

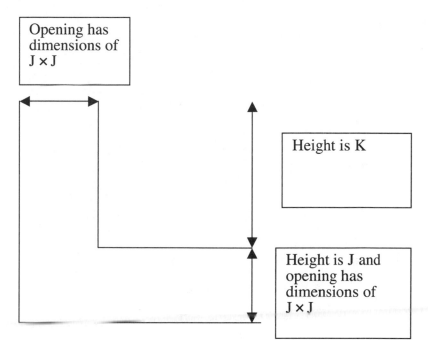

Figure 25-8. Dimensions of the Rocket Stove Combustion Chamber, Corresponding to Table 25-6.

the clay–sawdust mix up to the top of the pot. Widen the hole left by the cooking pot area an additional 2.5 cm. At the bottom of the cooking pot area, add three small clay–sawdust mixture bumps that measure 2.5 cm high and are located 120° from each other. The cooking pot will eventually rest on these supports (see left side of Fig. 25-6).

Then smooth the outer surface of the stove and allow it to dry for one month, after which you can remove the combustion chamber forms and smooth out the stove interior. Add small supports at the bottom of the combustion chamber to provide a small fuel shelf, which will facilitate air flow beneath the fire.

25.6.3 Combined Stoves: Rocket–Lorena and Improved Plancha Stoves

More advanced stove designs (such as the Rocket–Lorena and the Improved Plancha stoves) incorporate combustion chambers that increase fuel and cooking efficiency, larger cooking surfaces to allow cooking multiple items, and chimneys that direct air pollutants outdoors.

The *Rocket–Lorena stove* (as shown in Fig. 25-9) is a Lorena stove made with an insulating clay–organic material mixture as opposed to a clay–sand mixture. The firebox is modified to resemble a Rocket-type combustion chamber. The combustion chamber and

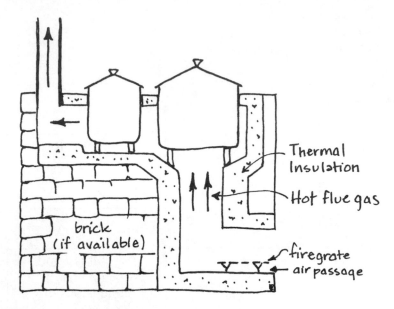

Figure 25-9. Sections of a Rocket–Lorena Improved Stove.

Source: Adapted from Ministry of Energy and Mineral Development, Republic of Uganda, and German Technical Cooperation (2004).

other tunnels are not formed by carving as in a Lorena-type construction but are made by using forms made of banana stems or other easily removable materials. The exterior dimensions of the Rocket–Lorena stove are similar to those of the regular Lorena stove.

The *Improved Plancha stove* is similar to a Rocket–Lorena stove but has a flat metal cooking surface (called the *plancha,* or griddle). The combustion chamber is typically made of stronger ceramic materials (see Fig. 25-10). The heated gas passes along the underside of the metal sheet, which is an excellent conductor and transfers heat to the cooking pots or directly to food laid on top of it. The hollow stove body can be made of brick, sheet metal, or Lorena mix. During construction, an insulating material is laid out below the cooking surface that covers the entire area. The combustion chamber is then placed on top of this material or at the front of the stove, and more insulating material is added to fill the interior of the stove to the top of the combustion chamber. The insulating material can be wood ash or similar material. The height of the stove is determined by the needs of the individual household, although the hot metal surface should be raised enough that small children cannot contact it.

⟫⟫⟫ 25.7 Ventilation and Personal Exposure

25.7.1 Ventilation

Many cooking spaces have insufficient ventilation and thus trap air pollutants in the cooking space and sometimes in other living spaces of a home. Two reasons for the lack

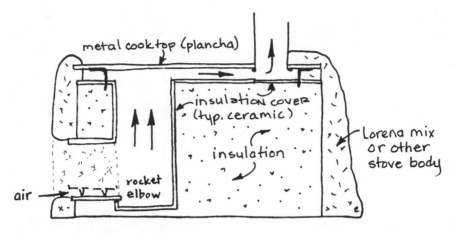

Figure 25-10. The Improved Plancha Stove.

Source: Adapted from Still and Winiarski (2001).

of ventilation in structures include the desire to protect the interior of the household from rain and wind in tropical areas and the desire to close off the home from the cold at higher elevations. In some windy areas, it is not necessarily the lack of ventilation that causes poor indoor air quality but rather the lack of controlled ventilation. Controlled air flow is required to provide sufficient air to burn fuel efficiently without undermining heat transfer to the cooking pot.

Even if a household continues to use the common three-stone fires or similar open fire, chimneys, elevated hoods, and complete fireplaces can usually be constructed around the fire, using scrap metal and local materials. The main purpose of such structures is to ventilate the room by directing smoke out of the household. Care should be taken not to cut off all air to the fire because it is needed for proper combustion.

Ventilation structures should be safe and fire resistant, should not impede the cook when preparing meals, and should take into account other geographic or cultural factors (e.g., preventing rain from penetrating a roof system with the addition of a chimney). Careful study of the wind patterns around the home may also provide information on how to improve ventilation by opening or closing existing windows or other openings.

25.7.2 Reducing Exposure

Changing personal behavior is perhaps the most difficult modification to implement, especially when it involves a practice as commonplace as cooking, watching children, or heating in colder climates. For example, having all non-cooks stay out of the cooking space during cooking could seem like a simple way of reducing exposure, but it could be complicated because of the social aspect of conversation between cooks and non-cooks or the care-giving responsibilities of mothers.

Many families reduce exposure by building their cooking space apart from the household living area. Although this solution can be effective to some extent, the problem of personal exposure within the cooking space remains. Because of convenience, food preparation areas are usually situated near the cooking area, increasing exposure of cooks to air pollutants. Simple modifications, such as changing the location of the main food preparation table, may decrease exposure to air pollutants.

Indoor smoke has some advantages. The presence of smoke is considered beneficial because it may keep away insects, which may be disease vectors. Smoke may also keep away insects that feed on certain housing materials (e.g., thatch).

))) References

Babcock, M. D. (2006). *The effects of traditional cooking technologies and small* control interventions on indoor air quality in Cayo Paloma, Panama, <http://www.cee.mtu.edu/peacecorps/resources.html> (Jan. 23, 2009).

Bruce, N., Perez-Padilla, R., and Albalak, R. (2002). "The health effects of indoor air pollution exposure in developing countries," World Health Organization, Geneva, Switzerland.

Bryden, M., Still, D., Scott, P., Hoffa, G., Ogle, D., Bailis, R., and Goyer, K. (2006). *Design principles for wood burning cook stoves.* Aprovecho Research Center (Eugene, Oregon), available through the Partnership for Clean Indoor Air, Washington, D.C.

Evans, I., and Boutette, M. (1981). *Lorena stoves: A manual for designing, building, and testing low-cost, wood-conserving cookstoves,* The Appropriate Technology Project of Volunteers in Asia, Stanford, Calif. <http://www.tecno-point.com/es/6/1734/lorena-ownerbuilt-stoves-estufas-de-lorena-de-construccion-propia-.html> (Jan. 23, 2009).

McGranahan, G., and Murray, F. (2003). *Air pollution and health in rapidly developing countries,* Earthscan Publications, London.

Ministry of Energy and Mineral Development, Republic of Uganda, and German Technical Cooperation. (2004). *How to build the improved household stoves: A construction manual for the Rocket–Lorena and shielded fire stoves.* Ministry of Energy and Mineral Development Energy Advisory Project, Kampala, Uganda, <http://www.energyandminerals.go.ug/DOCS/HOUSEHOLD%20Stoves%20Construction%20Manual.pdf> (Jan. 6, 2009).

Partnership for Clean Indoor Air website, <www.PCIAonline.org> (Jan. 6, 2009).

Smith, K. R. (1993). "Fuel combustion, air pollution exposure, and health: The situation in developing countries," *Annual Review of Energy and the Environment,* 18, 529–566.

Smith, K. R., and Mehta, S. (2003). "The burden of disease from indoor air pollution in developing countries: Comparison of estimates." International J. Hygiene and Environmental Health. 206, 279–289.

Smith, K. R., Apte, M. G., Yoqing, M., Wongsekiarttirat, W., and Kulkarni, A. (1994). "Air pollution and the energy ladder in Asian cities," Energy. 19(5), 587–600.

Smith-Sivertsen, T., Diaz, E., Bruce, N., Diaz, A., Khalakdina, A., Schei, M. A., McCracken, J., Arana, B., Klein, R., Thompson, L., and Smith, K. R. (2004). "Reducing indoor air pollution with a randomized intervention design—A presentation of the stove intervention study in the Guatemalan highlands." Norsk Epidemiologi, 14(2), 137–143.

Still, D., and Winiarski, L. (2001). "Increasing fuel efficiency and reducing harmful emissions in traditional cooking stoves," *Boiling Point,* Vol. 47, 36–39, <http://practicalactionpublishing.org/?id=bp47_abs > (Jan. 6, 2009).

Sundell, J. (2004). "On the history of indoor air quality and health," Indoor Air, 14(s7), 51–58.

World Health Organization (WHO). (2005). *WHO air quality guidelines for particulate matter, ozone, nitrogen dioxide and sulfur dioxide.* Global Update 2005: Summary of Risk Assessment. World Health Organization, Geneva, Switzerland.

))) Further Reading

Partnership for Clean Indoor Air, <http://www.pciaonline.org> (May 21, 2009).

Smith, K. R., Samet, J. M., Romieu, I., and Nigel, B. (2000). "Indoor air pollution in developing countries and acute lower respiratory infections in children." *Thorax*, 55, 518–532.

Still, D., Pinnell, M., Ogle, D., and van Appel, B. (2003). "Insulative ceramics for improved cooking stoves," *Boiling Point*, Vol. 49, 7–10, <http://practicalactionpublishing.org/docs/energy/boiling point49/pdf> (Jan. 6, 2009).

Appendix A

*Tabulated Values for
Designing Pipeline Crossings*

Appendix A-1. Pipe Properties

Nominal Tube Size	Outside Diameter (in.)	Specification	Inside Diameter (in.)	Wall Thickness (in.)	Area (in.²)	Moment of Inertia (in.⁴)	Radius of Gyration (in.)	Weight per Foot (lb)	Weight of Water per Foot (lb)	Critical Bending Moment (in.-lb)
1/2	0.840	5S	0.710	0.065	0.158	0.012	0.275	0.53	0.014	1,026
		10S	0.674	0.083	0.197	0.014	0.269	0.67	0.013	1,227
		40-Std-40S	0.622	0.109	0.250	0.017	0.261	0.85	0.011	1,465
		80-XS-80S	0.546	0.147	0.320	0.020	0.250	1.08	0.008	1,721
		160	0.466	0.187	0.384	0.022	0.240	1.30	0.006	1,896
		XXS	0.252	0.294	0.504	0.024	0.219	1.71	0.002	2,078
3/4	1.050	5S	0.920	0.065	0.201	0.025	0.349	0.68	0.024	1,680
		10S	0.884	0.083	0.252	0.030	0.343	0.85	0.022	2,036
		40-Std-40S	0.824	0.113	0.333	0.037	0.334	1.13	0.019	2,540
		80-XS-80S	0.742	0.154	0.433	0.045	0.321	1.47	0.016	3,071
		160	0.614	0.218	0.570	0.053	0.304	1.93	0.011	3,613
		XXS	0.434	0.308	0.718	0.058	0.284	2.44	0.005	3,972
1	1.315	5S	1.185	0.065	0.255	0.050	0.443	0.86	0.040	2,737
		10S	1.097	0.109	0.413	0.076	0.428	1.40	0.034	4,144
		40-Std-40S	1.049	0.133	0.494	0.087	0.421	1.67	0.031	4,782
		80-XS-80S	0.957	0.179	0.639	0.106	0.407	2.17	0.026	5,782
		160	0.815	0.250	0.836	0.125	0.387	2.84	0.019	6,851
		XXS	0.599	0.358	1.076	0.140	0.361	3.65	0.010	7,691
1 1/4	1.660	5S	1.530	0.065	0.326	0.104	0.564	1.10	0.066	4,500
		10S	1.442	0.109	0.531	0.160	0.550	1.80	0.059	6,961
		40-Std-40S	1.380	0.140	0.669	0.195	0.540	2.27	0.054	8,445
		80-XS-80S	1.278	0.191	0.881	0.242	0.524	2.99	0.046	10,487
		160	1.160	0.250	1.107	0.284	0.506	3.76	0.038	12,312
		XXS	0.896	0.382	1.534	0.341	0.472	5.21	0.023	14,795

Nominal	OD	Schedule								
1½	1.900	5S	1.770	0.065	0.375	0.158	0.649	1.27	0.089	5,984
		10S	1.682	0.109	0.613	0.247	0.634	2.08	0.080	9,353
		40-Std-40S	1.610	0.145	0.799	0.310	0.623	2.71	0.074	11,743
		80-XS-80S	1.500	0.200	1.068	0.391	0.605	3.63	0.064	14,825
		160	1.338	0.281	1.429	0.482	0.581	4.85	0.051	18,280
		XXS	1.100	0.400	1.885	0.568	0.549	6.40	0.034	21,518
2	2.375	5S	2.245	0.065	0.472	0.315	0.817	1.60	0.143	9,546
		10S	2.157	0.109	0.776	0.499	0.802	2.63	0.132	15,133
		40-Std-40S	2.067	0.154	1.075	0.666	0.787	3.65	0.121	20,183
		80-XS-80S	1.939	0.218	1.477	0.868	0.766	5.02	0.107	26,312
		160	1.689	0.343	2.190	1.162	0.729	7.44	0.081	35,237
		XXS	1.503	0.436	2.656	1.311	0.703	9.02	0.064	39,753
2½	2.875	5S	2.709	0.083	0.728	0.710	0.988	2.00	0.208	17,781
		10S	2.635	0.120	1.039	0.987	0.975	3.50	0.197	24,724
		40-Std-40S	2.469	0.203	1.704	1.530	0.947	5.80	0.173	38,305
		80-XS-80S	2.323	0.276	2.254	1.924	0.924	7.70	0.153	48,190
		160	2.125	0.375	2.945	2.353	0.894	10.00	0.128	58,921
		XXS	1.771	0.552	4.028	2.871	0.844	14.00	0.089	71,895
3	3.500	5S	3.334	0.083	0.891	1.301	1.208	3.00	0.315	26,767
		10S	3.260	0.120	1.274	1.822	1.196	4.30	0.301	37,480
		40-Std-40S	3.068	0.216	2.228	3.017	1.164	7.60	0.267	62,067
		80-XS-80S	2.900	0.300	3.015	3.894	1.136	10.20	0.239	80,112
		160	2.626	0.437	4.205	5.032	1.094	14.30	0.196	103,514
		XXS	2.300	0.600	5.466	5.993	1.047	19.00	0.150	123,274
4	4.500	5S	4.334	0.083	1.152	2.810	1.562	3.90	0.533	44,957
		10S	4.260	0.120	1.651	3.963	1.549	5.60	0.515	63,403
		40-Std-40S	4.026	0.237	3.174	7.233	1.510	11.00	0.460	115,722
		80-XS-80S	3.826	0.337	4.407	9.510	1.477	15.00	0.415	153,768
		120	3.626	0.437	5.578	11.643	1.445	19.00	0.373	186,293
		160	3.438	0.531	6.621	13.271	1.416	23.00	0.335	212,335
		XXS	3.152	0.674	8.101	15.284	1.374	28.00	0.282	244,539

Appendix A-2. Cable Properties

Construction	Diameter (in.)	Weight per Unit (lb/ft)	Minimum Breaking Strength (lb)	Area of Steel (in.^{2})
1×19	1/8	0.035	2,100	0.0092
	5/32	0.055	3,300	0.0144
	3/16	0.077	4,700	0.0207
	7/32	0.102	6,300	0.0282
	1/4	0.135	8,200	0.0369
	9/32	0.170	10,300	0.0467
	5/16	0.210	12,500	0.0576
	3/8	0.301	17,500	0.0830
7×7	1/8	0.028	1,700	0.0074
	5/32	0.043	2,600	0.0115
	3/16	0.062	3,700	0.0166
	7/32	0.083	4,800	0.0225
	1/4	0.106	6,100	0.0294
	9/32	0.134	7,600	0.0373
	5/16	0.167	9,200	0.0460
	11/32	0.201	11,100	0.0557
	3/8	0.236	13,100	0.0662
7×19	1/8	0.029	2,000	0.0079
	5/32	0.045	2,800	0.0123
	3/16	0.065	4,200	0.0178
	7/32	0.086	5,600	0.0242
	1/4	0.110	7,000	0.0316
	9/32	0.139	8,000	0.0399
	5/16	0.173	9,800	0.0493
	11/32	0.207	12,500	0.0597
	3/8	0.243	14,400	0.0710
6×19	1/4	0.120	5,880	0.0316
	5/16	0.180	9,160	0.0493
	3/8	0.260	13,120	0.0710
6×36	1/4	0.120	5,340	0.0303
	5/16	0.180	8,320	0.0474
	3/8	0.260	11,900	0.0682

Appendix A-3. Critical Axial Loads (in lb) for Steel Pipe of Various Effective Heights

Normal Tube Size	Specification	Effective Height in Feet								
		3	3.5	4	4.5	5	5.5	6	6.5	7
¹/₂	5S	2162	1523	1017	644	385	219	118	60	29
	10S	2586	1795	1178	731	429	238	125	62	29
	40-Std-40S	3082	2092	1338	806	457	245	123	58	26
	80-XS-80S	3583	2350	1445	832	449	227	108	48	20
	160	3877	2450	1443	792	405	193	86	35	14
	XXS	3954	2280	1208	588	263	108	41	14	5
³/₄	5S	4418	3246	2485	1871	1352	959	653	430	274
	10S	5354	3933	3003	2238	1622	1121	753	489	307
	40-Std-40S	6679	4907	3717	2724	1925	1311	861	545	333
	80-XS-80S	8076	5944	4423	3165	2177	1439	915	559	328
	160	9501	6977	5015	3450	2231	1430	862	497	274
	XXS	10445	7509	5143	3350	2024	1221	684	364	184
1	5S	9014	6623	5071	4006	3231	2613	2058	1587	1199
	10S	13650	10026	7676	6066	4898	3879	3005	2276	1667
	40-Std-40S	15750	11572	8859	7000	5625	4418	3390	2543	1864
	80-XS-80S	19044	13991	10712	8484	6715	5185	3906	2872	2059
	160	22563	16577	12692	10003	7725	5805	4245	3021	2092
	XXS	25329	18609	14282	10956	8145	5871	4102	2777	1823
1¹/₄	5S	11725	11725	10523	8315	6735	5566	4677	3983	3352
	10S	19120	19120	16280	12863	10419	8611	7249	6126	5107
	40-Std-40S	24067	24067	19750	15605	12640	10446	8799	7388	6118
	80-XS-80S	31733	31733	24526	19378	15646	12972	9056	9056	7412
	160	39867	37607	28793	22750	18427	15254	12708	10420	8409
	XXS	55214	45190	34599	27337	22123	18246	14783	11760	9185

Appendix A-3. Critical Axial Loads (in lb) for Steel Pipe of Various Effective Heights (*Continued*)

Normal Tube Size	Specification	Effective Height in Feet								
		3	3.5	4	4.5	5	5.5	6	6.5	7
1½	5S	13490	13490	13490	12656	10251	8472	7119	6066	5235
	10S	22079	22079	22079	19781	16023	13242	11127	9481	8194
	40-Std-40S	28780	28780	28780	24836	20118	16626	13970	11904	10284
	80-XS-80S	38453	38453	38453	31353	25396	20989	17636	15027	12940
	160	51452	51452	48930	38661	31315	25880	21747	18571	15781
	XXS	67858	67858	57598	45510	36863	30465	25651	21665	18052
2	5S	16982	16982	16982	16982	16982	16894	14196	12096	10429
	10S	27934	27934	27934	27934	27934	26782	22504	19175	16534
	40-Std-40S	38683	38683	38683	38683	38683	35718	30013	25573	22050
	80-XS-80S	53181	53181	53181	53181	53181	46565	39127	33339	28746
	160	78826	78826	78826	78826	75455	62359	52399	44648	38497
	XXS	95613	95613	95613	95613	85126	70352	59115	50370	43432
2½	5S	26209	26209	26209	26209	26209	26209	26209	26209	23516
	10S	37390	37390	37390	37390	37390	37390	37390	37390	32699
	40-Std-40S	61346	61346	61346	61346	61346	61346	61346	58754	50660
	80-XS-80S	81127	81127	81127	81127	81127	81127	81127	73915	63733
	160	106029	106029	106029	106029	106029	106029	106029	90375	77925
	XXS	145024	145024	145024	145024	145024	145024	129420	110275	95084
3	5S	32076	32076	32076	32076	32076	32076	32076	32076	32076
	10S	45872	45872	45872	45872	45872	45872	45872	45872	45872
	40-Std-40S	80225	80225	80225	80225	80225	80225	80225	80225	80225
	80-XS-80S	108573	10573	108573	108573	108573	108573	108573	108573	108573
	160	151384	151384	151384	151384	151384	151384	151384	151384	151384
	XXS	196789	196789	196789	196789	196789	196789	196789	196789	196789

Normal Tube Size	Specification	Effective Height in Feet								
		7.5	8	8.5	9	9.5	10	10.5	11	11.5
4	5S	41463	41463	41463	41463	41463	41463	41463	41463	41463
	10S	59444	59444	59444	59444	59444	59444	59444	59444	59444
	40-Std-40S	114266	114266	114266	114266	114266	114266	114266	114266	114266
	80-XS-80S	158668	158668	158668	158668	158668	158668	158668	158668	158668
	120	200808	200808	200808	200808	200808	200808	200808	200808	200808
	160	238357	238357	238357	238357	238357	238357	238357	238357	238357
	XXS	291647	291647	291647	291647	291647	291647	291647	291647	291647
1	5S	887	643	456	317	216	144	94	60	38
	10S	1222	866	601	407	270	175	111	69	42
	40-Std-40S	1335	934	639	427	279	178	111	68	40
	80-XS-80S	1441	984	655	426	270	167	101	59	34
	160	1410	925	590	367	222	130	75	42	23
	XXS	1160	715	427	248	135	76	40	20	10
1 1/4	5S	2785	2284	1850	1479	1168	910	701	532	399
	10S	4201	3410	2730	2157	1681	1293	981	734	542
	40-Std-40S	4996	4023	3194	2501	1931	1471	1104	818	597
	80-XS-80S	5977	4749	3718	2867	2179	1632	1204	875	627
	160	6680	5222	4018	3044	2269	1665	1202	854	598
	XXS	7044	5304	3922	2847	2023	1420	976	658	436
1 1/2	5S	4551	3918	3341	2821	2360	1955	1604	1303	1048
	10S	7076	6050	5120	4289	3557	2921	2374	1910	1521
	40-Std-40S	8832	7505	6311	5252	4325	3524	2841	2267	1790
	80-XS-80S	11014	9271	7718	6354	5173	4165	3316	2611	2034
	160	13249	10990	9007	7293	5835	4612	3602	2779	2119
	XXS	14840	12036	9631	7603	5921	4550	3449	2579	1903

Appendix A-3. Critical Axial Loads (in lb) for Steel Pipe of Various Effective Heights (*Continued*)

Normal Tube Size	Specification	Effective Height in Feet								
		7.5	8	8.5	9	9.5	10	10.5	11	11.5
2	5S	9085	7985	7073	6324	5649	5016	4426	3883	3385
	10S	14403	12659	11213	10023	8916	7881	6922	6042	5240
	40-Std-40S	19208	16882	14970	13345	11817	10396	9086	7889	6805
	80-XS-80S	25041	22009	19543	17311	15229	13304	11543	9945	8510
	160	33535	29540	26029	22761	19751	17008	14534	12325	10372
	XXS	37834	33283	29050	25147	21590	18384	15525	13003	10802
2¹/₂	5S	20485	18005	15949	14226	12768	11523	10452	9546	8690
	10S	28484	25035	22176	19781	17753	16022	14544	13265	12047
	40-Std-40S	44131	38787	34358	30646	27505	24824	22570	20473	18487
	80-XS-80S	55518	48795	43224	38554	34603	31269	28358	25596	22992
	160	67882	59662	52849	47140	42309	38268	34472	30895	27549
	XXS	82829	72798	64486	57520	51724	46272	41159	36403	32013
3	5S	32076	32076	29228	26070	23398	21117	19154	17452	15967
	10S	45872	45872	40926	36505	32764	29569	26820	24437	22359
	40-Std-40S	80225	76510	67774	60452	54256	48966	44414	40468	37026
	80-XS-80S	108573	98753	87477	78027	70030	63202	57326	52233	47790
	160	145182	127601	113031	100821	90487	81665	74072	67491	61750
	XXS	172897	151960	134608	120067	107761	97254	88213	80376	73703
4	5S	41463	41463	41463	41463	41463	41463	41361	37687	34481
	10S	59444	59444	59444	59444	59444	59444	58333	53150	48629
	40-Std-40S	114266	114266	114266	114266	114266	114266	106467	97008	88756
	80-XS-80S	158668	158668	158668	158668	158668	155972	141471	128902	117937
	120	200808	200808	200808	200808	200808	188963	171395	156168	142883
	160	238357	238357	238357	238357	238357	215379	195355	177999	162857
	XXS	291647	291647	291647	291647	274841	248044	224983	204995	187557

Normal Tube Size	Specification	Effective Height in Feet								
		12	12.5	13	13.5	14	14.5	15	15.5	16
1 1/4	5S	296	216	156	111	78	54	37	25	17
	10S	395	284	201	141	57	66	45	30	19
	40-Std-40S	430	305	214	148	100	67	45	29	19
	80-XS-80S	442	307	211	142	94	62	40	25	16
	160	412	279	186	122	79	50	31	19	12
	XXS	284	181	114	70	42	25	15	8	5
1 1/2	5S	835	659	515	399	336	232	175	130	96
	10S	1199	936	723	553	449	314	233	171	124
	40-Std-40S	1399	1081	827	626	469	348	255	185	133
	80-XS-80S	1566	1193	899	669	453	359	259	184	130
	160	1596	1188	873	634	455	323	226	157	107
	XXS	1385	995	705	493	340	231	155	103	67
2	5S	2933	2526	2162	1840	1556	1307	1092	907	748
	10S	4516	3868	3291	2783	2339	1953	1620	1336	1095
	40-Std-40S	5832	4965	4199	3528	2945	2442	2012	1647	1339
	80-XS-80S	7231	6102	5114	4256	3518	2888	2354	1906	1532
	160	8661	7178	5903	4818	3902	3136	2501	1979	1555
	XXS	8899	7272	5893	4736	3776	2985	2341	1820	1404
2 1/2	5S	7878	7112	6394	5725	5134	4532	4007	3528	3093
	10S	10893	9808	8794	7850	6978	6177	5444	4777	4174
	40-Std-40S	16618	14871	13247	11747	10370	9113	7972	6942	6018
	80-XS-80S	20555	18290	16196	14274	12220	10930	9496	8211	7066
	160	24440	21572	18943	16550	14386	12442	10705	9165	7806
	XXS	27991	24336	21038	18083	15454	13132	11096	9322	7787

Appendix A-3. Critical Axial Loads (in lb) for Steel Pipe of Various Effective Heights (Continued)

Normal Tube Size	Specification	Effective Height in Feet								
		12	12.5	13	13.5	14	14.5	15	15.5	16
3	5S	14664	13515	12503	11613	10756	9935	9151	8405	7698
	10S	20534	18924	17526	16252	15028	13857	12741	11681	10680
	40-Std-40S	34004	31339	29040	26816	24688	22661	20738	18921	17211
	80-XS-80S	43890	40542	37410	34412	31555	28844	26282	23873	21616
	160	56836	52291	47946	43814	39901	36214	32757	29529	26528
	XXS	67544	61671	56100	50843	45908	41298	37013	33050	29402
4	5S	31667	29185	26983	25021	23266	21689	20267	18981	17813
	10S	44661	41159	38054	35288	32812	30588	28583	26769	25122
	40-Std-40S	81514	75123	69456	64406	59888	55829	52169	48858	45852
	80-XS-80S	108314	99822	92291	79577	79577	74184	69321	64921	61011
	120	131225	120937	111813	103684	96410	89876	83984	78653	73991
	160	149569	137842	127443	118178	109887	102439	95724	89836	84269
	XXS	172253	158748	146771	136101	126553	117976	110458	103424	96628

Normal Tube Size	Specification	Effective Height in Feet								
		16.5	17	17.5	18	18.5	19	19.5	20	20.5
2	5S	614	500	405	327	261	208	164	129	101
	10S	891	721	580	463	368	290	227	177	137
	40-Std-40S	1081	868	692	548	431	337	262	202	155
	80-XS-80S	1223	970	764	597	464	358	274	208	157
	160	1212	937	719	548	414	311	231	171	125
	XXS	1074	815	613	458	339	249	181	131	94

Tube Size	Specification	21	21.5	22	22.5	23	23.5	24	24.5	25
2½	5S	2701	2348	2033	1753	1506	1288	1096	930	785
	10S	3632	3147	2714	2332	1994	1698	1440	1216	1023
	40-Std-40S	5193	4461	3815	3248	2752	2322	1950	1630	1357
	80-XS-80S	6052	5159	4376	3695	3105	2597	2161	1790	1476
	160	6614	5576	4677	3903	3241	2677	2200	1799	1464
	XXS	6468	5341	4386	3581	2907	2347	1883	1503	1193
3	5S	7032	6405	5818	5270	4760	4287	3851	3450	3081
	10S	9736	8850	8022	7251	6536	5874	5264	4704	4192
	40-Std-40S	15609	14114	12723	11435	10247	9154	8054	7241	6411
	80-XS-80S	19511	17556	15747	14080	12550	11150	9876	8720	7675
	160	23752	21194	18847	16703	14753	12986	11392	9960	8678
	XXS	26060	23012	20245	17745	15496	13482	11686	10092	8683
4	5S	16750	15806	14922	14064	13233	12431	11657	10914	10201
	10S	23622	22302	21035	19805	18617	17469	16365	15305	14289
	40-Std-40S	43202	40632	38260	35909	33643	31463	29372	27371	25461
	80-XS-80S	57420	53939	50575	47332	44214	41225	38366	35639	33044
	120	69447	65055	60823	56755	52856	49128	45575	42196	38992
	160	78887	73699	68712	63933	59365	55012	50875	46953	43246
	XXS	90084	83803	77792	72055	66599	61423	56527	519209	47566

| Normal Tube Size | Specification | Effective Height in Feet | | | | | | | | |
		21	21.5	22	22.5	23	23.5	24	24.5	25
2½	5S	660	553	461	383	317	261	214	175	142
	10S	856	714	592	490	403	330	269	219	177
	40-Std-40S	1124	927	761	622	506	410	330	265	212
	80-XS-80S	1211	989	804	650	523	419	334	265	210
	160	1185	954	764	609	483	381	299	234	182
	XXS	941	738	576	447	344	264	201	153	115

Appendix A-3. Critical Axial Loads (in lb) for Steel Pipe of Various Effective Heights (*Continued*)

| Normal Tube Size | Specification | Effective Height in Feet | | | | | | | | | |
		21	21.5	22	22.5	23	23.5	24	24.5	25
3	5S	2745	2438	2160	1908	1681	1476	1293	1130	984
	10S	3725	3300	2916	2569	2257	1977	1727	1504	1307
	40-Std-40S	5659	4980	4369	3822	3334	2899	2513	2172	1872
	80-XS-80S	6734	5889	5135	4463	3866	3339	2875	2467	2111
	160	7535	6521	5624	4834	4141	3535	3008	2550	2155
	XXS	7443	6356	5408	4585	3872	3258	2731	2281	1898
4	5S	9518	8867	8246	7656	7096	6566	6066	5595	5151
	10S	13317	12391	11510	10673	9880	9131	8424	7759	7134
	40-Std-40S	23642	21913	20275	18726	17264	15888	14595	13384	12251
	80-XS-80S	30581	18148	26045	23969	22017	20187	18474	16875	15385
	120	35961	33101	30409	27881	25514	23302	21241	19324	17546
	160	39751	36464	33381	30497	27805	25300	22973	20818	18827
	XXS	43492	39681	36126	32819	29750	26910	24289	21876	19660

Normal Tube Size	Specification	Effective Height in Feet					
		25.5	26	26.5	27	27.5	28
3	5S	855	741	640	551	473	406
	10S	1132	977	842	723	619	528
	40-Std-40S	1608	1377	1176	1002	850	720
	80-XS-80S	1800	1530	1297	1096	923	775
	160	1815	1523	1274	1062	882	731
	XXS	1574	1300	1070	877	716	583
4	5S	4735	4345	3981	3641	3324	3030
	10S	6549	6001	5490	5013	4571	4160
	40-Std-40S	11195	10211	9297	8449	7666	6942
	80-XS-80S	14001	12718	11531	10435	9425	8498
	120	15901	14381	12982	11696	10517	9438
	160	16992	15305	13757	12341	11048	9870
	XXS	17630	15776	14087	12551	11159	9899

Appendix A-4. C, X1, and L for Different Sag Ratios and Drop Ratios

Sag Ratio		0.00	0.01	0.02	0.03	0.04	0.05	0.06	0.07	0.08	0.09	0.10	0.11
0.030	C	4.171657	5.055271	6.702506	16.667389								
	X1	0.500	0.550	0.634	1.000								
	L	1.002395982	1.001681137	1.001127554	1.00059825								
0.032	C	3.911572	4.674587	6.014267	10.004330								
	X1	0.500	0.547	0.620	0.800								
	L	1.002725467	1.001957782	1.001352071	1.00086607								
0.034	C	3.682123	4.347760	5.460797	8.157825								
	X1	0.500	0.543	0.609	0.745								
	L	1.003076044	1.002255581	1.001597545	1.00107583								
0.036	C	3.478206	4.064053	5.004659	7.007947								
	X1	0.500	0.541	0.600	0.710								
	L	1.00344768	1.002574509	1.001864043	1.001298146								
0.038	C	3.295788	3.815422	4.621481	6.187425								
	X1	0.500	0.538	0.592	0.685								
	L	1.003840344	1.002914539	1.002151594	1.001538116								
0.040	C	3.131644	3.595716	4.294597	5.560547	12.504116							
	X1	0.500	0.536	0.586	0.667	1.000							
	L	1.004254	1.003275641	1.002460205	1.001797416	1.00106598							
0.042	C	2.983164	3.400149	4.012166	5.060865	8.027578							
	X1	0.500	0.534	0.580	0.652	0.821							
	L	1.004688613	1.003657784	1.002789871	1.002076781	1.00145865							
0.044	C	2.848212	3.224940	3.765517	4.650688	6.714217							
	X1	0.500	0.532	0.575	0.639	0.768							
	L	1.005144145	1.004060935	1.003140578	1.002376583	1.001723467							
0.046	C	2.725024	3.067060	3.548135	4.306553	5.872607							
	X1	0.500	0.531	0.571	0.629	0.735							
	L	1.005620555	1.004485059	1.003512306	1.002697022	1.002007321							
0.048	C	2.612128	2.924057	3.355021	4.012888	5.258605							
	X1	0.500	0.529	0.567	0.620	0.710							
	L	1.006117804	1.004930119	1.003905029	1.003038213	1.00230593							
0.050	C	2.508289	2.793918	3.182270	3.758847	4.780833	10.008313						
	X1	0.500	0.528	0.563	0.612	0.691	1.000						
	L	1.006635846	1.005396076	1.004318719	1.003400217	1.002622202	1.001664728						
0.052	C	2.412463	2.674983	3.026783	3.536597	4.393990	6.728070						
	X1	0.500	0.527	0.560	0.606	0.675	0.836						
	L	1.007174639	1.00588289	1.004753344	1.003783065	1.002957451	1.002168792						

0.054	C	2.323759	2.565863	2.886066	3.340308	4.072071	5.728425	
	X1	0.500	0.525	0.557	0.600	0.662	0.786	
	L	1.007734134	1.006390518	1.00520887	1.004186767	1.003312372	1.002517871	
0.056	C	2.241415	2.465392	2.758092	3.165535	3.798700	5.075137	
	X1	0.500	0.524	0.555	0.595	0.651	0.753	
	L	1.008314285	1.006918917	1.005685253	1.004611319	1.003687359	1.002855671	
0.058	C	2.164771	2.372581	2.641188	3.006824	3.562886	4.591123	
	X1	0.500	0.524	0.553	0.590	0.590	0.729	
	L	1.008915041	1.007468042	1.00618247	1.005056709	1.004082647	1.003224675	
0.060	C	2.093258	2.286588	2.533968	2.867439	3.356894	4.209599	8.343⬚8
	X1	0.500	0.523	0.550	0.586	0.634	0.710	1.000
	L	1.009536352	1.008037844	1.006700464	1.005522914	1.004498383	1.003599246	1.002⬚5982
0.062	C	2.026379	2.206688	2.435268	2.739184	3.175078	3.897212	5.8044⬚3
	X1	0.500	0.522	0.548	0.582	0.626	0.694	0.848
	L	1.010178164	1.008628276	1.007239197	1.006009909	1.004934654	1.003991409	1.0030⬚332
0.064	C	1.963700	2.132258	2.344106	2.622272	3.013198	3.634674	5.0086⬚8
	X1	0.500	0.521	0.547	0.578	0.620	0.681	0.800
	L	1.010840423	1.009239287	1.007798624	1.00651766	1.005391511	1.004402251	1.00345⬚155
0.066	C	1.904839	2.062754	2.259645	2.515234	2.867991	3.409751	4.4809⬚
	X1	0.500	0.520	0.545	0.575	0.614	0.670	0.768
	L	1.011523073	1.009870824	1.008378697	1.007046133	1.005868978	1.004832411	1.00387 107
0.068	C	1.849459	1.997706	2.181171	2.416847	2.736898	3.214176	4.08546
	X1	0.500	0.520	0.543	0.572	0.609	0.660	0.744
	L	1.012226057	1.010522835	1.008979367	1.007595289	1.006367063	1.005282283	1.00429⬚133
0.070	C	1.797261	1.936699	2.108066	2.326086	2.617876	3.042087	3.77068⬚	7.154489
	X1	0.500	0.519	0.542	0.569	0.604	0.651	0.725	1.000
	L	1.012949315	1.011195264	1.009600583	1.008165086	1.00688576	1.005752119	1.004726⬚54	1.003259232
0.072	C	1.747981	1.879368	2.039796	2.242087	2.509273	2.889182	3.510757	5.112336
	X1	0.500	0.519	0.540	0.567	0.600	0.644	0.710	0.857
	L	1.013692788	1.011888055	1.010242294	1.00875548	1.007425054	1.006242084	1.005176⬚08	1.004038108
0.074	C	1.701381	⬚.825393	1.975896	2.164110	2.409732	2.752206	3.290653	4.458120
	X1	0.500	0.518	0.539	0.564	0.596	0.637	0.697	0.811
	L	1.014456414	1.012601149	1.010904446	1.009366422	1.00798492	1.006752282	1.005643⬚4	1.004539668
0.076	C	1.657251	1.774489	1.915959	2.091523	2.318133	2.628639	3.100801	4.019415
	X1	0.500	0.518	0.538	0.562	0.592	0.631	0.685	0.780
	L	1.015240129	1.013334486	1.011586982	1.009997866	1.008565329	1.007282779	1.006129⬚5	1.005021787
0.078	C	1.615399	1.726400	1.859626	2.023780	2.233535	2.516493	2.934697	3.687669
	X1	0.500	0.517	0.537	0.560	0.589	0.625	0.675	0.757
	L	1.016043869	1.014088007	1.012289846	1.010649759	1.009166247	1.007833361⬚	1.006347⬚1	1.005506333

Appendix A-4. C, X1, and L for Different Sag Ratios and Drop Ratios *(Continued)*

Sag Ratio		0.00	0.01	0.02	0.03	0.04	0.05	0.06	0.07	0.08	0.09	0.10	0.11
0.080	C	1.575656	1.680903	1.806582	1.960407	2.155144	2.414174	2.787710	3.421600	6.263285			
	X1	0.500	0.517	0.536	0.558	0.585	0.620	0.666	0.738	1.000			
	L	1.016867567	1.014861647	1.013012979	1.011322049	1.009787633	1.008404048	1.007159013	1.006001161	1.004254001			
0.082	C	1.537866	1.637792	1.756548	1.900991	2.082287	2.320381	2.656424	3.200419	4.573334			
	X1	0.500	0.516	0.535	0.556	0.582	0.615	0.658	0.723	0.865			
	L	1.017711158	1.015655344	1.013756322	1.012014681	1.010429446	1.008996342	1.00770294	1.006510016	1.005181905			
0.084	C	1.501890	1.596887	1.709274	1.845171	2.014385	2.234045	2.538244	3.012004	4.022440			
	X1	0.500	0.516	0.534	0.555	0.580	0.611	0.651	0.710	0.821			
	L	1.018574571	1.016469033	1.014519812	1.012727597	1.011091638	1.009608229	1.008266668	1.007034965	1.005763892			
0.086	C	1.467602	1.558023	1.664538	1.792627	1.950939	2.154275	2.431152	2.848601	3.649661			
	X1	0.500	0.515	0.533	0.553	0.577	0.607	0.645	0.698	0.791			
	L	1.019457738	1.017302645	1.015303386	1.013460741	1.011774161	1.010240441	1.008850313	1.007577254	1.006316015			
0.088	C	1.434886	1.521052	1.622142	1.743077	1.891518	2.080322	2.333549	2.704928	3.365721			
	X1	0.500	0.515	0.532	0.552	0.575	0.603	0.639	0.688	0.768			
	L	1.020360587	1.018156114	1.016106981	1.014214052	1.012476963	1.01089295	1.009453951	1.008137683	1.006865449			
0.090	C	1.403638	1.485840	1.581908	1.696272	1.835744	2.011549	2.244145	2.577208	3.136584	5.570488		
	X1	0.500	0.515	0.531	0.550	0.573	0.600	0.633	0.679	0.750	1.000		
	L	1.021283045	1.019029371	1.01693053	1.014987468	1.013199991	1.011565723	1.010077628	1.008716784	1.007422043	1.005379744		
0.092	C	1.373761	1.452265	1.543674	1.651989	1.783285	1.947413	2.161888	2.462648	2.945065	4.141149		
	X1	0.500	0.514	0.530	0.549	0.570	0.596	0.628	0.671	0.734	0.871		
	L	1.02222504	1.019922345	1.017773966	1.015780927	1.013943188	1.012258721	1.010721371	1.009314922	1.007990532	1.006463508		
0.094	C	1.345168	1.420217	1.507295	1.610028	1.733852	1.887447	2.085907	2.359117	2.781119	3.668482		
	X1	0.500	0.514	0.530	0.548	0.568	0.593	0.624	0.664	0.721	0.829		
	L	1.023186496	1.020834963	1.018637221	1.016594364	1.014706496	1.012971901	1.011385185	1.00993235	1.008573551	1.007128378		
0.096	C	1.317780	1.389593	1.472641	1.570212	1.687187	1.831246	2.015472	2.264953	2.638304	3.346239		
	X1	0.500	0.514	0.529	0.546	0.566	0.590	0.620	0.657	0.710	0.800		
	L	1.024167337	1.021767154	1.019520225	1.017427714	1.015489855	1.013705216	1.012069065	1.010569246	1.009172705	1.007752177		
0.098	C	1.291521	1.360303	1.439592	1.532379	1.643061	1.778457	1.949969	2.178835	2.512218	3.099317		
	X1	0.500	0.513	0.528	0.545	0.565	0.588	0.616	0.651	0.699	0.777		
	L	1.025167487	1.022718843	1.020422908	1.018280908	1.016293204	1.014458615	1.012772995	1.011225731	1.009789035	1.008367744		
0.100	C	1.266324	1.332261	1.408039	1.496386	1.601270	1.728772	1.888874	2.099695	2.399709	2.899028	5.016576	
	X1	0.500	0.513	0.528	0.544	0.563	0.585	0.612	0.645	0.690	0.759	1.000	
	L	1.026186868	1.023689956	1.021345196	1.019153879	1.017116479	1.015232047	1.013496951	1.011901889	1.01042324	1.008987065	1.006635848	

0.102	C	1.242127	1.305391	1.377884	1.462100	1.561632	1.681918	1.83-740	2.026658	2.298438	2.730861	3.786563	
	X1	0.500	0.513	0.527	0.543	0.561	0.583	0.608	0.640	0.682	0.744	0.877	
	L	1.0272254	1.024680414	1.022287017	1.020046557	1.019959615	1.016625453	1.01240899	1.012597772	1.011075806	1.009615939	1.007881722	
0.104	C	1.218873	1.279620	1.349036	1.429404	1.523984	1.637655	1.77E177	1.958997	2.206615	2.586316	3.374878	
	X1	0.500	0.512	0.526	0.542	0.560	0.581	0.605	0.636	0.675	0.731	0.836	
	L	1.028283004	1.025690142	1.023248295	1.020958869	1.018822544	1.016838777	1.015204804	1.013313411	1.01174708	1.01025762	1.008631714	
0.106	C	1.196506	1.254883	1.321413	1.398190	1.488179	1.595771	1.727350	1.896102	2.122842	2.459931	3.092434	
	X1	0.500	0.512	0.526	0.541	0.558	0.578	0.602	0.631	0.668	0.720	0.807	
	L	1.029359597	1.026719059	1.024228956	1.021890745	1.0197052	1.017671958	1.015-8862	1.014048818	1.012437305	1.010914101	1.009328707	
0.108	C	1.174979	1.231120	1.294939	1.368358	1.454083	1.556076	1.680-65	1.837460	2.046006	2.347967	2.874910	
	X1	0.500	0.512	0.525	0.540	0.557	0.576	0.599	0.627	0.662	0.709	0.786	
	L	1.030455099	1.027767088	1.025228922	1.022342109	1.020607512	1.01852-4932	1.016592302	1.01480399	1.013146659	1.011586682	1.010011532	
0.110	C	1.154245	1.208276	1.269543	1.339820	1.421577	1.518359	1.635761	1.782628	1.975202	2.247742	2.697700	4.563669
	X1	0.500	0.512	0.525	0.539	0.556	0.575	0.597	0.623	0.656	0.700	0.768	1.000
	L	1.031569425	1.028834147	1.026248115	1.023812887	1.021252941	1.019397636	1.0174-5798	1.015578916	1.013875267	1.01227625	1.010694443	1.008021632
0.112	C	1.134262	1.186297	1.245163	1.312494	1.390552	1.482589	1.593513	1.731229	1.909689	2.157257	2.548338	3.490201
	X1	0.500	0.512	0.524	0.538	0.554	0.573	0.594	0.619	0.651	0.692	0.753	0.882
	L	1.032702492	1.029920154	1.027286457	1.024803004	1.022470821	1.020290003	1.0182-9054	1.016373572	1.014623214	1.012983426	1.011384376	1.009435347
0.114	C	1.114990	1.165137	1.221738	1.286303	1.360907	1.448508	1.5535-7	1.682935	1.848849	2.074987	2.419508	3.127184
	X1	0.500	0.511	0.524	0.538	0.553	0.571	0.592	0.616	0.646	0.685	0.740	0.842
	L	1.033854215	1.031025027	1.028343866	1.025812381	1.023431672	1.021201964	1.019112013	1.017187927	1.015390558	1.013708654	1.01208524	1.010272496
0.116	C	1.096392	1.144751	1.199215	1.261180	1.332554	1.416032	1.515553	1.637460	1.792163	1.999733	2.306506	2.876781
	X1	0.500	0.511	0.523	0.537	0.552	0.569	0.589	0.613	0.641	0.678	0.728	0.814
	L	1.035024508	1.032148682	1.029420263	1.026840941	1.024411888	1.022133451	1.020004614	1.018021947	1.016177332	1.014452259	1.012799448	1.011044056
0.118	C	1.078434	1.125098	1.177542	1.237060	1.305409	1.385050	1.479582	1.594556	1.739192	1.930538	2.206104	2.683093
	X1	0.500	0.511	0.523	0.536	0.551	0.568	0.587	0.610	0.637	0.672	0.718	0.793
	L	1.036213285	1.033291035	1.030515565	1.027888605	1.025411393	1.023084392	1.02090-795	1.018875587	1.016983553	1.015214475	1.013528582	1.011795152
0.120	C	1.061085	1.106140	1.156673	1.213885	1.279397	1.355460	1.44533-	1.554000	1.689558	1.866627	2.115982	2.524712
	X1	0.500	0.511	0.522	0.535	0.550	0.566	0.585	0.607	0.633	0.666	0.709	0.775
	L	1.037420458	1.034452001	1.031629688	1.028955291	1.026430109	1.024054714	1.021825492	1.019748488	1.017809222	1.015995477	1.014273727	1.012542424

Note: The sag is defined as the vertical distance between the highest points of the cable's arc (at the fixture points) and the lowest point (the apex). The drop is the vertical distance between one fixture point and the other. The sag ratio and drop ratio are defined as the sag and the drop, respectively, divided by the span. In a crossing where both fixture points are at the same height, the drop and the drop ratio are zero. The point where the cable first touches the tower (and is no longer hanging freely) is known as the fixture point. For a given sag ratio and drop ratio the length of cable (not accounting for the cable stretching under tension) can be estimated (length of span times L) and the horizontal distance to the apex can be located from the highest fixture (length of span times X1). C is a constant to be used with the equations in Chapter 13 to define the shape of the curve and therefore estimate the location of the cable at any point along the span.

Index

NOTE: Page locators in *italics* refer to figures; locators in **bold** refer to tables.

About the Authors

James R. Mihelcic is a professor of civil and environmental engineering and a State of Florida 21st Century World Class Scholar at the University of South Florida. Dr. Mihelcic is a past president and board member of the Association of Environmental Engineering and Science Professors (AEESP). He is a recipient of the AEESP-Wiley Interscience Award for Outstanding Contributions to Environmental Engineering and Science Education and other environmental engineering education awards. Dr. Mihelcic is a member of the U.S. Environmental Protection Agency Science Advisory Board Environmental Engineering Committee and a board-certified member of the American Academy of Environmental Engineers (AAEE). His teaching and research interests are in green engineering, sustainable development, reform of engineering education, and the impact of global stressors such as population, land use, and climate on water and sanitation. He founded the first Master's International program in civil and environmental engineering in 1997 and now directs such a program at the University of South Florida (http://cee.eng.usf.edu/peacecorps). Dr. Mihelcic is a contributor to the 2009 UNESCO report *Engineering: Issues and Challenges for Development* and is the lead author of two other books: *Fundamentals of Environmental Engineering* (John Wiley & Sons, 1999) and *Environmental Engineering: Fundamentals, Design, Sustainability* (John Wiley & Sons, 2009). He has studied environmental policy as an AAAS-U.S. Environmental Protection Agency Environmental Fellow and has traveled extensively in the developing world to serve and conduct research on development issues related to water, sanitation, and global health.

Lauren M. Fry is a doctoral candidate in environmental engineering at Michigan Technological University, with research interests in the area of water resources and sustainable sanitation in developing countries. As a doctoral student, she has been involved in forming Michigan Tech's D80 Center. She earned a B.A. in physics from Gustavus Adolphus College and is a graduate of the Master's International program, which involved two years of engineering service as a Peace Corps volunteer in Cameroon. While in Cameroon, she conducted springbox and VIP latrine construction projects and facilitated participatory project management of community projects. She has also worked with the Office of Prevention, Pesticides, and Toxic Substances of the U.S. Environmental Protection Agency as a Pollution Prevention/Environmental Justice Fellow.

Elizabeth A. Myre is an international water and sanitation program manager. Her most recent employment took place in Haiti. She received an M.S. in environmental engineering from Michigan Technological University. As a graduate student, her involvement with Engineers Without Borders and an internship with UNICEF took her to Guatemala, Bolivia, and India. Elizabeth's multi-disciplinary approach to engineering stems from her undergraduate education in Linguistics, Foreign Languages, and International Development from Ohio State University.

Linda D. Phillips is a lecturer and Patel Associate in the Department of Civil and Environmental Engineering at the University of South Florida. Linda has a B.S. and an M.S. in civil engineering, specializing in construction management. She has over 20 years of practical experience working as a project engineer. In 2000, at the request of her students, Linda started the International Capstone Design course, now taking close to 170 students to developing countries to do their capstone design projects (http://cee.eng.usf.edu/ICD). The International Capstone Design program offers students a challenging and rewarding experience to learn engineering project design from inception to completion of construction contract documents, in a developing world setting. Students use their engineering skills, providing a service to communities in need, while satisfying ABET engineering accreditation requirements.

Brian D. Barkdoll is an associate professor of civil and environmental engineering at Michigan Technological University. Dr. Barkdoll is a Diplomate of the Academy of Water Resources Engineers (D.WRE). He is currently chair of the ASCE Sedimentation Technical Committee and a member of the ASCE Environmental Hydraulics Committee. He has won the ASCE Daniel W. Mead Award for Younger Members and the Chi Epsilon James M. Robbins Excellence-in-Teaching Award for the Southwest District. Dr. Barkdoll's teaching interests include fluid mechanics, hydraulics, hydrology, sediment transport, contaminant transport, and water collection and distribution. His research interests are in sedimentation, scour, oxygen transfer, clay permeability, vortices, acoustics, stream restoration, dams and reservoirs, intakes, water distribution systems, international development, and environmental sustainability. He served four years in Nepal as an engineer with the Peace Corps.